NEUROMETHODS ☐ 24

Animal Models of Drug Addiction

NEUROMETHODS

Program Editors: Alan A. Boulton and Glen B. Baker

NEUROMETHODS ☐ 24

Animal Models of Drug Addiction

Edited by

Alan A. Boulton

University of Saskatchewan, Saskatoon, Canada

Glen B. Baker

University of Alberta, Edmonton, Canada

and

Peter H. Wu

University of Toronto, Toronto, Canada

Humana Press • Totowa, New Jersey

© 1992 The Humana Press Inc.
999 Riverview Drive, Suite 208
Totowa, New Jersey 07512

Printed in the United States of America.

Library of Congress Cataloging-in-Publication Data

Main entry under title:

Animal models of drug addiction / edited by Alan A. Boulton, Glen B.
   Baker, and Peter Wu.
         p. cm. -- (Neuromethods ; 24)
      Includes bibliographical references and index.
      ISBN 0-89603-217-5
      1. Drug abuse—Animal models. 2. Alcoholism—Animal models.
I. Boulton, A. A. (Alan A.) II. Baker, Glen B., 1947– . III. Wu,
Peter. IV. Series.
         [DNLM: 1. Alcohol, Ethyl—pharmacology. 2. Disease Models,
Animal. 3. Drug Tolerance. 4. Models, Genetic. 5. Substance
Dependence—physiopathology. W1 NE337G v. 24 / QV 77 A5992]
RC564.A56 1992
616.86'27—dc20
DNLM/DLC
for Library of Congress   94-49180
                                                                CIP

Preface to the Series

When the President of Humana Press first suggested that a series on methods in the neurosciences might be useful, one of us (AAB) was quite skeptical; only after discussions with GBB and some searching both of memory and library shelves did it seem that perhaps the publisher was right. Although some excellent methods books have recently appeared, notably in neuroanatomy, it is a fact that there is a dearth in this particular field, a fact attested to by the alacrity and enthusiasm with which most of the contributors to this series accepted our invitations and suggested additional topics and areas. After a somewhat hesitant start, essentially in the neurochemistry section, the series has grown and will encompass neurochemistry, neuropsychiatry, neurology, neuropathology, neurogenetics, neuroethology, molecular neurobiology, animal models of nervous disease, and no doubt many more "neuros." Although we have tried to include adequate methodological detail and in many cases detailed protocols, we have also tried to include wherever possible a short introductory review of the methods and/or related substances, comparisons with other methods, and the relationship of the substances being analyzed to neurological and psychiatric disorders. Recognizing our own limitations, we have invited a guest editor to join with us on most volumes in order to ensure complete coverage of the field. These editors will add their specialized knowledge and competencies. We anticipate that this series will fill a gap; we can only hope that it will be filled appropriately and with the right amount of expertise with respect to each method, substance or group of substances, and area treated.

Alan A. Boulton
Glen B. Baker

Preface
to the Animal Models Volumes

This volume describes animal models of drug addiction. Because of increasing public concern over the ethical treatment of animals in research, we felt it incumbent upon us to include this general preface in order to indicate why we think further research using animals is necessary.

Animals should only be used when suitable alternatives are not available, and humans can only be experimented upon in severely proscribed circumstances. Alternative procedures using cell or tissue culture are inadequate in any models requiring assessments of behavioral change or of complex in vivo processes. However, when the distress, discomfort, or pain to the animals outweighs the anticipated gains for human welfare, the research is not ethical and should not be carried out.

It is imperative that each individual researcher examine his/her own research from a critical moral standpoint before engaging in it, and take into consideration the animals' welfare as well as the anticipated gains. Furthermore, once a decision to proceed with research is made, it is the researcher's responsibility to ensure that the animals' welfare is of prime concern in terms of appropriate housing, feeding, and maximum reduction of any uncomfortable or distressing effects of the experimental conditions. *Frequent formalized monitoring of these conditions must be conducted.*

It is essential to conform to national and local animal welfare regulations, whether codified in law or set down by self-regulatory bodies. We urge readers who wish to adopt any of the procedures described to follow closely both the letter of their own national and local regulations and the spirit of these guidelines.

The Editors

Preface

The theme of this volume, the use of animal models in the study of addiction, is one that has found wide acceptance in behavioral pharmacology. Yet it is also one that carries the inherent risk of ambiguity, because it contains two concepts that are frequently ill-defined or, perhaps worse, left undefined. These are the concepts of *addiction* and of *model*. The first is important because it tells us exactly what it is that we are trying to model. The second is equally important, because it tells us exactly what we are trying to do in relation to addiction. Unfortunately, neither is easy to define in a way that will be universally acceptable.

Corrigall's chapter in this volume comes squarely to grips with the first question, by citing the definitions of addiction that are offered in the 1988 Report of the US Surgeon General on Nicotine Addiction, and the 1989 Committee Report of the Royal Society of Canada, on Tobacco, Nicotine and Addiction. Both of these reports define the central problem of addiction as drug-seeking and drug-taking behavior that is very strongly motivated by the reinforcing psychoactive effects of a drug, and that is very difficult to give up, even when the user is also strongly motivated to do so.

The phenomena of tolerance and physical dependence (identifiable only as a withdrawal reaction when drug use is abruptly terminated) are recognized as frequent, or even characteristic, *consequences* of addiction, but not intrinsic parts of it. Similarly, the harm that may result from addiction is a legitimate social concern and a major reason for studying addiction, but not an essential defining feature of it.

The definition of "model" is perhaps more difficult to agree on, yet it is essential for an understanding of what we can and cannot hope to learn from the use of animal models of addiction. In everyday language, the word "model" has various meanings. It may mean an ideal example to be copied, such as a model student, or a role model. Sometimes it means a prototype, such as a new model of car. It may mean a miniature but perfect copy, such as a model airplane, and in that sense a rat model of addiction might be a tiny perfect copy of addiction in humans. The most relevant meaning, however, is a description or analogy that will help us to visualize the structure or function of something that we cannot otherwise study.

An animal model of addiction, in that sense, should b e one that replicates all the factors contributing to the production of addiction in humans, that reproduces in the animals the essential components of the picture of addiction as seen in humans, and that therefore permits a valid and detailed analysis of the underlying mechanisms, and their possible prevention or treatment, by interventions that are not physically or ethically possible in humans.

Extensive study of addiction in humans, however, has revealed many important aspects that are not included in the definition given above. In alcoholism, for example, there is an important genetic component of risk, or predisposition, that has been the subject of a great deal of recent research in animal models. But there are also very important social, financial, occupational, and psychiatric factors contributing to individual risk of alcoholism, especially in those who do not have an apparent genetic predisposition. Moreover, the genetic predisposition does not invariably lead to alcoholism in humans, so that these same external factors can also exert important modulatory influences that may facilitate or deter the actual occurrence of alcoholism in those at genetic risk.

What, then, can we reasonably expect to learn from the study of animal "models"? Clearly, we are unlikely ever to have a complete animal model of alcoholism, because no experimental animals, even primates, are subject to *all* of the social influences that act on humans and their drug use. However, it is possible to model many of these influences.

For example, the demonstrated effect of variation in price on the level of consumption of alcohol by humans could be modeled by systematic variation of the amount of work the animal must perform to obtain the drug, as has been done with other reinforcers. The role of social environmental stresses has to some extent been modeled by studies of drug consumption in animals living in group cages, and occupying different rungs of the dominance ladder. Much more could be done to examine the effects of other external influences, but it may be an impossible task to build a sufficiently complex model to test all the recognized external factors and study their interactions with endogenous or genetic determinants.

It is much more probable that animal models will yield valuable information about individual components of the overall picture. The various chapters of the present volume make clear that animal models can be used fruitfully to examine the reinforcing and discriminative stimulus properties of psychoactive drugs, the production of tolerance and various components of physical dependence, and the risk of organic complications from heavy use. Animal models may therefore also help to explore the neural circuitry and the neuronal and other cellular mechanisms underlying these various aspects of addiction. They can permit studies of the interactions among behavioral, genetic, and environmental influences acting on these components of addiction.

But it is important to recognize that these are not models of *addiction*. As Harris and Crabbe point out, it is certain that a great many genes contribute to the total risk of addiction, and to the expression of its individual features. Moreover, the role of each gene is not specific for addiction, but for a biological function that may participate in addiction. For example, animals genetically selected for proneness to seizures induced by alcohol withdrawal are also prone to seizures induced by withdrawal of other depressant drugs, or by administration of seizure-provoking drugs. Therefore, there is no guarantee that the overall risk of addiction, even to a specific drug, is determined by exactly the same set of genes in all cases. Indeed, it has been proposed that there are multiple forms of human alcoholism, for example, that reflect different combinations of genetic and environmental influences.

It is thus essential, both for the clarity of our concepts and for the appropriate formulation of our research objectives and protocols, to be very precise in defining what we attempt to model. This will help to avoid inappropriate explorations and excessive claims on the one hand and, on the other, any unwarranted frustration or pessimism about the applicability of the results to humans.

There is also another potential benefit of defining with precision the extent and objectives of our modeling. If we are clear that we model only specific limited components of addiction, we are more likely to be alert to the fact that these models may also yield valuable information about the fundamental biological processes underlying these separate components. For example, what we learn about tolerance may contribute to knowledge about neuroadaptive processes in general. If we extract such added benefit from the use of animal models, we are more likely to meet the ethical requirements laid out in the Series Editors' preface to the volumes on Animal Models.

H. Kalant

Contents

Genetic Animal Models: *A User's Guide*
R. Adron Harris and John C. Crabbe

Contents

Alcohol Tolerance:
Methodological and Experimental Issues
Anh Dzung Lê, S. John Mihic, and Peter H. Wu

Animal Models of Drug Addiction: *Barbiturates*
Michiko Okamoto

Opiate Withdrawal-Produced Dysphoria:
A Taste Preference Conditioning Model
Ronald F. Mucha

A Rodent Model for Nicotine Self-Administration
William A. Corrigall

Animal Models for Assessing Hallucinogenic Agents
Richard A. Glennon

Animal Models for Caffeine Exposure in the Perinatal Period
Ronnie Guillet

Contributors

WILLIAM A. CORRIGALL • *Addiction Research Foundation, Toronto, Ontario, Canada*

JOHN C. CRABBE • *Departments of Medical Psychology and Pharmacology, Oregon Health Sciences University, Portland, OR*

RICHARD A. GLENNON • *Department of Medicinal Chemistry, Virginia Commonwealth University, Richmond, VA*

LARRY A. GRUPP • *Addiction Research Foundation, Toronto, and Department of Pharmacology, University of Toronto, Toronto, Ontario, Canada*

RONNIE GUILLET • *Departments of Pediatrics and Psychology, University of Rochester, Rochester, NY*

R. ADRON HARRIS • *Department of Pharmacology, University of Colorado School of Medicine, Denver, CO*

PETRI HYYTIÄ • *Research Laboratories, State Alcohol Co. (Alko Ltd.), Helsinki, Finland*

KALERVO KIIANMAA • *Research Laboratories, State Alcohol Co. (Alko Ltd.), Helsinki, Finland*

ANH DZUNG LÊ • *Addiction Research Foundation, Toronto, Ontario, Canada*

RICHARD G. LISTER • *Laboratory of Clinical Studies, DICBR, NIAAA, Bethesda, MD*

S. JOHN MIHIC • *Department of Pharmacology, University of Colorado Health Sciences Center, Denver, CO*

RONALD F. MUCHA • *Psychological Institute I, University of Cologne, Cologne, Germany*

MICHIKO OKAMOTO • *Department of Pharmacology, Cornell University Medical College, New York, NY*

NICOLE R. RICHARDSON • *Department of Psychology, Carleton University, Ottawa, Ontario, Canada*

DAVID C. S. ROBERTS • *Department of Psychology, Carleton University, Ottawa, Ontario, Canada*

xxi

JOHN DAVID SINCLAIR • *Research Laboratories, State Alcohol Co. (Alko Ltd.), Helsinki, Finland*

ROBERT B. STEWART • *Department of Psychiatry and Behavioral Sciences, University of Texas Mental Sciences Institute, Houston, TX*

PETER H. WU • *Department of Pharmacology, University of Toronto, Toronto, Ontario, Canada*

Models of Alcohol Consumption Using the Laboratory Rat

Robert B. Stewart and Larry A. Grupp

1. Introduction

For 50 years, since the original work by Richter and Campbell (1940), rats have been used in experiments to model human alcohol abuse. The use of a nonhuman species, such as the rat, to model a behavior that seems to be inherently human has been justified in terms of the relatively economical and convenient provision of large, genetically uniform subject populations, and the degree of control that is possible over such factors as housing, diet, and previous drug experience. In addition, certain physiological or pharmacological manipulations can be administered to rats that would not be possible with human subjects.

Much effort has been expended in devising procedures to increase ethanol consumption by rats. One rationale for this endeavor is the assumption that a higher baseline of ethanol self-administration in an animal model will provide a superior testing ground for treatment interventions designed to reduce alcohol drinking by addicts. A second rationale is the assumption that something can be learned about the etiology of alcohol dependence in humans by the examination of experimental methods that successfully increase ethanol consumption by animals, such as rats, that may be reluctant to drink significant amounts of the drug (Deitrich and Melchior, 1985). Thus, not only must such methods bring about the desired result of an

From: *Neuromethods, Vol. 24: Animal Models of Drug Addiction*
Eds: A. Boulton, G. Baker, and P. H. Wu ©1992 The Humana Press Inc.

increase in ethanol self-administration, but they also must have relevance to the real world of human alcohol abuse. For example, the limitation of available drinking fluids to ethanol solutions alone can shed little light on the causal factors of alcohol drinking in humans, i.e., the manipulation lacks relevance to drug use by people since human alcohol drinkers invariably have other fluids available. Similarly, sweetening the ethanol solution also results in increased ethanol intake by rats (Eriksson, 1969), but for reasons that may not be operative in human alcohol abuse. This issue of relevance has been a source of much debate, as shall become apparent in the following review of some of the main approaches used to model human alcohol consumption using rats.

2. The Continuous-Access, Two-Bottle Choice Model

The oldest and most straightforward approach to an animal model has been to offer rats a continuous choice between an ethanol solution and water or other drug-free fluids (Richter and Campbell, 1940; Richter, 1941). This procedure is also referred to as the free-choice or preference model. In the most basic form of this method, rats are housed individually in a cage with food, water, and an ethanol solution freely available. Results are expressed in terms of both the absolute and relative amounts of the two fluids that are consumed during consecutive 24-h periods. The highest ethanol concentration preferred to water varies considerably among individual rats, but tends to occur within the range of 1–6% w/v when rats are tested that are not selectively bred for ethanol preference (Richter and Campbell, 1940).

The continuous-access, two-bottle choice method has been widely used, but has been severely criticized (Lester and Freed, 1973; Mello, 1973; Cicero, 1979; Kalant, 1983; Meisch, 1984) because the mean daily dose consumed by groups of rats is usually well below the rat's metabolic capacity of approx 7.2 g ethanol/kg body wt/d (Wallgren and Barry, 1970). With such low levels of ethanol consumption, tolerance, physical dependence, and organic damage are not produced. Any intoxication that may occur is hard to detect. The necessity for an animal model to show these physical consequences of high ethanol intake is debatable (Cappell and LeBlanc, 1981), but the most serious criticism of this model is the doubt over whether the rats are consuming the ethanol for its pharmacological effects or for other reasons, such as the calories it provides or its taste and gustatory properties.

The comparison of daily intake to daily metabolic capacity may be misleading if the temporal distribution of drinking within each 24-h period is not considered. Closer examination of patterns of ethanol drinking has shown that rats consume ethanol mainly during the dark 12-h period of their light/dark cycle (Aalto and Kiianmaa, 1984; Gill et al., 1986). Furthermore, the drinking occurs in discrete bouts such that in a duration of 1 or 2 min, a rat might drink enough of an 8% w/v ethanol solution to achieve doses in the range of 0.13–0.6 g/kg body wt (Gill et al., 1986). Given such a pattern of ethanol intake, it is possible that at certain times, the drinking may exceed the rate of metabolism for ethanol. However, drinking enough ethanol so that blood ethanol levels can be detected may be a necessary condition to assume that rats are taking in the drug for its postingestive pharmacological (presumably central nervous system [CNS]) effects, but it is not a sufficient condition because, for example, the calories or taste of the ethanol may motivate its intake, whereas the resulting level of ethanol in the blood occurs as a "side effect" of its consumption as a food.

Insofar as there has been a general unease and dissatisfaction with the two-bottle choice model because of the low levels of ethanol consumption usually achieved, various strategies have been employed to develop a more satisfactory model of alcohol abuse. These strategies include: the selective breeding of rats for high ethanol intake; the use of nonoral routes of ethanol administration; and the testing of various techniques to increase the oral self-administration of ethanol.

3. Selective Breeding of Rats for High Ethanol Intake

Concern over the low levels of ethanol consumption seen with the two-bottle choice model is based on the measurement of the mean daily ethanol intake by groups of genetically heterogeneous rats. However, there is much variability in the amount of ethanol that is consumed by individual rats, as was first noted by Richter and Campell (1940). Low mean daily ethanol intake is a reflection of the fact that most rats within a given population avoid ethanol. However, focusing on mean intake masks the fact that a small percentage of rats within a given population may drink large amounts of ethanol. Selective breeding of rats that show a high or low preference for ethanol has resulted in the development of rat lines that consistently self-administer very high or very low amounts of ethanol.

Three such lines that have been well characterized are the UChA (low alcohol preference) and UChB (high alcohol preference) line (Mardones and Segozia-Riquelene, 1983), the AA (Alko Alcohol) and ANA (Alko-Nonalcohol) lines (Eriksson, 1968), and the P (alcohol-preferring) and NP (alcohol-non-preferring) lines (Li et al., 1979). The P rats, for example, have been shown to drink about 7 g/kg body wt/d of an 8% w/v ethanol solution that results in blood ethanol levels as high as 152 mg/dL (Murphy et al., 1986). Ethanol tolerance (Gatto et al., 1987a) and physical dependence (Waller et al., 1982) have been demonstrated in P rats after chronic ethanol self-administration. The P line of rats satisfies most of the rigorous criteria set for animal models of alcohol abuse by such investigators as Lester and Freed (1973) and Cicero (1979), yet these rats were derived from a genetically heterogeneous Wistar strain using the standard continuous-access, two-bottle choice procedure that was condemned as "essentially worthless" (Cicero, 1979, p. 540) by the same individuals.

The demonstration that selective breeding can produce a line of rats with high ethanol consumption provides experimental confirmation of the importance of genetic factors in determining the risk for alcohol abuse already noted in humans (Cloninger, 1987). Such selection studies also afford an opportunity to determine if there is a true genetic correlation between high alcohol consumption and certain neurochemical, anatomical, and behavioral traits. The usefulness of genetically selected lines, however, need not be restricted to questions of heritability. These rat lines are equally useful for the examination of environmental factors, for example, using behavioral techniques that may increase ethanol intake by rats selectively bred for low ethanol preference (Samson et al., 1989).

A related area of research has been the evaluation of ethanol drinking by rodent strains or lines selectively bred for characteristics other than preference for ethanol. Again, the goal is to determine if there is a genetic correlation between certain physiological characteristics and ethanol consumption. The C57BL and DBA mice strains, for example, were originally bred for use in cancer research and only later were found to respond differentially to ethanol and to differ with respect to ethanol intake (Staats, 1966; Kakihana et al., 1966). The Dahl salt-sensitive (SS) line of rats was selectively bred to develop a sustained increase in blood pressure when fed a high salt diet, whereas the salt-resistant (SR) line was selected to remain normotensive (Dahl et al., 1962). Using a continuous-access, two-bottle choice procedure, Grupp et al. (1986a) found that rats of the SS line drank more ethanol than the SR rats. Low plasma renin activity has been postulated to be an inherited characteristic associated with ethanol preference because both the SS line (Iwai et al., 1973) and the alcohol-preferring P line (Grupp et al., 1989) have low plasma renin activity relative to the SR and alcohol-nonpreferring NP lines, respectively.

4. The Use of Nonoral Routes of Administration in Self-Administration Studies

4.1. Intravenous Route

One postulated explanation for the low levels of ethanol consumption shown by most rats is simply that the animals find the taste of the ethanol solution to be aversive. In addition, the time lag between the act of consuming the drug and the achievement of an appreciable CNS effect may be overly long when the oral route is used (Samson et al., 1988). Such a delay may hamper the learning of an association between drinking and the subsequent consequences of that drinking. For these reasons, the use of other, nonoral routes of administration has been incorporated into ethanol self-administration models. Initial reports (Smith and Davis, 1974; Smith et al., 1976) indicated that rats will self-administer ethanol by the intravenous (iv) route if infusions are contingent upon lever-presses. However, only very low unit doses of ethanol were effective (0.1–3.0 mg ethanol/kg body wt/infusion) and only very low rates of self-administration are typically seen. For example, Sinden and LeMagnen (1982) reported a mean of 62.5 mg ethanol/kg body wt/d for their most effective infusion dose.

On the other hand, even such modest results have not always been replicated. Oei and Singer (1979) compared iv ethanol self-administration in rats that were food-deprived and maintained at 80% of their free-feeding body wt to self-administration in rats that were allowed *ad libitum* access to food and were not weight-reduced. Only the hungry, weight-reduced rats responded for ethanol infusions at a rate that exceeded that for saline infusions by control groups. Grupp (1981) demonstrated the sensitive nature of iv ethanol self-administration by showing that rats would not sustain responding on even very low fixed ratio schedules for ethanol infusions. Numan (1981) found that rats exposed to a forced regimen of ethanol infusions (9–16 g/kg/d, for an average of 30 d) would subsequently self-administer ethanol when iv infusions were contingent upon lever-presses, but he found that rats not given the forced ethanol treatment would not self-administer ethanol. Numan et al. (1984) later attempted to replicate previously published positive results (Smith and Davis, 1974; Sinden and LeMagnen, 1982) using the same very low doses of ethanol. They reported a failure to achieve iv ethanol self-administration. Similarly, Collins et al. (1984), in a paper that recommended an iv self-administration model using rats as a screening technique to predict the abuse liability of new drugs, reported a failure to show iv ethanol self-administration by rats under the same conditions as those under which rats self-administered other drugs. This raises the interesting possibility that if ethanol was evaluated as a *new* drug using this method, it would have been deemed to have no abuse liability. DeNoble et al. (1985) also found that rats would not self-administer ethanol intravenously even when they used rats that had a history of iv pentobarbital self-administration. The success rate for iv ethanol self-administration appears to be much better for non-human primates (Deneau et al., 1969; Woods et al., 1971) than for rats.

4.2. Intragastric Route

The intragastric (ig) route of administration has also been tested in studies in which lever-presses produce infusions directly into the stomach. Smith et al. (1976) compared the same range of doses (0.1–3.0 mg/kg body wt/infusion) using both the iv and ig routes, and found the iv route to be more effective. Again, the amounts self-administered were low. Deutsch and Hardy (1976) devised a variation of the ig self-administration technique whereby the drinking of a drug-free flavored liquid activates a pump that injects a proportional amount of ethanol solution directly into the rat's stomach via a chronically implanted catheter. A control fluid with a different flavor is also available, the drinking of which produces an intragastric infusion of water. They reported some success with this method but, as was the case with Numan's (1981) iv study described above, it was necessary for the rats to have first undergone a period of forced ethanol administration before they would self-administer the drug. Naive controls not previously exposed to ethanol avoided the ethanol-paired flavor. The selectively bred alcohol-preferring P line of rats has also been tested using this technique. Rats of the P line will self-administer even more ethanol intragastrically than they do orally in the standard two-bottle choice procedure, without the necessity of pretreatment with forced intubation with ethanol (Waller et al., 1984).

5. Techniques to Increase Oral Ethanol Self-Administration by Rats

The two-bottle choice model results in very low ethanol intake when rats that are not selected for ethanol preference are tested. On the other hand, iv and ig self-administration studies have not produced robust rates of ethanol self-infusion. Attention has therefore focused on the genetic models described above and on the development of behavioral manipulations to increase ethanol drinking by rats that normally do not drink very much ethanol.

5.1. Methods Using Body Weight Reduction

5.1.1. Schedule-Induced Polydipsia

Probably the most powerful manipulation for increasing oral ethanol intake by rats is the polydipsia technique (Falk, 1961; Falk et al., 1972). Rats are partially deprived of food and maintained at a fixed percentage (typically, 80%) of their free-feeding body wt. When presented with small portions of food on an intermittent schedule (for example, one 45-mg food pellet every 1–2 min), rats will drink large volumes of water in the absence of any obvious physiological need (Falk, 1961). Falk called this drinking *schedule-induced polydipsia*. The substitution of an ethanol solution for the water results in substantial intake of the drug (Lester, 1961; Freed and Lester, 1970). When offered a choice between water and solutions of ethanol and water, rats will prefer concentrations up to approx 5% ethanol and water (Samson and Falk, 1974). This finding is similar to that obtained with the standard two-bottle choice model previously described, except that the total amount of fluid consumed is of a much higher order. By spacing six 1-h polydipsia sessions throughout each day, Falk et al. (1972) found that rats would drink ethanol doses (13.1 g/kg/d) that exceeded the animals' 24-h metabolic capacity. Sustained blood ethanol concentrations of 100–300 mg% were measured, and when the ethanol was withdrawn after 3 mo of such drinking, severe physical dependence was confirmed by the observation of tonic-clonic seizures in which some of the animals died.

There is a wide range of opinions concerning the relationship of the polydipsia model to alcohol abuse by humans. Falk and coworkers (Falk and Samson, 1976; Falk and Tang, 1977; Falk, 1983) have argued in favor of the proposition that excessive alcohol use in humans may be generated by intermittent schedules of reinforcement. This theory of addiction stresses the role of environmental factors and polydipsia is presented as an example of how situational variables may induce excessive behaviors of many kinds, including the abuse of alcohol. On the other hand, Freed and Lester (1970) have observed that the model may be flawed since the rats are weight-reduced and, therefore, may be consuming the ethanol for the calories it provides. Another concern, not shared by Falk and coworkers (*vida supra*), is the lack of drug-specificity with the polydipsia model. Many different behaviors can be exaggerated by intermittent schedules of reinforcement, not all of them related to the self-administration of ethanol or other drugs or even ingestive in nature (Falk, 1971). In a review of animal models, Cicero (1979) classified the polydipsia model as a technique for *forced* ethanol administration—efficacious in terms of producing intoxication, but irrelevant in terms of learning the causes of alcohol abuse. The debate concerning the polydipsia model often reveals two fundamentally different camps regarding the factors considered to be operating to produce drug addiction. One side views addiction as an abnormal physiological process, i.e., as a disease state. The other views addiction as a psychosocial phenomenon in which otherwise normal behaviors are exaggerated by environmental factors.

Leaving aside this theoretical debate, the desired end point of some of the original polydipsia experiments was the production of high and sustained ethanol intake under conditions of intermittent food reinforcement in order to produce tolerance and physical dependence (Falk et al., 1972; Samson and Falk, 1975). However, a related line of research was concerned with the level of ethanol intake that remained when rats had initially self-administered ethanol under conditions of schedule-induced polydipsia, but then later had the intermittent food reinforcement schedule discontinued. The general finding (Freed et al., 1970; Freed and Lester, 1970) was that ethanol self-administration persisted, but that this ethanol intake was reduced compared to that observed during the schedule-induced polydipsia.

Since this level of drinking was higher than what would be expected by naive rats, Meisch and Thompson (1971, 1974a) began to investigate the polydipsia phenomenon. They were not interested in the excessive schedule-induced ethanol intake *per se*, but in the development of methods for establishing ethanol drinking subsequent to the removal of the intermittent food presentation schedule. In the course of the parametric analysis of polydipsia, Meisch (1976) substituted a single feeding of the rats' daily ration of food for the intermittent presentation of food pellets. Since rats have a propensity to eat and drink at the same time, if water was made available, the rats would drink substantial amounts. This pattern was not disrupted when solutions with increasingly higher concentrations of ethanol were substituted for the water. When ethanol solutions of relatively high concentrations (8–32% w/v) were later offered in the absence of the availability of food, the ethanol drinking remained elevated. This modification and simplification of the schedule-induced polydipsia procedure is referred to as the prandial drinking method.

5.1.2. Prandial Drinking Method

Most of the studies using the prandial drinking method have been operant conditioning experiments. The rats must make a response, such as a lever-press, in order to activate a liquid delivery system that presents a small volume of ethanol solution (e.g., 0.1 mL). However, the use of operant conditioning techniques is not necessary because similar results have also been obtained using graduated drinking tubes attached to standard wire cages (MacDonnell and Marcucella, 1979; Stewart and Grupp, 1984). The prandial drinking method has been used to study the effects of diet (Grupp et al., 1986b), drugs (Samson and Doyle, 1985; Samson and Grant, 1985; Pfeffer and Samson, 1985; Grupp et al., 1986b), schedules of ethanol reinforcement (Beardsley et al., 1983), and genetic selection (Ritz et al., 1986; Suzuki et al., 1988) on ethanol self-administration.

Figure 1 shows data from an experiment in which the prandial drinking method was used to establish ethanol drinking. A group of eight male Long Evans hooded rats were food-restricted and maintained at 80% of their free-feeding body wt. Water was continuously available in the rats' home cages. The experiment was divided into two phases. The Acquisition phase was the period during which the rats were trained to self-administer ethanol and had 26 daily trials. The Test phase was the period during which the ethanol intake was evaluated after the training was completed and had 30 daily trials. Each drinking trial con-

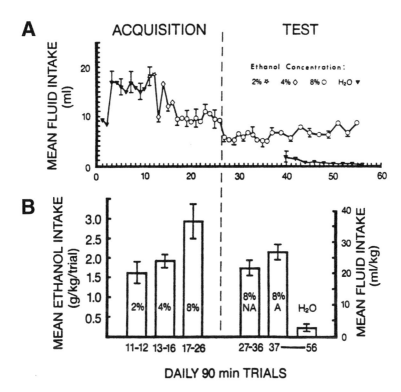

Fig. 1. Mean ± SEM ethanol and water intake for 8 rats in a prandial drinking method experiment. The upper panel shows the mean fluid intake (mL) for each daily 90-min trial during the Acquisition phase (trials 1–26), when the drinking was induced by the presentation of food within each trial, and during the Test phase (trials 27–56), when food was no longer presented to induce the drinking. The lower panel shows the mean ± SEM ethanol and water intake for the combined trials indicated under each bar graph. The left ordinate scale measures the ethanol dose (g/kg body wt/90 min) consumed during the ethanol trials. The right ordinate scale measures the volume of fluid consumed (mL/kg body wt/90 min) and is applicable to the trials during which either H_2O or 8% w/v ethanol solution was available. During the Test phase, the 8% A (Alternating) trials alternated on a daily basis with H_2O trials, whereas the 8% NA (nonalternating) trials were given consecutively and did not alternate with H_2O trials.

sisted of removing the rats from the home cages and placing them individually for 90 min into another test cage that was equipped with a single 100-mL graduated drinking tube. The amount of fluid (mL) consumed during each trial was measured.

Figure 1A (left panel) shows the mean fluid intake (mL) for each daily trial of the Acquisition phase. Approximately 8 g of pelleted rat chow (20 g of chow/kg body wt) were made available during each trial and the eating of this food elevated drinking. The first 10 trials were done to habituate the rats to the procedure and only water was presented. Beginning with trial 11, the water was replaced by a 2% w/v ethanol solution for two trials, then a 4% w/v ethanol solution for four trials, and finally, an 8% w/v ethanol solution for 10 trials. The mean ethanol dose (g/kg body wt) consumed by each rat was calculated for the 2, 4, and 8% ethanol trials. These individual-rat means were used to calculate the mean ethanol intake (g/kg body wt) for the entire group, which is shown by the bar graphs in Fig. 1B (left panel).

Figure 1A (right panel) shows the mean fluid intake (mL) for each of the 30 daily 90-min trials of the Test phase. Food was no longer placed in the test cages during these trials. Ten consecutive trials with 8% w/v ethanol solution available were followed by 10 additional 8% ethanol trials that took place every other day. On intervening days, water was the available liquid. In the absence of the within-trial presentation of food, the ethanol solution intake was elevated compared to the water intake, which was reduced to very low levels. The mean volume of fluid consumed (mL/kg body wt) by each rat was calculated for the first 10 8% ethanol trials that did not alternate with water trials (8% Not Alternating [NA] trials), for the 10 8% ethanol trials that did alternate with water trials (8% Alternating [A] trials), and for the 10 trials in which water was the only available fluid (H_2O trials). These individual-rat means were used to calculate the group means, shown as bar graphs in Fig. 1B (right panel).

To examine blood ethanol levels achieved during the Test phase of a prandial drinking experiment, 12 rats were trained as previously described. Once trained, the rats were given a 15-min trial without food, with access to an 8% w/v ethanol solution. The mean ethanol dose consumed during that trial was 1.44 g/ kg body wt/15 min. After the drinking trial, 50-µL blood samples were taken from the tip of the tail at intervals for 4 h. Analysis of the blood samples was done by gas chromatography as described

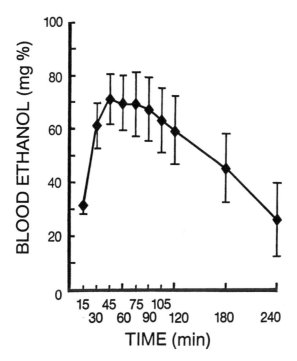

Fig. 2. Mean ± SEM blood ethanol levels (mg %) for 12 rats following a 15-min trial during which they drank an 8% w/v ethanol solution. The first (15 min) blood sampling time coincided with the cessation of drinking and additional samples were taken at the times shown.

elsewhere (Stewart and Grupp, 1988). The resulting blood ethanol time curve is shown in Fig 2.

Figure 3 shows the peak blood ethanol levels for the individual rats plotted as a function of dose (mL of 8% ethanol consumed converted to g/kg body wt). Dose and blood ethanol level were positively correlated ($r[10] = 0.89$, $p < 0.01$), indicating that the measurement of the volume of alcohol consumed is a reliable indicator of the blood ethanol level achieved.

After training using the prandial drinking method, rats will drink greater amounts of ethanol solution than water at concentrations as high as 32% w/v, either when the ethanol solution and the water are presented separately as the only available liquid during the drinking trials (Meisch and Thompson, 1974b),

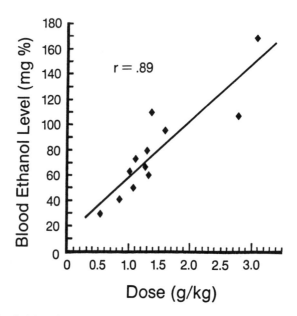

Fig. 3. Peak blood ethanol levels (mg%) in samples obtained after a 15-min drinking trial plotted as a function of dose (mL 8% w/v ethanol solution consumed converted to g ethanol/kg body wt). The points represent individual samples obtained from each of 12 rats. Dose and blood ethanol levels were positively correlated ($r = 0.89$).

or when both ethanol solution and water are presented concurrently (Meisch and Beardsley, 1975). As the ethanol concentration is raised from 8 to 32% w/v, the volume of fluid consumed decreases, but the ethanol dose received increases (Meisch and Thompson, 1974b). The temporal distribution of the drinking within each trial has a distinctive pattern, with most of the ethanol intake occurring near the beginning of the trial (Henningfield and Meisch, 1975; Stewart and Grupp, 1984).

Restoring *ad libitum* access to food in the home cages and allowing the rats to regain body wt usually results in a marked decrease in ethanol intake (Meisch and Thompson, 1974b; Stewart and Grupp, 1984; but *see* Beardsley et al., 1978 for an exception). However, the measurement of ethanol in the blood indicates that the intake of ethanol continues to exceed the drug's elimination rate (Stewart and Grupp, 1984). Since weight reduction has been repeatedly shown to increase ethanol intake (Westerfield and

Lawrow, 1953; *see* Meisch, 1984 for review), it has been argued that under these conditions, caloric restoration is the motivation for the ethanol self-administration (Lester and Freed, 1972). It is presumed that hungry rats take in ethanol as a food and not for the drug's pharmacological effects. On the other hand, weight reduction also increases the self-administration of many other drugs that do not have caloric content or any obvious anorectic properties (Carroll and Meisch, 1984), and these increases also occur when nonoral routes of drug administration are used (Carroll et al., 1981). This raises the possibility that the increases in ethanol intake observed in weight-reduced rats may represent an example of a more general phenomenon that may be unrelated to ethanol's food-like properties.

Additional evidence that weight-reduced rats may self-administer ethanol as a drug and not as a food comes from an experiment in which weight-reduced rats were trained to lever-press for the delivery of an ethanol solution using the prandial drinking method (Beardsley et al., 1983). The effects of schedules of ethanol reinforcement were examined. An intermittent reinforcement schedule was used in which a specified period of time must elapse after a lever-press has been reinforced before another lever-press is reinforced. As the time between reinforcements (the minimum-interreinforcer interval) was systematically changed from 0 to 480 s, the number of ethanol reinforcements obtained decreased. However, this reduction was not owing solely to the ceiling imposed by the maximum number of ethanol presentations obtainable within a trial. The quantity of reinforcements was always less than the maximum permitted by the value of the minimum-interreinforcer interval. It should be noted that when access to ethanol is unrestricted (a continuous reinforcement schedule), most of the drinking occurs at the very beginning of each drinking trial (Meisch and Thompson, 1974b; Stewart and Grupp, 1984). This initial burst of ethanol intake would be expected if the rats were, in effect, consuming a "loading dose" in order to achieve quickly an optimal drug effect that may correspond to a certain level of ethanol in the blood. The imposition of an interreinforcement interval may, if it is sufficiently long, make it impossible for the rats to achieve such a loading dose and the putative optimal drug effect. Thus, under these conditions, ethanol becomes a weaker reinforcer and fewer ethanol reinforcements are obtained than the maximum possible under the interval schedule. Similar findings have been reported in a clinical study (Gottheil et al., 1972) using human alcoholic subjects, many of whom would rather abstain than receive drinks that were spaced 60 min apart. Thus, it is possible that this interval schedule effect may indicate that the ethanol intake seen using the prandial drinking method is an example of drug-taking and not eating behavior.

5.2. Methods that Do Not Use Body Weight Reduction

Despite the fact that methods that include food deprivation are very effective, in recent years, attention has turned to procedures that increase ethanol intake without weight reduction. The "rule of thumb" for judging the success of these procedures is that ethanol self-administration should exceed the rats' ethanol metabolic rate of about 0.3 g ethanol/kg body wt/h (Wallgren and Barry, 1970) so that ethanol can be detected in the blood. Preferences for ethanol concentrations of 8% w/v or greater and blood ethanol levels higher than 100 mg% are causes for celebration.

5.2.1. Schedule-Induced Polydipsia and Prandial Drinking Method Without Weight Reduction

Since schedule-induced polydipsa and the prandial drinking method were effective for producing increased ethanol consumption in weight-reduced rats, attempts have been made to utilize these techniques using rats that are not weight-reduced. Samson and Pfeffer (1987) presented nonweight-reduced rats with small amounts of sucrose solution on an intermittant FI90-s schedule. They found that the polydipsia procedure was not as effective with full-weight rats and with liquid food reinforcers rather than dry food pellets. Only about half of their animals were induced to consume ethanol, in contrast with an almost 100% success rate for weight-reduced rats indicated in the literature. The amount of ethanol consumed by these rats was low compared to the impressive intake previously reported for weight-reduced rats (e.g., Falk et al., 1972). Along similar lines, Stewart et al. (1988) attempted to train rats that were not weight-reduced to self-administer ethanol using the prandial drinking method. The procedure used was identical to that described for the prandial drinking method except that the rats had *ad libitum* access to food at all times in their home cages, including the Acquisition (training) and Test phases of the experiment. A sweetened dry breakfast cereal was used in place of pelleted chow to induce prandial thirst during drinking trials of the Acquisition phase. Ethanol intake during the Test phase trials was only about one-third of the consumption shown by controls that were weight-reduced.

5.2.2. Habituation/Acclimation

Making the assumption that ethanol has an aversive taste that is a major impediment to oral self-administration by rats, several investigators have attempted to habituate nonweight-reduced rats to the taste of ethanol while exposing the rats to the drug's presumably reinforcing postingestional effects. In an experiment in which rats had continuous access to both ethanol solution and water, Veale and Myers (1969) found that if the concentration of the ethanol was systematically increased over a number of days, the animals would subseqently consume higher doses and concentrations of ethanol than controls not exposed to the regimen. Veale and Myers (1969) called this an "acclimation effect" and suggested that tolerance to ethanol may account for the increased ethanol intake observed. However, no independent behavioral or physiological measures were made to confirm that tolerance had indeed developed. In the same study (Veale and Myers, 1969), a separate control group was forced to drink a high-concentration ethanol solution as the sole source of fluid, but these rats showed a decrease in ethanol consumption when later offered a choice between ethanol solution and water. This finding illustrated the necessity for a gradual exposure to ascending concentrations and for the concurrent availability of water in order to bring about an increased preference for ethanol.

Other investigators have examined the contribution of taste to ethanol self-administration. One approach has been to make the ethanol solution more palatable by adulteration with sucrose or other sweeteners (Cullen et al., 1973; Kulkosky, 1979). However, removing the sweetener resulted in a return to low levels of ethanol intake. Samson (1986) hypothesized that this lack of maintained ethanol drinking may have been because the change from a preferred taste to a concentrated ethanol solution was too abrupt. He carried out an experiment in which rats were first trained to make a lever-pressing response for access to a 20% sucrose solution. Over the course of several daily 30-min sessions, the sucrose concentration was gradually reduced while ethanol was added at increasingly higher concentrations. Finally, the fluid consisted of an ethanol solution with no sucrose. With this "sucrose-fading" procedure (Samson, 1986), the rats had come to prefer ethanol concentrations of up to 32% w/v and doses on the order of 1 g/kg/30-min trial were self-administered.

5.2.3. Limitation of Temporal Access

Sinclair and Senter (1967,1968) gave free-feeding rats several weeks' access to an ethanol solution and water using a standard two-bottle choice procedure. The ethanol was then removed and reintroduced a few days later. A substantial but temporary increase in ethanol drinking was observed. A sustained increase in the amount of ethanol consumed per day develops if the ethanol solution is offered on alternate days with water available every day (Wayner et al., 1972; Wise, 1973; Pinel and Huang, 1976). Such increases may not be specific to ethanol solutions since presenting saccharin or quinine solutions every other day also increases the intake of those fluids (Wayner et al., 1972; Pinel and Huang, 1976).

MacDonnell and Marcucella (1979) and Linseman (1987) have investigated a technique in which rats are induced to drink ethanol in large, single bouts by limiting temporal access to a short interval (e.g., 1 h) each day. Food and water are continuously available in the rats' home cages. The rats are placed for 1 h each day into separate test cages that are equipped with two graduated drinking tubes. One of the tubes always contains water and the other contains a solution of ethanol and water. No food is available in the test cages. The concentration of the ethanol solution is systematically increased by offering a 3% w/v ethanol solution for 14 consecutive days, a 6% w/v solution for 14 d, and finally, a 12% w/v solution for 14 d (Linesman, 1988). Thus, this procedure also incorporates acclimation or habituation to the taste of ethanol as well as the limitation of temporal access to the drug. The rats maintain a preference for ethanol solution over water even at the highest ethanol concentration. They consume an ethanol dose of about 1 g/kg body wt/h, which results in blood ethanol concentrations of approx 50 mg% (Linesman, 1987). Nost of the ethanol intake occurs as a single bout at the beginning of the drinking trial.

Noting that alternate-day access to nonalcoholic solutions increases the rate of intake of those fluids (Wayner et al., 1972; Pinel and Huang, 1976), Linseman (1988) examined the possibility that the increases in ethanol intake seen with 1-h daily access also may not be specific to ethanol. She found that 1-h daily access to a mildly bitter sucrose octaacetate solution did not result in increased intake of that solution. Since there is some discrepancy, the drug-specificity of these limited-access procedures should be further investigated by comparing alternate-day and 1-h access using identical nonalcoholic flavored solutions.

The relevance or relationship of the limited access or "alcohol deprivation" (Sinclair and Senter, 1968) effect to human drug use patterns is not clear, except that life's contingencies, work responsibilities, and social norms certainly impose periods during which alcohol drinking is either not possible or unacceptable. It could be hypothesized that such limitations on the availability of alcohol may contribute directly to increase the amount consumed once alcohol is made available. This idea has gained some support from a study in which college students were observed to drink more alcohol after a short period of abstinence than if they had been drinking the previous evening (Burish et al., 1981). However, temporal access to ethanol is an important experimental variable since many of the techniques aimed at increasing ethanol intake in rats incorporate schedules of limited access to ethanol.

6. Conclusions

Until recently, an overview of this field would have engendered a great deal of pessimism concerning the usefulness of rats to model human alcohol consumption. The low levels of oral intake shown by unselected rats in the continuous-access, two-bottle choice model, plus the failure to demonstrate robust ethanol self-administration by rats using nonoral routes of administration (e.g., Numan et al., 1984) suggested that ethanol was, at best, only a weak reinforcer for this species. Schedule-induced polydipsia (Falk et al., 1972) and the prandial drinking method (Meisch, 1976) successfully produced higher rates of ethanol self-administration, but these procedures involved maintaining the rats at reduced body wt. There was serious doubt concerning whether rats ever drink ethanol to experience the drug's pharmacological effects (Lester and Freed, 1972).

However, the characterization of rats that have been selectively bred for high ethanol intake has produced evidence that ethanol's pharmacological effects may be reinforcing for some rats. The ethanol-preferring (P) rats (Li et al., 1979), for example, drink sufficient quantities of ethanol in the presence of food and water to achieve high blood ethanol levels (Murphy et al., 1986) and to produce ethanol tolerance and physical dependence (Gatto et al., 1987a; Waller et al., 1982). High oral intake alone, however, does not constitute evidence that the ethanol is being consumed as a drug and not as a food, since the selective breeding may have resulted in an aberrant preference for the taste of ethanol or for some other quality that ethanol may possess as a food. The P rats also will self-administer even higher ethanol doses via the ig route of administration than they consume orally (Waller et al., 1984) and will show a conditioned taste preference for a saccharin solution that has been paired with a low dose of ethanol given by ip injection (Froehlich et al., 1988). These results are more difficult to interpret as being artifacts of an enhanced preference for the taste, smell, or other food-like qualities of ethanol. The most convincing evidence comes from studies in which P rats have been compared to alcohol-nonpreferring (NP) rats for responses to ethanol in addition to ethanol self-administration. Low-dose injections of ethanol increase spontaneous motor activity in the P, but not in the NP rats, whereas moderate doses of ethanol produce motor impairment in the NP, but not the P rats (Waller et al., 1986). The P rats develop tolerance to high-dose ethanol effects, such as motor impairment, more quickly than NP rats (Waller et al., 1983), and this tolerance persists for a longer period of time in the P compared to the NP rats (Gatto et al., 1987b). These findings suggest that the selection for ethanol preference may be associated with differential sensitivities to the pharmacological effects of the drug. Such an association would not be expected to accompany selection for an exaggerated preference for a particular food.

The P line was the result of the selective breeding of the highest drinkers in a foundation stock of genetically heterogeneous rats. Therefore, it is probable that at least the highest drinkers, which constitute a small portion of unselected rat populations, must possess characteristics in common with P rats and probably self-administer ethanol for its pharmacological effects. However, there is no reason to rule out the possibility that rats that drink moderate or even low amounts of ethanol are doing so in order to experience the pharmacological effects of the drug. It also follows that the individual differences in the propensity to self-administer ethanol shown by unselected rats are, at least in part, genetically transmitted. In this respect, the ethanol drinking shown by laboratory rats may mirror the variability in preference for alcohol seen in "unselected" populations of humans.

The fact that most unselected rats do not self-administer large quantities of ethanol may reflect genetic variability, but some of this variability is likely to be the result of environmental factors. The demonstration that certain behavioral manipulations can increase the self-administration of ethanol by rats may also illustrate the importance of environmental factors in the etiology of alcohol abuse by humans (Falk, 1983). Many different techniques have been used with rats, but almost all involve a procedure to habituate the animals to the taste of the ethanol solution, plus a schedule of limited access to the ethanol. An important question is whether these diverse methods result in similar levels and/or patterns of ethanol intake once the rats have been trained using these techniques and peak ethanol consumption is reached (Meisch, 1976; Samson et al., 1988; Stewart et al., 1988).

Unfortunately, the manipulations that produce the greatest increases in ethanol intake by rats, schedule-induced polydipsia and the prandial drinking method, also utilize body wt reduction. Thus, the issue of whether the rats are consuming the ethanol as a food or for its pharmacological effects has been raised with regard to these methods (Freed and Lester, 1970). As previously discussed, increased self-administration of ethanol by weight-reduced rats does not prove that the rats are consuming the ethanol for calories (Carroll and Meisch, 1984). On the other hand, ethanol self-administration by freely feeding rats does not eliminate the possibility that the ethanol is being consumed as a food, since rats will readily consume certain palatable foods (e.g., a sucrose solution; Samson 1986) even when they are not hungry. Since the relevance of weight reduction to drug-taking behavior is not fully understood, weight reduction probably should not be included in an animal model for the purpose of studying other variables that may affect ethanol intake. However, the weight reduction effect should be investigated as an interesting phenomenon in its own right. The idea that food reinforcement may share some common neural pathways with other reinforcers, such as alcohol and other abused drugs, has some empirical support (Wise, 1980). Therefore, it should not be surprising that a single manipulation, such as weight reduction, can increase responding for many different classes of reinforcers. It is possible that at some level of neural organization, the physiological processes involved in food, alcohol, and drug reinforcement may be inseparable.

References

Aalto J. and Kiianmaa K. (1984) Circadian rhythms of water and alcohol intake: Effect of REM-sleep deprivation and lesion of the suprachiasmatic nucleus. *Alcohol* **1**, 403–407.

Beardsley P. M., Lemaire G. A., and Meisch R. A. (1978) Ethanol-reinforced behavior of rats with concurrent access to food and water. *Psychopharmacology* **59**, 7–11.

Beardsley P. M., Lemaire G. A., and Meisch R. A. (1983) Effects of minimum-interreinforcer interval on ethanol-maintained performance of rats. *Pharmacol. Biochem. Behav.* **19**, 843–847.

Burish T. G., Maisto S. A., Cooper A. M., and Sobell M. B. (1981) Effects of voluntary short-term abstinence from alcohol on subsequent drinking patterns of college students. *J. Stud. Alcohol* **42**, 1013–1020.

Cappell H. and LeBlanc A. E. (1981) Tolerance and physical dependence: Do they play a role in alcohol and drug self-administration? in *Research Advances in Alcohol and Drug Problems, vol.6.* Isreal Y., Glaser F. B., Kalant H., Popham R. E., Schmidt W., and Smart R. G., eds., Plenum, New York, pp. 159–196.

Carroll M. E., France C. P., and Meisch R.A. (1981) Intravenous self-administration of etonitazene, cocaine, and phencyclidine in rats during food deprivation and satiation. *J. Pharmacol. Exp. Ther.* **217**, 241–247.

Carroll M. E. and Meisch R. A. (1984) Increased drug-reinforced behavior due to food deprivation, in *Advances in Behavioral Pharmacology,* vol. 4. Thompson T., Dews P. B., and Barrett J. E., eds., Academic, New York, pp. 47–88.

Cicero T. J. (1979) A critique of animal analogues of alcoholism, in *Biochemistry and Pharmacology of Ethanol,* vol. 2, Majchrowicz E. and Noble E. P., eds., Plenum, New York, pp. 533–560.

Cloninger C. R. (1987) Neurogenetic adaptive mechanisms in alcoholism. *Science* **236**, 410–416.

Collins R. J., Weeks J. R., Cooper M. M., Good P. I., and Russel R. R. (1984) Prediction of abuse liability of drugs using IV self-administration by rats. *Psychopharmacology* **82**, 6–13.

Cullen J. W., Croes R. A., and Gillis R. D. (1973) Alcohol selection by rats after experience with a sapid alcohol-sucrose solution. *Q. J. Stud. Alcohol* **34**, 769–773.

Dahl L. K., Heine M., and Tassinari L. (1962) Effects of chronic salt ingestion. Evidence that genetic factors play an important role in susceptibility to experimental hypertension. *J. Exp. Med.* **115**, 1173–1190.

Deitrich R. A. and Melchior C. L. (1985) A critical assessment of animal models for testing new drugs for altering ethanol intake, in *Research Advances in New Psychopharmacological Treatments for Alcoholism* , Naranjo C. A. and Sellers E. M., eds.), *Excerpta Medica*, Amsterdam, pp. 23–43.

Deneau G., Yanagita T., and Seever M. H. (1969) Self-administration of psychoactive substances by the monkey. *Psychopharmacologia (Berl.)* **16**, 30–84.

DeNoble V. J., Mele P. C., and Poter J. H. (1985) Intravenous self-administration of pentobarbital and ethanol in rats. *Pharmacol. Biochem. Behav.* **23**, 759–763.

Deutsch J. A. and Hardy W. T. (1976) Ethanol tolerance in the rat measured by the untasted intake of alcohol. *Behav. Biol.* **17**, 379–389.

Eriksson K. (1968) Genetic selection for voluntary alcohol consumption in the albino rat. *Science* **159**, 739–741.

Eriksson K. (1969) Factors affecting voluntary alcohol consumption in the albino rat. *Ann. Zool. Fennici.* **6**, 227–265.

Falk J. L. (1961) Production of polydipsia in normal rats by an intermittent food schedule. *Science* **133**, 195,196.

Falk J. L. (1971) The nature and determinants of adjunctive behavior. *Physiol. Behav.* **6**, 577–588.

Falk J. L. (1983) Drug dependence: Myth or motive? *Pharmacol. Biochem. Behav.* **19**, 385–391.

Falk J. L. and Samson H. H. (1976) Schedule-induced physical dependence on ethanol. *Pharmacol. Rev.* **27**, 449–464.

Falk J. L., Samson H. H., and Winger G. (1972) Behavioral maintenance of high concentrations of blood ethanol and physical dependence in the rat. *Science* **177**, 811–813.

Falk J. L. and Tang M. (1977) Animal model of alcoholism: Critique and progress in *Alcohol Intoxication and Withdrawal,* vol. 3B, Gross M. M., ed. Plenum, New York, pp. 465–493.

Freed E. X., Carpenter J. A., and Hymowitz N. (1970) Acquisition and extinction of schedule-induced polydipsic consumption of alcohol and water. *Psychol. Rep.* **26**, 915–922.

Freed E. X. and Lester D. (1970) Schedule-induced consumption of ethanol: Calories or chemotherapy? *Physiol. Behav.* **5**, 555–560.

Froehlich J. C., Harts J., Lumeng L., and Li T.-K. (1988) Differences in response to the aversive properties of ethanol in rats selectively bred for oral ethanol preference. *Pharmacol. Biochem. Behav.* **32**, 215–222.

Gatto G. J., Murphy J. M., Waller M. B., McBride W. J., Lumeng L., and Li T.-K. (1987a) Chronic ethanol tolerance through free-choice drinking in the P line of alcohol preferring rats. *Pharmacol. Biochem. Behav.* **28**, 111–115.

Gatto G. J., Murphy J. M., Waller M. B., McBride W. J., Lumeng L., and Li T.-K. (1987b) Persistence of tolerance to a single dose of ethanol in the selectively bred alcohol-preferring P rats. *Pharmacol. Biochem. Behav.* **28**, 105–110.

Gill K., France S., and Amit Z. (1986) Voluntary ethanol consumption in rats: An examination of blood-brain levels and behavior. *Alcoholism: Clin. Exp. Res.* **10**, 457–462.

Gottheil E., Murphy G. F., Skoloda T. E., and Corbit L. O. (1972) Fixed interval drinking decision: II. Drinking and discomfort in 25 alcoholics. *Q. J. Stud. Alcohol* **33**, 325–340.

Grupp L. A. (1981) An investigation of intravenous ethanol self-administration in rats using a fixed ratio schedule of reinforcement. *Physiol. Psychol.* **9**, 359–363.

Grupp L. A., Kalant H., and Leenen F. H. H. (1989) Alcohol intake is inversely related to plasma renin activity in the genetically selected alcohol-preferring and -non-preferring lines of rats. *Pharmacol. Biochem. Behav.* **32**, 1061–1063.

Grupp L. A., Perlanski E., Wanless I. R., and Stewart R. B. (1986a) Voluntary alcohol intake in the hypertension prone Dahl rat. *Pharmacol. Biochem. Behav.* **24,** 1167–1174.

Grupp L. A., Perlanski E., and Stewart R. B. (1986b) Diet and diuretics in the reduction of voluntary alcohol drinking in rats. *Alcohol and Alcoholism* **21,** 75–79.

Henningfield J. E. and Meisch R. A. (1975) Ethanol-reinforced responding and intake as a function of volume per reinforcement. *Pharmacol. Biochem. Behav.* **3,** 437–441.

Iwai J., Dahl L. K., and Knudsen K. D. (1973), Genetic influences on the renin-angiotensin system. Low renin activities in hypertension-prone rats. *Circ. Res.* **32,** 678–684.

Kakihana R., Brown D. R., McClearn G. E., and Tabershaw I. R. (1966) Brain sensitivity to alcohol in inbred mouse strains. *Science* **154,** 1574,1575.

Kalant H. (1983) Animal models of alcohol and drug dependence: Some questions, answers, and clinical implications, in *Etiologic Aspects of Alcohol and Drug Abuse* Gottheil E., Druley K. A., Skododa T. E., and Waxman H. M., eds., C. C. Thomas, Springfield, IL, pp. 14–29.

Kulkosky P. J. (1979) Effect of addition of ethanol and NaCl on saccharin + glucose polydipsia. *Pharmacol. Biochem. Behav.* **10,** 277–283.

Lester D. (1961) Self-maintenance of intoxication in the rat. *Q. J. Stud. Alcohol* **22,** 223–231.

Lester D. and Freed E. (1972) The rat views alcohol—Nutrition or nirvana? in *Biological Aspects of Alcohol Consumption*, Forsander O. and Eriksson K., eds., The Finnish Foundation for Alcohol Studies, Helsinki, pp. 27–29.

Lester D. and Freed E. (1973) Criteria for an animal model of alcoholism. *Pharmacol. Biochem. Behav.* **1,** 103–107.

Li T.-K., Lumeng L., McBride W. J., and Waller M. B. (1979) Progress toward a voluntary oral consumption model of alcoholism. *Drug Alcohol Depend.* **4,** 45–60.

Linseman M. A. (1987) Alcohol consumption in free-feeding rats: Procedural, genetic and pharmacokinetic factors. *Psychopharmacology* **92,** 254–261.

Linseman M. A. (1988) Consumption of alcohol compared to another bitter solution in a limited access drinking paradigm. *Alcohol* **5,** 301–303.

MacDonnell J. S. and Marcucella H. (1979) Increasing the rate of ethanol consumption in food- and water-satiated rats. *Pharmacol. Biochem. Behav.* **10,** 211–216.

Mardones J. and Segozia-Riquelene N. (1983) Thirty-two years of selection of rats by ethanol preference: UChA and UChB strains. *Neurobehav. Toxicol. Teratol.* **5,** 171–178.

Meisch R. A. (1976) The function of schedule-induced-polydipsia in establishing ethanol as a positive reinforcer. *Pharmacol. Rev.* **27,** 465–473.

Meisch R. A. (1984) Alcohol self-administration by experimental animals, in *Research Advances in Alcohol and Drug Problems*, vol. 8, Smart R. G., Glaser F. B., Isreal Y., Cappell H., Kalant H. Schmidt W., and Sellers E. M., eds., Plenum, New York, pp. 23–45.

Meisch R. A. and Beardsley P. (1975) Ethanol as a reinforcer for rats: Effects of concurrent access to water and alternate positions of water and ethanol. *Psychopharmacologia (Berl.)* **43**, 19–23.

Meisch R. A. and Thompson T. (1971) Ethanol intake in the absence of concurrent food reinforcement. *Psychopharmacologia (Berl.)* **22**, 72–79.

Meisch R. A. and Thompson T. (1974a) Rapid establishment of ethanol as a reinforcer for rats. *Psychopharmacologia (Berl.)* **37**, 311–321.

Meisch R. A. and Thompson T. (1974b) Ethanol intake as a function of concentration during food deprivation and satiation. *Pharmacol. Biochem. Behav.* **2**, 589–596.

Mello N. K. (1973) A review of methods to induce alcohol addiction in animals. *Pharmacol. Biochem. Behav.* **1**, 89–101.

Murphy J. M., Gatto G. J., Waller M. B., McBride W. J., Lumeng L., and Li T.-K. (1986) Effects of scheduled access on ethanol intake by the alcohol preferring P line of rats. *Alcohol* **3**, 331–336.

Numan R. (1981) Multiple exposures to ethanol facilitate intravenous self-administration by rats. *Pharmacol. Biochem. Behav.* **15**, 101–108.

Numan R., Naparzewska A. M., and Adler C. M. (1984) Absence of reinforcement with low dose intravenous ethanol self-administration in rats. *Pharmacol. Biochem. Behav.* **21**, 609–615.

Oei T. P. S. and Singer G. (1979) Effects of a fixed time schedule and body weight on ethanol self-administration. *Pharmacol. Biochem. Behav.* **10**, 767–770.

Pfeffer A. O. and Samson H. H. (1985) Oral ethanol reinforcement in the rat: Effects of acute amphetamine. *Alcohol* **2**, 693–697.

Pinel J. P. J and Huang E. (1976) Effects of periodic withdrawal on ethanol and saccharin selection in rats. *Physiol. Behav.* **16**, 693–698.

Richter C. P. (1941) Alcohol as food. *Q. J. Stud. Alcohol* **1**, 650–662.

Richter C. P. and Campbell K. H. (1940) Alcohol taste thresholds and concentrations of solutions preferred by rats. *Science* **9**, 507,508.

Ritz M. C., George F. R., deFiebre C. M., and Meisch R. A. (1986) Genetic differences in the establishment of ethanol as a reinforcer. *Pharmacol. Biochem. Behav.* **24**, 1089–1094.

Samson H. H. (1986) Initiation of ethanol reinforcement using a sucrose substitution procedure in food- and water-sated rats. *Alcoholism: Clin. Exp. Res.* **10**, 436–442.

Samson H. H. and Doyle T. F. (1985) Oral ethanol self-administration in the rat: Effect of naloxone. *Pharmacol. Biochem. Behav.* **22**, 91–99.

Samson H. H. and Falk J. L. (1974) Alternation of fluid preference in ethanol-dependent animals. *J. Pharmacol. Behav. Ther.* **190**, 365–376.

Samson H. H. and Falk J. L. (1975) Pattern of daily blood ethanol elevation and the development of physical dependence. *Pharmacol. Biochem. Behav.* **3**, 1119–1123.

Samson H. H. and Grant K. A. (1985) Chlordiazepoxide effects on ethanol self-administration: Dependence on concurrent conditions. *J. Exp. Anal. Behav.* **43**, 353–364.

Samson H. H. and Pfeffer A. O. (1987) Initiation of ethanol-maintained responding using a schedule-induction procedure in free feeding rats. *Alcohol Drug Res.* **7**, 461–469.

Samson H. H., Pfeffer A. O., and Tolliver G. A. (1988) Oral ethanol self-administration in rats: Models of alcohol-seeking behavior. *Alcoholism: Clin. Exp. Res.* **12**, 591–598.

Samson H. H., Tolliver G. A., Lumeng L., and Li T.-K. (1989) Ethanol reinforcement in the alcohol-nonpreferring (NP) rat: Initiation using behavioral techniques without food restriction. *Alcoholism: Clin. Exp. Res.* **13**, 378–385.

Sinclair J. D. and Senter R. J. (1967) Increased preference for ethanol in rats following alcohol deprivation. *Psychon. Sci.* **8**, 11,12.

Sinclair J. D. and Senter R. J. (1968) Development of an alcohol deprivation effect in rats. *Q. J. Stud. Alc.* **29**, 863–867.

Sinden J. D. and LeMagnen J. (1982) Parameters of low-dose ethanol intravenous self-administration in the rat. *Pharmacol. Biochem. Behav.* **16**, 181–183.

Smith S. G. and Davis W. N. (1974) Intravenous alcohol self-administration in the rat. *Pharmacol. Res. Commun.* **6**, 397–401.

Smith S. G., Werner T. E., and Davis W. M. (1976) Comparison between intravenous and intragastric alcohol self-administration. *Physiol. Psychol.* **4**, 91–93.

Staats J. (1966) The inbred mouse, in *Biology of the Laboratory Mouse*, Green E. L., ed., McGraw Hill, New York, pp. 1–11.

Stewart R. B. and Grupp L. A. (1984) A simplified procedure for producing ethanol self-selection in rats. *Pharmacol. Biochem. Behav.* **21**, 255–258.

Stewart R. B. and Grupp L. A. (1988) Conditioned place aversion mediated by orally self-administered ethanol in the rat: A consideration of blood ethanol levels. *Pharmacol. Biochem. Behav.* **32**, 331–371.

Stewart R. B., Perlanski E., and Grupp L. A. (1988) Ethanol as a reinforcer for rats: Factors of facilitation and constraint. *Alcoholism: Clin. Exp. Res.* **12**, 599–608.

Suzuki T., George, F. R., and Meisch R. A. (1988) Differential establishment and maintenance of oral ethanol reinforced behavior in Lewis and Fischer 344 inbred rat strains. *J. Pharmacol. Exp. Ther.* **245**, 164–170.

Veale W. L. and Myers R. D. (1969) Increased alcohol preference in rats following repeated exposures to alcohol. *Psychopharmacologia (Berl.)* **15**, 361–372.

Waller M. B., McBride W. J., Gatto G. J., Lumeng L., and Li T.-K. (1984) Intragastric self-infusion of ethanol by the P and NP (alcohol-preferring and -nonpreferring) lines of rats. *Science* **225**, 78–80.

Waller M. B., McBride W. J., Lumeng L., and Li T.-K. (1982) Induction of dependence on ethanol by free-choice drinking in alcohol preferring rats. *Pharmacol. Biochem. Behav.* **16**, 501–507.

Waller M. B., McBride W. J., Lumeng L., and Li T.-K. (1983) Initial sensitivity and acute tolerance to ethanol in the P and NP lines of rats. *Pharmacol. Biochem. Behav.* **19**, 683–686.

Waller M. B., Murphy J. M., McBride W. J., Lumeng L., and Li T.-K. (1986) Effect of low dose ethanol on spontaneous motor activity in alcohol-preferring and -nonpreferring lines of rats. *Pharmacol. Biochem. Behav.* **24,** 617–625.

Wallgren H. and Barry H. III (1970) *Actions of Alcohol.* Elsevier, Amsterdam.

Wayner M. J., Greenberg I., Tartaglione R., Nolley D., Fraley S., and Cott A. (1972) A new factor affecting the consumption of ethyl alcohol and other sapid fluids. *Physiol. Behav.* **8,** 345–362.

Westerfield W. W. and Lawrow J. (1953) The effect of caloric restriction and thiamin deficiency on the voluntary consumption of alcohol by rats. *Q. J. Stud. Alc.* **14,** 378–384.

Wise R. A. (1973) Voluntary ethanol intake in rats following exposure to ethanol on various schedules. *Psychopharmacologia (Berl.)* **29,** 203–210.

Wise R. A. (1980) Action of drugs of abuse on brain reward systems. *Pharmacol. Biochem. Behav.* **13,** 213–223.

Woods J. H., Young A. M., and Winger G. (1971) The reinforcing property of ethanol, in *Biological Aspects of Alcohol. Advances in Mental Science,* vol. 3, Roach M. K., McIsaac W., and Creaven R. T., eds., University of Texas Press, Austin, pp. 371–387.

Development of an Animal Model of Ethanol Abuse

Genetic Approach

Kalervo Kiianmaa, Petri Hyytiä, and John David Sinclair

1. Introduction

Ethanol abuse is usually understood to include heavy ethanol use as well as the development of medical, legal, social, and/or family problems from drinking. The various problems associated with the excessive use of alcoholic beverages have made ethanol abuse and alcoholism a dominant medical issue in modern society, causing huge costs in health and social care.

Improved medical treatment of ethanol abuse and alcoholism coud potentially reduce the problems related to drinking substantially. For instance, many of the health problems related to alcohol, such as liver diseases, are results of its pathological use and could be reduced or eliminated by an effective treatment for alcohol abuse. The development of such treatments, however, is dependent on an understanding of the mechanisms responsible, and this understanding can only be obtained through research.

Obvious ethical and methodological problems limit research with human subjects on ethanol abuse and alcoholism regardless of whether it is conducted in normal volunteers or alcoholics. Therefore, the development of animal models is important

From: *Neuromethods, Vol. 24: Animal Models of Drug Addiction*
Eds: A. Boulton, G. Baker, and P. H. Wu ©1992 The Humana Press Inc.

for studies on ethanol-related health problems. Rodents and primates have been the most widely used animals in alcohol research, but probably all species are susceptible to alcohol intoxication, acquire tolerance to alcohol, and exhibit withdrawal reactions after the termination of alcohol administration. The use of animal models for studying these effects has been readily accepted (cf Eriksson et al., 1980). There has been increasing evidence that alcohol also produces positive reinforcement for many animals and that animal studies can be used with high predictive validity for screening pharmacological agents that alter alcohol consumption (Sinclair, 1987). Thus, useful animal models of ethanol abuse are available.

Although most laboratory animals voluntarily consume ethanol (Fuller, 1985), the amount of alcohol selected varies among species, among strains within a species, and among individuals within a strain. The variability among individual rats within most strains is rather similar to the variability seen in human alcohol consumption. The variability among strains and among species, on the other hand, was one of the initial indications for a genetic influence on alcohol drinking.

2. Ethanol Abuse and Genetics

Although the basic effects of ethanol consumption on the body are well known, the mechanisms of ethanol's various actions, including etiology of alcoholism, have not been solved. Biomedical research in the field, however, has contributed greatly to our understanding of these phenomena. For instance, it is now generally accepted that an important factor in determining one's liability to abuse ethanol and develop alcoholism is one's genetic constitution, and that genetic factors are responsible for a predisposition of some people to become alcoholics.

2.1. Human Studies

The fundamental issue in studies on the genetic basis of human ethanol abuse and alcoholism is the separation of environmental and genetic effects. Various methods have been used to attempt to solve it. The main approaches have been studies involving families, half-siblings, twins, and adoptees. This work has been covered in several recent books and reviews (Goedde and Agarval, 1987; Kiianmaa et al., 1989; Deitrich and Spuhler, 1984; Schuckit et al., 1985; Fuller et al., 1985; Cloninger, 1987).

2.1.1. Family and Half-Sibling Studies

Family studies have shown that alcoholics are more likely than nonalcoholics to have an alcoholic father, mother, or other relative (Cotton, 1979; Schuckit, 1988,1989). Although these results might be caused by a genetic predisposition to alcoholism, they could also be caused by environmental factors, such as having an alcoholic as a role model or the child-rearing practices of alcoholic parents. More evidence for a genetic risk was obtained when half-sibs with a biological alcoholic parent and a comparative group of half-sibs without an alcoholic biological parent were compared (Schuckit et al., 1972). Those with an alcoholic biological parent exhibited more than fivefold greater incidence of alcoholism compared to controls.

2.1.2. Twin Studies

Studies with identical (monozygous) and fraternal (dizygous) twins have contributed greatly to the clarification of the genetic and environmental influences on alcoholism. The most common approach has been to compare identical and fraternal twins with regard to what percentage of the twins have both members being the same—either both being alcoholics or both not being alcoholics. These studies are based on the hypothesis that any observed differences between identical twins is attributable to environmental effects, whereas a difference between dizygotic twins is presumed to be attributable to both environmental and genetic effects. The study by Kaij (1960) demonstrated that the concordance rate of alcoholism was 54% in identical twins, but only 32% in fraternal twins, whereas in a later study, the figures were 26% in identical twins and 12% in fraternal twins (Hrubec and Omenn, 1981). The greater similarity of identical twins compared to fraternal twins suggests that there is a genetic influence in alcoholism.

Other studies have further demonstrated genetic influences on the alcohol drinking patterns and the effects of ethanol

(Partanen et al., 1966). The Finnish Twin Cohort study examined 879 pairs of identical and 1940 pairs of fraternal twin brothers and found greater similarity among identical twin brothers in the frequency, quantity, and density of drinking than among fraternal twin brothers. Furthermore, this could not be explained by the greater social contact of the identical twin brothers with one another (Kaprio et al., 1987). It has also been shown that identical twins demonstrate almost identical responses to ethanol, whereas fraternal twins usually differ in this respect (Propping, 1977).

2.1.3. Adoption Studies

An even better separation of heredity and environment is obtained by measuring the rate of alcoholism among children of alcoholics and nonalcoholics who were adopted and raised by alcoholic and nonalcoholic foster parents. A large study by Goodwin and coworkers (1973,1977) revealed that adopted-out sons with a biological parent who was an alcoholic were three or four times more likely to be alcoholic than the adopted-out sons of nonalcoholic biological parents. A significant increase was not found, however, among adopted-out daughters.

These findings were confirmed in the Stockholm Adoption Study (Bohman et al., 1981; Cloninger et al., 1981; *see also* Cloninger, 1987; Schuckit et al., 1985), which evaluated the inheritance of ethanol abuse in 862 men and 913 women of known paternity who were adopted at an early age by nonrelatives. Most subjects were separated from their biological relatives in the first few months of life and adopted by nonrelatives before the age of three years. Analysis of ethanol abuse and other parameters obtained from official sources revealed that ethanol abuse was significantly greater in adopted-out sons of alcoholic biological fathers than in sons of nonalcoholic parents. In addition, twice as many of the adopted-out sons of alcoholic biological mothers were ethanol abusers than the sons of nonalcoholic parents. The effect of parental alcoholism on adopted-out daughters was more complex. Ethanol abuse was over three times more frequent in the adopted-away daughters of alcoholic mothers than in the daughters of nonalcoholic parents. The excess of ethanol abusers in the daughters of alcoholic fathers was not significant. Ethanol abuse in adoptive parents was not associated with a greater risk of ethanol abuse in adoptees, indicating that imitation of parental ethanol abuse was not an important cause of alcoholism.

Further analysis of the data demonstrated that the population of alcoholics being studied could be segregated into two prototypic groups (Cloninger, 1987,1989). The two groups were distinguished in terms of ethanol-related symptoms, patterns of inheritance, and personality traits, which have been described in detail elsewhere (Bohman et al., 1987; Cloninger, 1987,1989). The combination of both genetic and environmental factors was necessary for the development of alcoholism in the more common, milieu-limited type of alcoholism (type 1). In the less common, male-limited form (type 2), ethanol abuse was highly heritable in men. This study, therefore, showed that risk for alcoholism is complex and heterogeneous, with both genetic and environmental factors playing a role in its onset.

One can conclude from the human studies described that there is a significant genetic component to ethanol abuse, and that some individuals are at greater risk for becoming alcoholic because of genetic factors.

2.2. Animal Studies

Much of our knowledge about the role of heredity in ethanol abuse comes from animal studies, many of which have been focused on ethanol consumption. In these studies, the animal is presented with a choice between water and an ethanol solution. The initial evidence that there is a genetic component in ethanol preference comes from numerous studies using this procedure with inbred mice.

Inbred strains represent populations of genetically identical individuals that have been produced by more than 20 generations of mating of closely related animals, such as siblings (Belknap, 1980). Theoretically, this has resulted in random fixation of alleles with the animals being homozygotic at all gene loci.

The studies clearly established that various inbred strains of mice widely differ in ethanol intake. The C57BL mice voluntarily consume high amounts of ethanol, whereas C3H/2, A/2,

BALB/c, and DBA/2N strains prefer water (McClearn and Rodgers, 1959; Rodgers and McClearn, 1962; Yoshimoto and Komura, 1987). Demonstration of significant strain differences is itself presumptive evidence of involvement of genotype in ethanol preference. Accordingly, studies using inbred rat strains have also shown differences in ethanol preference (Brewster, 1968; Satinder, 1970), and provided further evidence on the inheritance of ethanol preference.

These early studies provided the stimulus to use selective breeding to study the genetics of ethanol drinking. In this technique, selected individuals are intermated to produce offspring of a desired phenotype.

Selective breeding is an ancient technique, for example, used to develop varieties of dogs and domestic cattle with specific characteristics. Although cattle breeders mostly are interested in selection in one direction, in basic research, selection is usually performed bidirectionally, i.e., selecting for high and low extremes. This was done with respect to the alcohol drinking by rats almost 30 years ago by Eriksson (1968,1969,1971) in the Alko Research Laboratories in Helsinki, Finland, and resulted in the establishment of the AA (Alko, Alcohol) line that voluntarily consumes high amounts of ethanol, and the ANA (Alko, Non-Alcohol) line that chooses water to the virtual exclusion of ethanol. Similar programs were started elsewhere, and led to the UCHA and UCHB rat lines at the University of Chile (Mardones, 1960,1972), and to the P and NP lines, as well as the HAD and LAD lines at the Indiana School of Medicine in Indianapolis (Li et al., 1981,1987; Lumeng et al., 1986). The results of these selection programs clearly indicate that genetic factors influence the animals' ethanol drinking; otherwise, it would not have been possible to separate the traits.

Thus, the animal studies and the research with humans are in agreement in showing that genetic factors influence the amount of alcohol individuals consume. This conclusion has had a major impact on the entire field of alcohol research. It has helped to change the concept of alcoholism held both by people treating alcoholics and by alcoholics themselves. It also has stimulated the drive for prevention of alcoholism and aimed it toward the early identification of high-risk individuals.

From the vantage point of basic research, however, these results have only succeeded in establishing that there is a real phenomenon to be studied—the genetic contribution to the development of alcohol abuse. The next step then must be the delineation of which inherited factors are responsible for a high risk for alcoholism, how are they inherited, and how they interact with environmental, experiential factors to cause some (but not all) high-risk individuals to develop the disease. So far, there have been many suggestions but no solid answers produced for these questions. Nevertheless, much effort is currently being devoted to these questions in alcohol research institutions around the world. Again, the work is proceeding with a combination of studies on humans and on experimental animals.

3. Animal Models in Studies on Ethanol Abuse

An animal model provides an ideal basis for studying a genetically transmitted trait. With an animal model, both the genotype and the environment can be manipulated in a controlled manner, thus allowing conclusions to be made about both forms of influence and about their interactions.

Historically, the development of animal models of ethanol abuse has proceeded by first the publication of criteria for such a model (cf Lester and Freed, 1973; Cicero, 1980) and then by studies showing that a particular model meets these criteria (Li et al., 1987). The proposed criteria basically are:

1. The animal must voluntarily self-administer ethanol in pharmacologically significant amounts;
2. Tolerance to ethanol should be demonstrable as a result of the alcohol intake; and
3. Dependence on ethanol should develop from the intake as evidenced by withdrawal during subsequent abstinence.

Although the establishment of a set of criteria for an animal model of alcoholism has generated much useful research and discussion, this approach in general can be criticized. First, human alcoholism is a complex, heterogeneous disorder, and therefore, it is not justified to have only a single set of criteria for animals to meet without further specifications. Second, the published criteria emphasize tolerance and dependence, assuming that they contribute to the abuse of ethanol. Our knowledge of the mechanisms that are primary in the development of ethanol abuse and alcoholism is sparse, however, and we do not yet know if tolerance and dependence on ethanol play a role in the initiation and maintenance of excessive ethanol intake (Tabakoff, 1991). Animal models of ethanol abuse should not be limited by adherence to unproven assumptions, but rather should be made so that proposed hypotheses about the biological factors contributing to the problem can be tested. An animal model should make it possible to separate different variables, such as ethanol-seeking behavior and voluntary ethanol consumption, sensitivity, tolerance, and dependence, for genetic analysis and determination of which variables contribute to alcohol abuse and how they interact with environmental factors.

Finally, the requirements for an animal model depend on the use for which it is intended. An animal model intended for screening pharmacological agents with potential usefulness in the treatment of alcoholism does not require face validity; i.e., the drinking by the animals does not necessarily need to share characteristics found in human drinking, such as the development of tolerance and dependence, as specified by the criteria. Instead, it must show high predictive validity, i.e., the ability to distinguish those drugs that will suppress drinking by human alcoholics.

3.1. Genetic Correlations

Numerous methods are available for performing genetic studies to answer the stated questions concerning the pharmacogenetics of ethanol abuse. The principal methods currently available are the use of heterogeneous stocks, inbred mice, recombinant inbred strains, and genetically selected lines of animals. The studies are usually planned to determine the extent to which a measured variable—behavioral, biochemical, or neurochemical—correlates with the selected trait, usually a behavioral response to ethanol. A lack of correlation across genotypes is evidence that the selected trait is not caused by the measured variable. A significant correlation does not prove there is a causal link between the variable and the selected trait, but it does allow that a causal link may exist. Therefore, genotypes differing in an ethanol-related trait are powerful tools to study mechanisms of ethanol's effects. For estimation of genetic correlations and interpretation of the experiments, it is important to know the principles of the correlational analysis method (cf Crabbe et al., 1990b).

3.2. Heterogeneous Populations

Probably the simplest way to study genetic correlations is to choose individual animals from a suitable heterogeneous stock that score high or low on an ethanol-related trait and then examine the scores of the same individual animals for another trait. In this method, the same animals are used to test for both traits, and thus, the first test cannot have a permanent effect that could influence the second test (Goldstein, 1989). It has also been pointed out that since, in the case of ethanol-related traits, heritabilities are of moderate magnitude, environmental variability is likely to be more important than genetic variability in determining an individual's phenotype (Deitrich and Spuhler, 1984; Crabbe et al., 1990b). Therefore, demonstration of genetic correlations in heterogeneous stocks may be difficult because the "signal" (i.e., a genetic influence) is obscured by environmental "noise."

3.3. Inbred Mice

A wide selection of inbred mice has been developed for different purposes, such as cancer and genetic research. Inbred mice can provide a powerful research tool; they also are readily available and economical to use (Crabbe, 1989). Differences among inbred strains have been found in their ethanol consumption (McClearn and Rodgers, 1959; Rodgers and McClearn, 1962; Yoshimoto and Komura, 1987) and other ethanol-related traits (Belknap, 1980). Demonstration of differences among the strains has provided evidence for a genetic influence on the phenotypic traits in question.

Inbred mice are not, however, particularly suitable for research on the mechanisms causing ethanol abuse. They can be used to demonstrate genetic correlations between ethanol consumption or other ethanol-related traits and various behavioral, physiological, and biochemical measures. One should keep in mind, however, that because all the genes of a particular strain have become homozygotically fixed in a random manner, there is a high probability that two strains differing in, e.g., ethanol preference, will also differ on any other measured variable even if there is no causal link. Many strains, therefore, are, needed to demonstrate a meaningful genetic correlation between ethanol consumption and a biochemical mechanism (Deitrich and Spuhler, 1984; Crabbe et al., 1990b), e.g., eight inbred strains were used to demonstrate a relationship between ethanol sleep time and Purkinje neuron sensitivity to ethanol (Spuhler et al., 1982).

3.4. Recombinant Inbred Strains

Genetic correlations can also be assessed by using recombinant strains derived from two inbred strains or two selectively bred lines that widely differ in a trait of interest (Deitrich and Spuhler, 1984; Crabbe, 1989). The two strains are first crossed to produce an F_1 generation, and an F_2 generation is then produced by randomly intercrossing individuals of the F_1 generation. Separate inbred strains are then started by brother–sister mating from the F_2 generation. The resulting inbred strains should represent a gradient of the extreme responses of the parental strains. For instance, 16 recombinant inbred strains of mice derived from long-sleep (LS) and short-sleep (SS), selected for differential sensitivity to narcosis from high ethanol doses, were used to study genetic correlations between different behavioral effects of acute ethanol administration (DeFries et al., 1989; Erwin et al., 1990). In these studies, only small correlations were found between the effects of high doses of ethanol and the effects of low doses, suggesting an independence of the underlying inherited mechanisms.

3.5. Lines Selected for Ethanol-Related Traits

As previously described, numerous lines of rats differing in voluntary ethanol intake (UCHA/UCHB; AA/ANA; P/NP; HAD/LAD) have been successfully developed in different laboratories and provide evidence that genetic factors influence ethanol consumption. Selected lines have been widely used to study mechanisms of ethanol abuse because they also provide a convenient means for obtaining animals with extreme reactions to ethanol. Bidirectional selections for numerous generations usually produce two lines with opposite extreme responses to ethanol, whereas the responses of most individuals from the base population fall in between the extremes. Separation of the selected lines is usually complete by 10–20 generations of breeding. Insofar as there is no overlap between the high and low lines on the selected trait, it is possible to use the animals in studies without having to confirm the phenotype of the animals by exposing them to ethanol, which is an important advantage in pharmacological studies.

Selected lines are useful in studies on the mechanisms of ethanol abuse and ethanol's effects. Since the lines are produced by selectively breeding animals from a heterogeneous base population for a specific ethanol-related trait, the selected lines should theoretically differ from each other only in the trait upon which selection has been applied, and in traits that are related to the selected trait either causally or through genetic linkage. Although this theoretical ideal is only reached with an infinitely large breeding population (as discussed below), selected lines are valuable tools with which to search for the existence of genetic correlation between the selected trait and a specific biochemical, neurochemical, or behavioral trait, and to test hypotheses regarding underlying causes of ethanol abuse (Deitrich and Spuhler, 1984; Crabbe et al., 1985, 1990b; Schuckit et al., 1985; Deitrich, 1990).

Early studies used selected lines differing in voluntary ethanol consumption in the investigation of ethanol abuse. Subsequently, lines differing in many other characteristics related to the effects of ethanol, such as sensitivity to ethanol, capacity to develop tolerance, liability to develop physical dependence, and withdrawal severity after chronic ethanol administration, have been developed, partly to test the possible contribution of these factors to the development of ethanol abuse and partly because of interest in the factors themselves. The lines and related work have been recently discussed elsewhere (McClearn et al., 1981;

Deitrich and Spuhler, 1984; Li et al., 1987; Crabbe, 1989; Kiianmaa et al., 1989; Phillips et al., 1989; Sinclair et al., 1989; Deitrich and Pawlowski, 1990), and the lines will be only listed here.

3.5.1. Voluntary Ethanol Consumption

UCHA and UCHB rats are the oldest lines selected for differential voluntary ethanol intake. They were developed by J. Mardones in Santiago, Chile (Mardones, 1960,1972).

AA (Alko, Alcohol) and ANA (Alko, Non-Alcohol) rats have been selectively outbred for their voluntary intake of 10% ethanol solution in a free-choice situation (Eriksson, 1968,1969,1971; Sinclair et al., 1989).

P (Preferring) and NP (Non-Preferring) rats, each line started originally from a single breeding pair, were also developed for differences in their preference for 10% ethanol solution (Li et al., 1981,1987).

HAD (High Alcohol Drinking) and LAD (Low Alcohol Drinking) rats are the result of replicating the development work of the P and NP lines (Lumeng et al., 1986).

SP (Sardinia Preferring) and SNP (Sardinia Non-Preferring) rats are rather recent lines developed in Cagliari, Italy, and differ in their preference for 10% ethanol solution (Fadda et al., 1989).

3.5.2. Sensitivity to Ethanol

LS (Long-Sleep) and SS (Short-Sleep) mice have been selectively outbred for the duration of ethanol-induced loss of righting reflex (sleep time), and are currently at the Institute for Behavioral Genetics in Boulder, Colorado (McClearn and Kakihana, 1981; Phillips et al., 1989; Deitrich, 1990).

FAST and SLOW mice differ in the stimulatory effect of ethanol (2 g/kg) in the open field apparatus (Crabbe et al., 1987b,1990a; Phillips et al., 1989,1991).

HOT and COLD mice differ in the hypothermic effect of an acute dose (3 g/kg) of ethanol (Crabbe et al., 1987a,1990a; Phillips et al., 1989). Both Fast/Slow and Hot/Cold mice are outbred lines.

AT (Alcohol Tolerant) and ANT (Alcohol Non-Tolerant) rats are outbred lines selected for differential ethanol-induced (2 g/kg) impairment of motor performance on the tilting-plane (Eriksson and Rusi, 1981; Eriksson, 1990).

HAS (High Alcohol Sensitive) and LAS (Low Alcohol Sensitive) rats have been selectively outbred for the duration of ethanol-induced (3 g/kg) loss of righting reflex in a test similar to that used in the breeding of LS and SS mice (Spuhler et al., 1990).

3.5.3. Ethanol Withdrawal

SEW (Severe Ethanol Withdrawal) and MEW (Mild Ethanol Withdrawal) mice have been selectively outbred for the severity of ethanol withdrawal syndrome (McClearn et al., 1982; Phillips et al., 1989). Physical dependence is produced by providing a liquid diet containing ethanol for 9 d, and withdrawal severity is scored on a battery of tests.

WSP (Withdrawal Seizure Prone) and WSR (Withdrawal Seizure Resistant) mice have been selected for severe and mild signs of withdrawal induced by handling after 3 d of chronic inhalation of ethanol vapor (Crabbe et al., 1985; Crabbe, 1989; Phillips et al., 1989). WSP and WSR mice are also outbred lines.

4. Development of an Animal Model

Various factors that have to be considered in the development of an animal model of ethanol abuse via bidirectional selection of outbred animal lines will be discussed below with special reference to our experience in the selection of both AA/ANA and AT/ANT rat lines (Eriksson and Rusi, 1981). Selection of other animal lines and related work have also been widely described in the literature (cf McClearn et al., 1981; Deitrich and Pawlowski, 1990). The theory of selection experiments has been discussed by Falconer (1989), but *see also* DeFries (1981) and Roberts (1981).

4.1. Bidirectional Selection

Traditionally, unidirectional selective breeding has been used in agriculture to produce a single line with extreme characteristics in a desired direction, e.g., high yield. Although this approach could also be used in basic research, the bidirectional selective breeding procedure has typically been followed, producing a pair of lines, one with high scores on the phenotypic measure and one with low scores (Hyde, 1981). One reason for this in alcohol research is that it is important to determine not only the genetically determined factors contributing to high drinking, but also the genetic factors that provide protection against high drinking.

Another reason is that the "crop" in basic research is, at least initially, significant line differences, i.e., the discovery of factors correlated with and possibly causing differences in the selected trait. The yield of this crop is maximized by the bidirectional approach. The bidirectional approach can be fruitful if either the high characteristics, the low ones, or both are under genetic control. Furthermore, the variabilities within the high line and within the low line are both likely to be much smaller than those within the base population. The ability to detect significant differences from the base population in a single high line is obscured by this large variability in the base population. However, in comparisons between the high and low lines, this "noise" is eliminated and the experiment has much more power for disclosing correlates of the selected trait.

4.2. Selection Criteria

4.2.1. Choice of the Phenotype

Considering the central role of ethanol self-administration in the development of ethanol abuse and alcoholism, probably the most obvious basis for selecting animals would be voluntary ethanol consumption, and this has been the most commonly used phenotypic measure. Other responses to ethanol have also been used as phenotypic measures in order to test hypotheses about their relation to ethanol abuse. For instance, after the initial demonstration of differences among inbred mice and selected rat lines in ethanol consumption, several studies on correlates of these consumption differences found differences in ethanol sensitivity. This raised the hypothesis that the differences in voluntary ingestion might be related to the susceptibility of the animals to the effects of ethanol, and in turn, resulted in the development of the LS and SS mice (McClearn and Kakihana, 1981), and the AT and ANT rat lines (Eriksson and Rusi, 1981). The phenotypic measure selected could basically also be any other measurable response to ethanol, depending on the hypothesis to be tested (Goldstein, 1981), including traits related to ethanol metabolism, acute and acquired tolerance, dependence, or pathology (cf McClearn et al., 1981).

The selection process becomes complicated, however, if one decides to select for a trait, such as a neurochemical measure, that requires killing the animal. Obvious solutions in such a case would be

1. Continuing the line by rebreeding those parents whose first progeny showed the desired phenotype;
2. Producing the progeny before testing the parents, then continuing the line by breeding those progeny whose parents showed the phenotype (McClearn et al., 1981); or
3. Selecting breeders on the basis of measurements on their siblings (Falconer, 1989).

The AA and ANA rats have been selected for their voluntary ethanol consumption. The ethanol drinking is tested in individual cages when the animals are 3 mo old. Ethanol solution (10% v/v) is provided as the sole drinking fluid for the first 10 d, and then the animals are given a free choice between it and tap water for 3 wk, with the consumption during the third week being used in the selection of breeders (Eriksson, 1969). The positions of drinking tubes are reversed weekly. Food is available at all times.

The initial 10-d period with only alcohol available is probably not important, since several studies have shown that similar AA/ANA line differences develop if it is omitted. It is important, however, that the phenotypic measure is alcohol selection after prolonged (>3 wk) prior ethanol experience rather than the ethanol selection during the first day or so of access. In nonselected rats, the initial acceptance of alcohol during the first few days is closely related to taste reactions and is not highly correlated with the eventual levels chosen after 3 wk (Kampov-

Polevoy et al., 1990). The alcohol intake by the AAs, like that of most other rats and also that of hamsters and monkeys, starts out relatively low and then gradually increases to an asymptotic level during the first month of access; this increase probably involves the animals learning about the effects of alcohol. Lines developed on the basis of the initial acceptance of alcohol might have some relevance to the initial drinking by human teenagers, but lines developed on the basis of their drinking level after several weeks of experience probably have more relevance to human alcohol abuse after long-term prior experience.

From the point of genetic selection, the concentration of ethanol used to determine the phenotype does not seem to be a methodological factor of primary importance. Eriksson (1969) showed that rats consume roughly the same amount of ethanol over a certain concentration range (i.e., 5–10 %) depending on the strain to be tested. Some authors have preferred to present animals a series of concentrations for determination of individual selection threshold (Cicero and Myers, 1968). Subsequent work has shown, however, that one concentration (10%) gives roughly the same results as 6–8 concentrations (Drewek, 1980).

An important methodological factor, however, is the particular measure of ethanol consumption used for selecting breeders, since different phenotypic measures may lead to undesired results. Initially, the AA and ANA rats were selected on the basis of the amount of absolute ethanol consumed daily/U of body wt (g/kg), because this is the usual measure for administration of pharmacologically equivalent doses of ethanol. This phenotypic measure, however, applies selective pressure for differences in body wt, favoring selection of light AA animals and heavy ANA rats. Such a line difference in body wt did develop (Eriksson, 1969,1981; Eriksson and Närhi, 1973; Eriksson and Rusi, 1981). Consequently, after the 19th generation, the lines were selected on the basis of their energy preference, i.e., the energy derived from ethanol divided by the total energy from food and ethanol. This resulted initially in the disappearance of the difference in body wt and then, by the 29th generation, a reversal, with the AAs weighing more than the ANAs. After this, a combined measure of ethanol preference (the amount of

10% ethanol solution consumed as a percentage of the total fluid intake, E/T), the amount of ethanol consumed in g/kg body wt, and energy preference was used. Then, from the 39th generation on, the pharmacological role of ethanol was again emphasized, with the selection being based only on the g/kg measure, and the original line difference in body wt with the AAs being lighter was again gradually developed (Fig. 1). The differences in body wt have produced some confusion in interpreting the results, but they are not considered to have caused any major problems in most of the research conducted on the lines. It should be noted that the AAs have drunk markedly more alcohol than the ANAs, regardless of how alcohol consumption is measured and regardless of whether they weighed less, more, or the same as the ANA animals.

Another type of problem appeared during the development of SS and LS mice that were selected for the duration of ethanol-induced loss of righting reflex. Because the dose of ethanol used originally (3.3 g/kg) failed to block the righting reflex in the F_7 generation of the SS mice, the dose had to be increased to avoid zero scores, with the result that different amounts of ethanol were administered: The SS mice received 4.7 g/kg whereas the LS mice were given 3.8 g/kg (McClearn and Kakihana, 1981).

Future selective breeding programs might use different procedures and schedules of alcohol drinking in an attempt to produce animal models for specific forms of human drinking. For example, it has been suggested (Kampov-Polevoy et al., 1990) that it might be possible to develop one rat line with particular relevance to Cloninger's type 1 alcoholism and another for type 2 alcoholics. Another possibility would be an animal model of binge drinking. The AA, as well as the UChA, P, and HAD lines have all been developed for high levels of alcohol drinking when ethanol is available and is drunk every day. One might assume that these lines would have more relevance to that human drinking which also occurs on an everyday basis; i.e., they may be better models for French alcohol consumption than for that in Finland, where alcohol is typically consumed less frequently, but in larger amounts per drinking occasion. It has been found, however, that most rats (but not AAs; Sinclair and Li, 1989) with

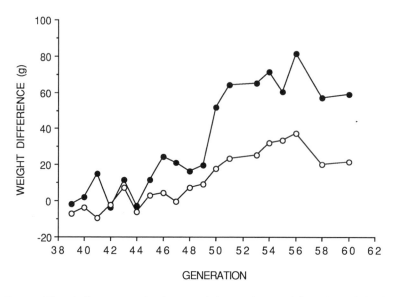

Fig. 1. The difference in body wt of the males and females of the high-drinking AA and low-drinking ANA rat lines (ANAs–AAs) during the first week of free choice between ethanol solution and water from generation F_{39} to F_{60}.—•— males; —o— females.

prolonged prior ethanol experience show a large temporary increase in alcohol drinking immediately after a week or more in which they have not had alcohol available. This "alcohol-deprivation effect" probably plays a role in human binge drinking, also. Consequently, a rat line selectively bred for especially high alcohol intake immediately after alcohol deprivation might have more relevance to the Finnish or Scandinavian styles of drinking and perhaps to the intake by binge alcoholics.

4.2.2. Selection for Multiple Traits

The problems related to the development of the AT and ANT rat lines were even more complex. The purpose of the selection was to produce two rat lines that show differential acute ethanol intoxication at the same level of blood ethanol (Eriksson and Rusi, 1981). Originally, the rats were tested for ethanol-induced impairment of motor performance on the rotarod and the tilting plane. The use of two tests on intoxication meant selecting the rats for multiple traits. Furthermore, the performance of the animals on the rotarod and tilting plane is also affected by body wt and the blood alcohol level (Arvola et al., 1958). Consequently, the selection criteria were aimed at producing maximal line differences on the two intoxication tests and no line differences in body wt or blood alcohol concentrations, with the major emphasis placed on the tilting plane and blood ethanol results. In later generations, the performance of the animals on the tilting plane was the only measure of ethanol intoxication. Ideally, the most efficient method for multiple traits would be index selection, in which all the measured traits are combined into a single numerical index (McClearn et al., 1981), but this procedure was not used in the development of the AT and ANT lines.

The tilting plane measure of intoxication used in the selection of AT and ANT rats is a "change score": the change in the sliding angle after ethanol administration, relative to the value obtained before ethanol administration. It has been argued that change scores result in increased unreliability of response measures (Nagoshi et al., 1986), but this problem has not been encountered with the AT and ANT lines, nor was it evident in the development of the Hot and Cold mouse lines developed for differential changes in ethanol-induced hypothermia (Crabbe and Weigel, 1987). If only the sliding angle after ethanol had been used as the phenotypic measure for the AT and ANT animals, it is possible that lines would have been developed differing in motor coordination, physical strength, and other factors affecting sober baseline performance rather than ones differing in their reaction to ethanol.

Other ways in which selection for multiple traits may be attempted include tandem selection, in which each trait is selected successively in turn in alternate generations, and by independent culling, where the animals to be bred must reach a certain, minimum standard on all traits to be tested (McClearn et al., 1981).

4.3. Foundation Stock

The foundation stock from which the selection is started must be a highly genetically heterogeneous population of animals to ensure sufficient genetic variability for the selection and to maximize the final gains from the selection. Earlier selection programs were impaired by the foundation stocks being of uncertain heterogeneity (McClearn, 1981). The foundation stock could be established for the particular project as was done, for instance, with the AT and ANT lines (Eriksson and Rusi, 1981). In this particular case, Wistar-, Sprague-Dawley-, and Long-Evans-originated animals were crossbred to produce a highly heterogeneous "Mixed" strain from which the AT and ANT lines were developed. The other possibility is to use one of the established heterogeneous stocks of known genetic origin. The heterogeneous stock of mice (HS/Ibg), which is maintained at the Institute for Behavioral Genetics, University of Colorado, was derived from intercrosses of eight inbred mice strains (McClearn et al., 1970). The genetically heterogeneous stock of rats (HS/NIH) was also developed from intercrosses of eight inbred rat strains (Hansen and Spuhler, 1984). It has been well characterized in terms of ethanol-related phenotypes (cf Spuhler et al., 1990), and is available from the Animal Resource Center at the National Institutes of Health (NIH) in Bethesda, MD.

4.4. Methods of Selection

4.4.1. Individual and Family Selection

In most cases, the method of selection is individual selection, but it is also possible to select families or to use within-family selection (cf Roberts, 1981; Falconer, 1989). In individual selection, a specific animal is chosen for breeding on the basis of its own phenotype. This is a simple method to execute and produces a rapid response to selection, but it may contribute to inbreeding since the selected individuals may come from only a few families. Family selection uses whole "good" families to provide the breeding individuals. This method may be valuable when heritability of the selected trait is low and the family mean is a more reliable guide than an individual's phenotype. In practice, family selection is of little importance in the case of laboratory animals because it demands much space and at the same time limits the stock to a few families, making inbreeding unavoidable.

4.4.2. Within-Family Selection

The method that has been most recommended is within-family selection (cf McClearn et al., 1981). In this method, every family contributes the same number of parents to the subsequent generation, and the breeding individuals are selected from the families on the basis of their deviation from their own family mean. The basis of this method is to select breeding individuals exposed to similar environmental conditions as are all the members of the same family. It also provides a reliable estimate of the phenotype, and it reduces inbreeding because every family contributes the same number of parents for the next generation. This method may be less efficient in the short-term than individual selection because the initial response to selection may be reduced, but it probably gives better long-term results because the longer retention of more alleles increases the eventual limits that may be reached.

4.5. Response to Selection

4.5.1. Factors Affecting the Response to Selection

The rate at which selected lines diverge from the original base population on the phenotypic measure—i.e., the response to selection—depends on numerous factors, most of which are, to some extent, under the control of the experimenter. The rate depends on the standard deviation of the phenotype, the heritability of the trait, and the intensity of the selection. If the individuals in the population vary much among themselves, which usually is the case in genetically heterogeneous populations, the phenotypic standard deviation is great, and one can expect a rapid response to selection. One should also keep in mind the benefits of reducing measurement error and of obtaining an accurate estimate of the phenotype produced by reducing environmental variance (Roberts, 1981). Therefore, unnecessary changes in the environment should be avoided by keeping the physical circumstances and diet constant and by minimizing disease. An example of the problems produced by environmental changes can be seen in our AA and ANA lines (Hilakivi et al., 1984). From the F_{20} to F_{28} generations, the commercial diet used was contaminated by an aldehyde-dehydrogenase inhibitor, cyanamide, which suppresses alcohol drinking. When this problem was corrected, the alcohol consumption of the AAs rose again. The alcohol intake by the ANAs also increased and did not return to the lower level seen prior to the cyanamide-contaminated diet, despite many generations of selection, until after a "revitalization" of the lines.

The heritability of the trait describes the relative importance of heredity in determining the observed variation in the population. It also predicts the similarity of parents and their offspring (Falconer, 1989; DeFries, 1981; Roberts, 1981). Therefore, the response to selection depends on the heritability of the character. Many ethanol-related traits have been estimated to have relatively low heritabilities (cf Crabbe et al., 1990b), but they still are responsive to genetic selection if there is a high phenotypic standard deviation between individuals.

The intensity of the selection means the proportion of the animals within the population selected for breeding. The response to selection is more rapid if a smaller proportion of the population is selected for breeding because the breeders then will be more extreme individuals. Low heritability can, thus, also be compensated by selecting more intensely. Very intense selection maximizes the initial rate of response, but this is achieved at the expense of the final gain. After the initial divergence, selection in later generations operates largely on recessive genes and ones with only a minor contribution to the desired phenotype, but these alleles, if relatively rare, may be lost at an early stage when only a few breeders are used.

The actual rate of divergence, in terms of time rather than generations, obviously also depends on the interval between generations. More divergence can be produced in, e.g., 2 yr if six generations are produced than if there are only three. Therefore, to save time in the selection, the phenotype should be determined as early as possible and the generations should follow each other as rapidly as possible (cf Roberts, 1981). On the other hand, it probably would be counterproductive to select for alcohol drinking in immature animals or to reduce the duration of prior ethanol drinking experience in order to decrease the generation interval.

Two opposing factors influence the optimal generations for testing for correlates of the selected trait. In order to include correlates of all factors influencing the phenotype, it is desirable to wait as many generations as possible so that all possible contributing alleles will have been segregated. Tests made after only a few generations when a significant phenotypic line difference first appears will miss correlates of recessive or less important alleles that have not yet been separated. On this basis, it is advisable to wait at least 10 generations (cf McClearn et al., 1981), or even 20 or 30 generations (cf DeFries, 1981) if further response to selection is still being observed.

The other factor is random genetic drift and the accidental loss or fixation of alleles, which can produce spurious line differences that are not causally or genetically linked to the phenotypic difference. The probability of significant, but spurious line differences arising is very low in the early generations, but increases rapidly after about the 10–15th generation, depending on the number of breeders maintaining each line (Sinclair et al., 1987). Therefore, tests for correlates after 20 or 30 generations (or 61 generations for the AA and ANA lines) are quite likely to be confounded by spurious line differences.

Considering both factors, the best time to test for correlates usually appears to be after a significant difference on the select trait has developed and before random genetic drift is likely to have produced many spurious line differences. Many factors, however, can influence the timing. Testing would have to be delayed for traits with low heritability. Intense selection shifts the optimal "window" downward, so that testing could begin at an earlier generation, but also becomes confounded by spurious line differences earlier. Using a larger number of breeders delays the effects of genetic drift. For additional information about issues that arise in estimation of genetic correlations and interpretation of the experiments, *see* Crabbe and coworkers (1990b) and Deitrich (1990).

4.5.2. Response to Selection for Ethanol Consumption

The selection of AA and ANA rats was initiated from a colony of 500–700 rats. The breeding animals were selected from 10–15 families to ensure large individual variation. From the paternal generation on, in every generation, the highest-drinking male and female from each AA litter were selected as AA breeders and mated across litters within the line following the procedure of within-family selection. Similarly, within the ANA line, the lowest-drinking male and female rat from each litter were selected as breeders. For each line, 20 breeding pairs were used. The population size tested in each generation for each line was 80–140 rats (Eriksson, 1969; Eriksson and Rusi, 1981).

The mean (± SD) alcohol intake in the F_{16} generation was in AA males, 7.5 ± 1.65; AA females, 9.87 ± 3.47: ANA males, 0.39 ± 0.23; and ANA females, 0.71 ± 0.39 g/kg (Eriksson, 1971). This appears to about the time when the lines had reached the selection plateau, i.e., the point at which there is no further response to selection. This, however, is obscured by the changes produced by the cyanamide-containing diet from the 20th to 28th generation. Subsequently, the mean alcohol consumption from the F_{31-37} generations was in AA males, 5.57; AA females; 7.20; ANA males, 1.49; and ANA females, 1.34 g/kg/d.

4.6. Problems Related to Continued Selection

4.6.1. Inbreeding

Loss of genetic diversity as a result of inbreeding and random genetic drift is a problem in selected lines maintained with a relatively small number of breeders. Inbreeding reduces genetic variance and, consequently, the response to selection within a selected line, and also the susceptibility to disease and fertility. Extensive inbreeding may result in the loss of lines owing to inability to breed (McClearn et al., 1981). Therefore, it is important to minimize inbreeding during the course of a selection experiment. To minimize inbreeding, one should avoid sib mating and keep the size of the population and the number of families as large as possible. It has been suggested that a minimum for a within-family selection is eight families per line (cf McClearn et al., 1981).

The problem of inbreeding became evident in the AA and ANA lines by about the 30th generation as a reduction in fertility, a decrease in the number of surviving progeny, and an increased susceptibility to disease (Hyytiä et al., 1987). Furthermore, a study estimating polymorphism at 25 enzyme loci of AA and ANA rat lines gave evidence suggesting that the outbreeding techniques used had not succeeded in preventing a decrease of heterozygosity (Eriksson et al., 1976). These changes were countered by a revitalization program to introduce new genetic variability into the lines by outcrossing (Hilakivi et al., 1984; Hyytiä et al., 1987). In this program, the F_{37} AA and ANA breeders were crossed with F_1 hybrids from Brown Norwegian and Lewis rats. The revitalization restored the fertility of both lines (Hyytiä et al., 1987). In the generations since revitalization, i.e., F_{39-60}, *the coefficient of inbreeding* (cf Falconer, 1989) can be estimated to have grown about 17%, i.e., 0.8% per generation, but so far there has been no significant reduction in litter size after revitalization (Fig. 2).

The line difference in ethanol consumption between the AA and ANA rats was reduced, but remained significant immediately after the revitalization (Hyytiä et al., 1987). Subsequently, the ethanol intake of the AA line slowly increased, going above the level just prior to revitalization and eventually reaching approximately the level observed in the F_{16} generation. There is currently no indication that a new plateau has been reached (Fig. 3). The large sex difference seen prior to revitalization, with females drinking more than males, has disappeared. The intake by the ANA line was soon reduced significantly below that seen just prior to revitalization and also somewhat lower than that in the F_{16} generation. They have shown no further response to selection after generation F_{46}, but whether this indicates a plateau in genetic terms or merely the bottom limit for measuring alcohol drinking cannot be determined. Introduction of new genetic variability into the lines via outcrossing has, thus, also resulted in a gain over the selection plateau observed immediately

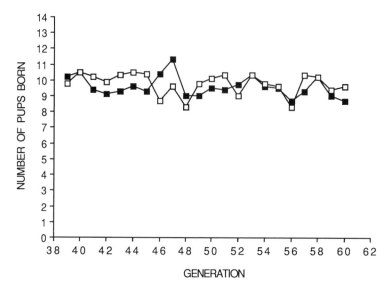

Fig. 2. The mean litter size of the AA and ANA rat lines after revitalization, in the F_{39-60} generations. —■— AA; —□— ANA.

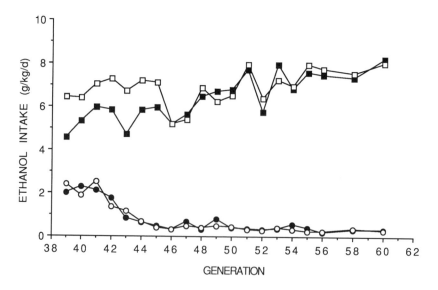

Fig. 3. Response to selection after revitalization of the AA and ANA rat lines. The data points represent the mean alcohol consumption, as g/kg body wt, during the third week of free choice between 10% (v/v) ethanol solution and water. —■— AA male; −□— AA female; —•— ANA male; —o— ANA female.

before revitalization, perhaps by restoring factors lost during the generations on cyanamide-contaminated food.

4.6.2. Spurious Line Differences

Another problem produced by inbreeding and random genetic drift when maintaining selected lines over many generations is the progressive increase in the probability of finding significant differences between the lines produced by chance rather than by selection (DeFries, 1981). For instance, by the F_{27} generation, the albino allele had been completely eliminated from the ANT line and the agouti allele from the AT line (Sinclair et al., 1987). Computer simulations showed that the observed loss of agouti coloring in the AT line was a very likely chance occurrence and, therefore, probably was only a spurious line difference, unrelated to ethanol sensitivity.

Random genetic drift thus caused the observation of false positive correlations between the selected trait and the measured variable, thus limiting the usefulness of the lines as a tool for research (Goldman et al., 1985; Sinclair et al., 1987). This is particularly a problem with variables determined by a single gene locus, such as many coat-color characteristics. With polygenically determined variables, drift at one locus may be counteracted by drift at other loci, making it less likely that significant spurious line differences will be observed. Nevertheless, they can still develop, and measures need to be taken to lessen their detrimental effect.

4.6.3. Control Line

In the previously described example, it was possible to estimate the probability of accidental loss of the agouti allele because the coat color of the F_1 generation had been recorded. Estimates can also be made if an unselected control line is maintained. Furthermore, a control line maintained at the same effective population size as the selected lines will accumulate inbreeding at the same rate, without the effects of selection, thus providing an indication of how much all of the lines have been affected by random genetic drift (cf McClearn et al., 1981). The control line also provides a means for monitoring the effects of environmental changes. In addition, since the control line is intermediate between the high and low line on the selected trait, the control line can be expected to lie between the high and low lines on any measured variable that is a true genetic correlate of the selected trait. Another reason for maintaining a control line is to give an indication of the direction of divergence. For example, with lines developed for differential alcohol consumption, it is important to know whether genetics contributes only to particularly high alcohol drinking, only to an aversion to alcohol, or to both. Similarly, when a correlate is discovered, it is important to know whether it contributes to high drinking, low drinking, or differentially to both. Some indication of the direction of divergence or influence can be obtained from records kept about the original base population or by measurements on animals similar to the base population, but a better method is to have a control line maintained under the same conditions as the selected lines.

4.6.4. Replicate Lines

Some form of replication is considered essential for the study of correlated responses in order to deal with the problem of spurious line differences. The most commonly suggested method is the maintenance of replicate lines. It has been suggested that an optimal bidirectional selection design has an unselected control line, in addition to high and low lines, and a replicate of all three lines, making a total of six lines (cf McClearn et al., 1981; Hyde, 1981).

Typically, earlier selection experiments for behaviors (cf Hyde, 1981) or ethanol-related traits (cf Crabbe, 1989) have not involved replicate lines. This probably has limited the conclusions derived from these populations, and may be at least a partial reason for conflicting results from different laboratories. It is possible to repeat tests made with one line pair in another line pair developed elsewhere with similar criteria; for example, we have replicated findings made on the AA and ANA lines by studying the P and NP lines. This is only a partial solution because a failure to replicate may be caused only by differences in procedure or baseline population and do not necessarily point to spurious results. More recent programs of selective breeding for ethanol-related traits have, however, used sophisticated selection designs: At least, the HAS/LAS, HAD/LAD, WSP/ WSR, SEW/MEW, Cold/Hot, and Fast/Slow lines have replicates and controls (Crabbe, 1989).

Replicate lines do not, however, provide as good a solution as is commonly believed and may produce a false sense of security, because the same spurious line difference can develop rather easily in both sets of high and low lines. For example, the probability of a spurious line difference concerning the agouti coat color having arisen in 27 generations of the AT and ANT lines was estimated to be nearly 0.50 (Sinclair et al., 1987). If there were replicate lines for the ATs and ANTs, the probability of the same line difference having developed in both sets, therefore, would be about 0.25. Thus, finding the same result in both sets is not grounds for concluding that it is not caused simply by random genetic drift. Furthermore, the rate of genetic drift increases rapidly as the number of breeders is decreased, so the probability of a spurious line difference developing in a set of replicate lines with 10 breeders each is much higher than the probability of the same difference developing in a pair of lines with 20 breeders each.

4.6.5. Restarting the Selection

As previously discussed, spurious line differences are not a major problem in the early generations, but become increasingly serious after about the 10–15th generation. Consequently, another way to cope with spurious line differences is to restart the selected lines periodically, and test for correlates only in those generations after the lines have diverged significantly on the phenotypic trait, when most line differences related to the phenotype can be observed, and before the time when spurious ones are likely to have developed (Sinclair et al., 1987). More specifically, one set of high and low lines is selectively bred. When they have diverged significantly; a second pair is started. The first pair is used in research looking for correlates until the second pair has diverged significantly, then the first pair is terminated and its laboratory space used for starting a third pair. The program thus continues with repeated restarting, either from the control foundation stock or from crosses of the high and low lines, always providing animals for correlate testing that are in the window when such testing is most efficient. A line difference observed in one set can be replicated in the next set and then again in the next, and so on, to provide any desired degree of confidence that it is not spurious.

References

Arvola A., Sammalisto L., and Wallgren H. (1958) A test for level of alcohol intoxication in the rat. *Q. J. Stud. Alcohol* **19**, 563–572.

Belknap J. K. (1980) Genetic factors in the effects of alcohol: neurosensitivity, functional tolerance and physical dependence. in *Alcohol Tolerance and Dependence* (Rigter H. and Crabbe J. C., eds.) Elsevier, Amsterdam, pp. 157–180.

Bohman M., Sigvardsson S., and Cloninger C. R. (1981) Maternal inheritance of alcohol abuse: cross fostering analysis of adopted women. *Arch. Gen. Psychiatry* **38**, 965–969.

Bohman M., Cloninger R., Sigvardsson S., and von Knorring A-L. (1987) The genetics of alcoholism and related disorders. *J. Psychiatr. Res.* **21**, 447–452.

Brewster D. J. (1968) Genetic analysis of ethanol preference in rats selected for emotional reactivity. *J. Hered.* **59**, 283–286.

Cicero T. J. (1980) Animal models of alcoholism? in *Animal Models in Alcohol Research* (Eriksson K., Sinclair J. D., and Kiianmaa K., eds.) Academic Press, London, pp. 99–117.

Cicero T. J. and Myers R. D. (1968) Selection of a sinle ethanol test solution in free-choice studies with animals. *Q. J. Stud. Alcohol* **29**, 446–448.

Cloninger C. R. (1987) Neurogenetic adaptive mechanisms in alcoholism. *Science* **236**, 410–416.

Cloninger C. R. (1989) Clinical heterogeneity in families of alcoholics, in *Genetic Aspects of Alcoholism* (Kiianmaa K., Tabakoff B., and Saito T., eds.) The Finnish Foundation for Alcohol Studies, vol. 37, Helsinki, pp. 55–65.

Cloninger C. R., Bohman M., and Sigvardsson S. (1981) Inheritance of alcohol abuse: cross fostering analysis of adopted men. *Arch. Gen. Psychiatry* **38**, 861–868.

Cotton N. S. (1979) The familial incidence of alcoholism. *J. Stud. Alcohol* **40**, 89–116.

Crabbe J. C. (1989) Genetic animal models in the study of alcoholism. *Alcohol: Clin. Exp. Res.* **13**, 120–127.

Crabbe J. C. and Weigel R. M. (1987) Quantification of individual sensitivities to ethanol in selective breeding experiments: Difference scores versus regression residuals. *Alcohol: Clin. Exp. Res.* **11**, 544–549.

Crabbe J. C., McSwigan J. D., and Belknap J. K. (1985) The role of genetics in substance abuse, in *Determinants of Substance Abuse* (Galizio M. and Maisto S. A., eds.), Plenum, New York, pp. 13–64.

Crabbe J. C., Kosobud A., Young E. R., Tam B. R., and McSwigan J. D. (1985) Bidirectional selection for susceptibility to ethanol withdrawal seizures in *Mus musculus*. *Behav. Genet.* **15**, 521–536.

Crabbe J. C., Kosobud A., Tam B. R., Young E. R., and Deutsch C. M. (1987a) Genetic selection of mouse lines sensitive (COLD) and resistant (HOT) to acute ethanol hypothermia. *Alcohol Drug Res.* **7**, 163–174.

Crabbe J. C., Young E. R., Deutsch C. M., Tam B. R., and Kosobud A. (1987b) Mice genetically selected for differences in open-field activity after ethanol. *Pharmacol. Biochem. Behav.* **27**, 577–581.

Crabbe C., Feller D. J., and Phillips T. J. (1990a) Selective breeding for two measures of sensitivity to ethanol, in *Initial Sensitivity to Alcohol* (Deitrich R. A. and Pawlowski A. A., eds.) NIAAA Research Monograph 20, US Government Printing Office, Rockville, Md, pp. 123–150.

Crabbe J. C., Phillips T. J., Kosobud A., and Belknap J. K. (1990b) Estimation of genetic correlation of experiments using selectively bred and inbred animals. *Alcohol: Clin. Exp. Res.* **14**, 141–151.

DeFries J. C. (1981) Current perspectives on selective breeding: example and theory, in *Development of Animal Models as Pharmacogenetic Tools* (McClearn G. E., Deitrich R. A. and Erwin E., eds.), NIAAA Research Monograph 6, US Government Printing Office, Rockville, MD, pp. 11–35.

DeFries J. C., Wilson R., Erwin V. G., and Petersen D. R. (1989) LS X SS recombinant inbred strains of mice: Initial characterization. *Alcohol: Clin. Exp. Res.* **13**, 196–200.

Deitrich R. A. (1990) Selective breeding of mice and rats for initial sensitivity to ethanol: contributions to understanding of ethanol's actions, in *Initial Sensitivity to Alcohol* (Deitrich R. A. and Pawlowski A. A., eds.) NIAAA Research Monograph 20, US Government Printing Office, Rockville, MD, pp. 7–59.

Deitrich R. A. and Pawlowski A. A., eds. (1990) *Initial Sensitivity to Alcohol*, NIAAA Research Monograph 20, US Government Printing Office, Rockville, MD.

Deitrich R. A. and Spuhler K. (1985) Genetics of alcoholism and alcohol actions, in *Research Advances in Alcohol and Drug Problems*, vol. 8 (Smart K. and Sellers E. M., eds.) Plenum, New York, pp. 47–98.

Drewek K. J. (1980) Inherited drinking and its behavioral correlates, in *Animal Models in Alcohol Research* (Eriksson K., Sinclair J. D., and Kiianmaa K., eds.), Academic Press, London, pp. 35–49.

Eriksson C. J. P. (1981) Finnish selection studies on alcohol-related behaviors: Factors regulating voluntary alcohol consumption, in *Development of Animal Models as Pharmacogenetic Tools* (McClearn G. E., Deitrich R. A., and Erwin G., eds.) NIAAA Research Monograph 6, US Government Printing Office, Rockville, MD, pp. 119–145.

Eriksson C. J. P. (1990) Finnish selective breeding studies for initial sensitivity to ethanol: Update 1988 on the AT and ANT rat lines, in *Initial Sensitivity to Alcohol* (Deitrich R. A. and Pawlowski A. A., eds.) NIAAA Research Monograph 20, US Government Printing Office, Rockville, MD, pp. 61–86.

Eriksson K. (1968) Genetic selection for voluntary alcohol consumption in the albino rat. *Science* 159, 739–741.

Eriksson K. (1969) Factors affecting voluntary alcohol consumption in the albino rat. *Ann. Zool. Fenn.* 6, 227–265.

Eriksson K. (1971) Rat strains specifically selected for their voluntary alcohol consumption. *Ann. Med. Exp. Biol. Fenn.* 49, 67–72.

Eriksson K., Sinclair J. D., and Kiianmaa K. (1980) *Animal Models in Alcohol Research*. Academic Press, London.

Eriksson K. and Närhi M. (1973) Specially selected rat strains as a model of alcoholism, in *The Laboratory Animal in Drug Testing* (Spiegel A., ed.) Springer Verlag, Stuttgart, pp. 282–301.

Eriksson K. and Rusi M. (1981) Finnish selection studies on alcohol-related behaviors: General outline, in *Development of Animal Models as Pharmacogenetic Tools* (McClearn G. E., Deitrich R. A., and Erwin G., eds.) NIAAA Research Monograph 6, US Government Printing Office, Rockville, MD, pp. 87–117.

Eriksson K., Halkka O., Lokki J., and Saura A. (1976) Enzyme polymorphism in feral, outbred and inbred rats *(Rattus norvegicus)*. *Heredity* 37, 341–349.

Erwin V. G., Jones B. C., and Radcliffe R. (1990) Further characterization of LS X SS recombinant inbred strains of mice: Activating and hypothermic effects of ethanol. *Alcohol: Clin. Exp. Res.* 14, 200–204.

Fadda F., Mosca E., Colombo G., and Gessa G. L. (1989) Effect of spontaneous ingestion of ethanol on brain dopamine metabolism. *Life Sci.* 44, 281–287.

Falconer, D. S. (1989) *Introduction to Quantitative Genetics*. Longman, Hong Kong.

Fuller J. L. (1985) The genetics of alcohol consumption in animals. *Soc. Biol.* 32, 210–221.

Fuller J. L., McClearn G. E., Wilson J. R., and Crowe L. (1985) Genetics and Human Encounter with Alcohol. *Social Biol.* vol. 32. 327 pp.

Goedde H. W. and Agarwal D. P., eds. (1987) Genetics and Alcoholism. *Proc. Clin. Biol. Res.* vol 241. 353 pp.

Goldman D., Nelson R., Deitrich R. A., Baker R. C., Spuhler K., Marley H., Ebert M., and Merril C. R. (1985) Genetic brain polypeptide variants in inbred mice and in mouse strains with high and low sensitivity to alcohol. *Brain Res.* 341, 130–138.

Goldstein D. (1981) What traits to breed for and how to measure them, in *Development of Animal Models as Pharmacogenetic Tools* (McClearn G. E., Deitrich R. A., and Erwin G., eds.) NIAAA Research Monograph 6, US Government Printing Office, Rockville, MD, pp. 233–247.

Goldstein D. (1989) Animal models developed by selective breeding: Some questions raised and a few answered, in *Genetic Aspects of Alcoholism* (Kiianmaa K., Tabakoff B., and Saito T., eds.) The Finnish Foundation for Alcohol Studies, vol. 37, Helsinki, pp. 229–238.

Goodwin D. W., Schulsinger F., Hermansen L., Guze S. B., and Winokur G. (1973) Alcohol problems in adoptees raised apart from alcoholic biological parents. *Arch. Gen. Psychiatry* **28**, 238–243.

Goodwin D. W., Schulsinger F., Moller N., Hermansen L., Winokur G., and Guze S. B. (1977) Drinking problems in adopted and nonadopted sons of alcoholics. *Arch. Gen. Psychiatry* **31**, 164–169.

Hansen C. and Spuhler K. (1984) Development of National Institutes of Health genetically heterogenous rat stock. *Alcohol: Clin. Exp. Res.* **8**, 477–479.

Hilakivi L., Eriksson C. J. P., Sarviharju M., and Sinclair J. D. (1984) Revitalization of the AA and ANA rat lines: Effects on some line characteristics. *Alcohol* **1**, 71–75.

Hrubec Z. and Omenn G. S. (1981) Evidence of genetic predisposition to alcoholic cirrhosis and psychosis: Twin concordances for alcoholism and its biological end points by zygosity among male veterans. *Alcohol: Clin. Exp. Res.* **5**, 207–215.

Hyde J. S. (1981) A review of selective breeding programs, in *Development of Animal Models as Pharmacogenetic Tools* (McClearn G. E., Deitrich R. A., and Erwin G., eds.) NIAAA Research Monograph 6, US Government Printing Office, Rockville, MD, pp. 59–77.

Hyytiä P., Halkka O., and Eriksson K. (1987) Alcohol-preferring (AA) and alcohol-avoiding (ANA) lines of rats after intogression of alien genes, in *Advances in Biomedical Alcohol Research, Alcohol Alcohol.* (Lindros K. O., Ylikahri R., and Kiianmaa K., eds.) Suppl. 1, pp. 351–355.

Kaij L. (1960) *Alcoholism in Twins. Studies on the Etiology and Sequelae of Abuse of Alcohol.* Almquist and Wiksell, Stockholm.

Kampov-Polevoy, A. B., Kasheffskaya R. P., and Sinclair J. D. (1990) Initial acceptance of ethanol: Gustatory factors and patterns of alcohol drinking. *Alcohol* **7**, 83–85.

Kaprio J., Koskenvuo M., Langinvainio H., Romanov K., Sarna S., and Rose R. J. (1987) Genetic influences on use and abuse of alcohol: a study of 5638 adult Finnish twin brothers. *Alcohol: Clin. Exp. Res.* **11**, 349–356.

Kiianmaa K., Tabakoff B., and Saito T., eds, (1989) *Genetic Aspects of Alcoholism*, The Finnish Foundation for Alcohol Studies, vol, 37. Helsinki.

Lester D. and Freed E. X. (1973) Criteria for an animal model of alcoholism. *Pharmacol. Biochem. Behav.* **1**, 103–107.

Li T-K., Lumeng L., McBride W. J., and Waller B. M. (1981) Indiana selection studies on alcohol-related behaviors, in *Development of Animal Models as Pharmacogenetic Tools* (McClearn G. E, Deitrich R. A., and Ervin G., eds.) NIAAA Research Monograph 6, US Government Printing Office, Rockville, MD, pp. 171–191.

Li T-K., Lumeng L., McBride W. J., and Murphy, J. M. R. (1987) Rodent lines
    selected for factors affecting alcohol consumption, in *Advances in Bio-
    medical Alcohol Research* (Lindros K. O., Ylikahri R., and Kiianmaa K.,
    eds.) *Alcohol Alcohol.* Suppl. 1, 91–96.
Lumeng L., Doolittle D. P., and Li T. (1986) New duplicate lines of rats that
    differ in voluntary alcohol consumption. *Alcohol and Alcoholism* 21, A125.
Mardones J. (1960) Experimentally induced changes in the free selection of
    ethanol. *Int. Rev. Neurobiol.* 2, 41–76.
Mardones J. (1972) Experimentally induced changes in alcohol appetite, in
    Biological Aspects of Alcohol Consumption (Forsander O. and Eriksson
    K., eds.) The Finnish Foundation for Alcohol Studies, vol. 20, Helsinki,)
    pp. 15–23.
McClearn G. E. (1981) Introduction chapters, in *Development of Animal Mod-
    els as Pharmacogenetic Tools* (McClearn G. E., Deitrich R. A., and Erwin
    G., eds.) NIAAA Research Monograph 6. US Government Printing
    Office, Rockville, MD, pp. 3–10, 81–85.
McClearn G. E. and Kakihana R. (1981) Selective breeding for ethanol sensi-
    tivity: short-sleep and long-sleep mice, in *Development of Animal Mod-
    els as Pharmacogenetic Tools* (McClearn G. E., Deitrich R. A., and Erwin
    G., eds.) NIAAA Research Monograph 6, US Government Printing
    Office, Rockville, MD, pp. 147–159.
McClearn G. E. and Rodgers D. A., (1959) Differences in alcohol preference
    among inbred strains of mice. *Q. J. Stud. Alcohol* 20, 691–695.
McClearn G. E., Wilson J. R., and Meredith W. (1970) The use of isogenic
    and heterogenic mouse stocks in behavioral research, in *Contributions
    to Behavior-Genetic Analysis: The Mouse as a Prototype* (Lindzey G. and
    Thiessen D. D., eds.) Appleton-Century-Crofts, New York, pp. 3–22.
McClearn G. E., Deitrich R. A., and Erwin G., eds. (1981) *Development of
    Animal Models as Pharmacogenetic Tools*, NIAAA Research Monograph
    6, US Government Printing Office, Rockville, MD.
McClearn E., Wilson J. R., Petersen D. R., and Allen D. L. (1982) Selective
    breeding in mice for severity of ethanol withdrawal syndrome. *Subst.
    Alcohol Actions/Misuse* 3, 135–143.
Nagoshi C. T., Wilson J. R., and Plomin R. (1986) Use of regression residuals
    to quantify individual differences in acute sensitivity and tolerance to
    ethanol. *Alcohol: Clin. Exp. Res.* 10, 343–349.
Partanen J., Bruun K., and Markkanen T. (1966) *Inheritance of Drinking
    Behavior. A Study of Intelligence, Personality and Use of Alcohol of Adult
    Twins.* The Finnish Foundation for Alcohol Studies, vol. 14, Helsinki.
Phillips T. J., Feller D. J., and Crabbe J. C., (1989) Selected mouse lines, alco-
    hol and behavior. *Experientia* 45, 805–827.
Phillips T. J., Burkhart-Kasch S., Terdal E. S., and Crabbe J. C. (1991) Response
    to selection for ethanol-induced locomotor activation: genetic analyses
    and selection response characterization. *Psychopharmacology* 103, 557–
    566.

Propping P. (1977) Genetic control of ethanol action on the central nervous system: An EEG study in twins. *Hum. Genet.* **35,** 309–334.

Roberts R. C. (1981) Current perspectives on selective breeding: theoretical aspects, in *Development of Animal Models as Pharmacogenetic Tools* (McClearn G. E., Deitrich R. A., and Erwin G., eds.) NIAAA Research Monograph 6, US Government Printing Office, Rockville, MD, pp. 37–58.

Rodgers D. A and McClearn G. E. (1962) Mouse strain differences in preference for various concentrations of alcohol. *Q. J. Stud. Alcohol* **23,** 26–33.

Satinder K. P. (1970) Behavior-genetic-dependent self-selection of alcohol in rats. *J. Comp. Phvsiol. Psychol.* **80,** 422–434.

Schuckit M. A. (1988) Reactions to alcohol in sons of alcoholics and controls. *Alcohol: Clin. Exp. Res.* **12,** 465–470.

Schuckit M. A. (1989) Multiple markers of the response ethanol in sons of alcoholics and controls, in *Genetic Aspects of Alcoholism* (Kiianmaa K., Tabakoff B., and Saito T., eds.) The Finnish Foundation for Alcohol Studies, vol. 37, Helsinki, pp. 107-116.

Schuckit M. A., Goodwin D. A., and Winokur G. A. (1972) A study of alcoholism in half siblings. *Am. J. Psychiatry* **128,** 1132–1136.

Schuckit M. A., Li T-K., Cloninger C. R., and Deitrich R. A. (1985) Genetics of alcoholism. *Alcohol: Clin. Exp. Res.* **9,** 475–492.

Sinclair J. D. (1987) The feasibility of effective psychopharmacological treatments for alcoholism. *Br. J. Addict.* **82,** 1213–1223.

Sinclair, J. D. and Li, T.-K. (1989) Long and short alcohol deprivation: Effects on AA and P rats. *Alcohol* **6,** 505–509.

Sinclair J. D., Viitamaa T., and Hyytiä P. (1987) Behavioral and color variations between rat lines developed for differential alcohol sensitivity, in *Advances in Biomedical Alcohol Research* (Lindros K. O., Ylikahri R., and Kiianmaa K., eds.) *Alcohol Alcohol.* Suppl. 1, 449–453.

Sinclair J. D., Lê A. D., and Kiianmaa K. (1989) The AA and ANA rat lines, selected for differences in alcohol consumption. *Experientia* **45,** 798–805.

Spuhler K., Hoffer B., Weiner N., and Palmer M. (1982) Evidence for genetic correlation of hypnotic effects and cerebellar Purkinje neuron depression in response to ethanol in mice. *Pharmacol. Biochem. Behav.* **17,** 569–578.

Spuhler K., Deitrich R. A., and Baker R. C. (1990) Selective breeding of rats differing in sensitivity to the hypnotic effects of acute ethanol administration, in *Initial Sensitivity to Alcohol* (Deitrich R. A. and Pawlowski A. A., eds.) NIAAA Research Monograph 20, US Government Printing Office, Rockville, MD, pp. 87–102.

Tabakoff, B. (1991) One man's craving is another man's dependence. *Br. J. Addiction* **85,** 1253,1254.

Yoshimoto K. and Komura S. (1987) Reexamination of the relationship between alcohol preference and brain monoamines in inbred strains of mice including senescence-accelerated mice. *Pharmacol. Biochem. Behav.* **27,** 317–322.

Genetic Animal Models

A User's Guide

R. Adron Harris and John C. Crabbe

1. Overview

1.1. Introduction

The use of genetically specified animals to address fundamental questions in neuroscience continues to grow apace. In the early 1950s, Mardones (1951) reported the development by selective breeding of strains of rats preferring and not preferring to drink alcohol. McClearn and Rodgers (1959) reported that inbred mouse strains differed in their preference for 10% ethanol-in-tap water solutions. These pioneering studies represent the two methods still most frequently used today, selective breeding and the analysis of preexisting, defined genotypes (*see below*).

Although the term *pharmacogenetics* was popularized by Kalow (1962), he employed it in the context of analysis of genetically determined differences in drug metabolism. Vesell (1973) has reviewed much of the early work in this area, and a recent review discusses the future directions in genetic research on drug metabolism (Meyer, 1990). It is another sense of the term that we address in this chapter, namely the determination of drug responsiveness by genotype. Insofar as many of the pioneering pharmacogenetic experiments were performed to explore responses to ethanol, pharmacogenetic analysis of this drug is considerably more advanced than that for other drugs of abuse, and we

From: *Neuromethods, Vol. 24: Animal Models of Drug Addiction*
Eds: A. Boulton, G. Baker, and P. H. Wu ©1992 The Humana Press Inc.

will draw more frequently, but not exclusively, from alcohol research. Systematic reviews of the "state-of-the-art" of pharmacogenetic analyses for the different classes of drugs of abuse may be found in a recent volume (Crabbe and Harris, 1991). Currently, a major application of genetics is to test hypotheses about the relationship between different actions of a given drug. It appears particularly useful as a means of determining if an observed neurochemical action of a drug is related to its behavioral effects.

In this chapter, we will first discuss the general way in which genes may influence drug responses. We will then describe the various types of genetic animal models used historically and currently, their strengths and weaknesses, and we will project some future applications. Throughout, we will not attempt to review any aspect of this vast literature systematically, but we will provide examples that we hope illustrate the power and limits of genetic animal models.

1.2. Single Traits:
Genetic and Environmental Influences

The simplest application of pharmacogenetic principles is an attempt to analyze the probably mythical case where a single gene influences a drug response. Figure 1 schematically shows that Gene A, acting through the pharmacological/biochemical intervening Steps 1 and 2, leads to drug response X. It also shows that not all variability in drug response X is genetic: Environmental sources of influence (for example, availability of substrate, climate, and so forth), such as A', also influence X, either directly or perhaps through similar pharmacological/biochemical intermediates (1' and 2'). Practically, such a case might be approached where a single gene mutation leads to a defective gene product. As may seem obvious, examination of the literature suggests that all drug responses seem to be determined jointly by genetic and environmental influences.

Figure 2 introduces the next level of complexity. We know that in most cases, multiple genes, acting through intermediate steps, influence a single drug response. Such *polygenic* control may be mediated through convergence on a common pharma-

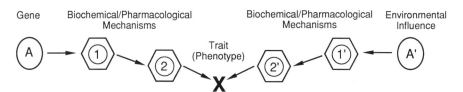

Fig. 1. Schematic representation of the pathways through which a gene (A), acting through biochemical/pharmacological mechanisms (1,2), influences a trait or phenotype (X). Also shown is the effect of an environmental influence (A') acting through intermediary mechanisms (1',2'). Adapted from Crabbe et al. (1990), with permission.

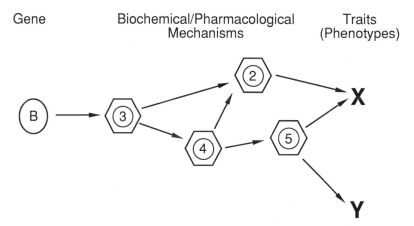

Fig. 2. Convergence of the influence of several genes on a single trait, a situation known as polygenic control. Environmental influences exist, but are not shown. *See* caption to Fig. 1.

cological mechanism (as in Genes A and B acting via mechanism 2 in Fig. 2), or divergence (as in Gene B acting through mechanisms 2 and 5). In this and subsequent figures, the role of environmental influences is deleted for clarity, but certainly remains in effect, and may be schematically conceptualized in parallel to the genetic paths. Figure 4 illustrates that the divergence of gene action may extend to the case where a single gene (B) influences both drug responses X and Y, a situation known as *pleiotropism*.

It can be seen that even attempts to trace a path from a single gene to a single drug response phenotype are necessarily com-

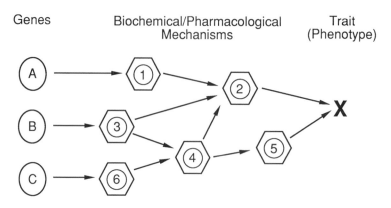

Fig. 3. Pleiotropic influence of a single gene on multiple traits. Environmental influences exist, but are not shown. *See* caption to Fig. 1.

plex and must be undertaken with care. Specifically, any phenotypes to be investigated must be as rigorously defined as possible. Great care must be taken to control environmental
conditions, because slight differences in environment may be
amplified through their interactions with multiple genes and
pathways.

1.3. Multiple Traits:
Genetic and Environmental Correlations

Most ventures into pharmacogenetic analysis have a broader
goal. The complexity of a number of genes influencing a number of drug-related phenotypes (as well as the environmental
influences) must be dissected. Figure 4 shows that in addition to
the notions of polygenic determination and pleiotropism introduced in the previous section, it is clear that genes may have
discrete or interactive effects. Genes A, B, C, and D all influence
Trait X, and genes B, C, and D all influence trait Y. An appropriate experiment would reveal the existence of a genetic correlation between the two traits. One of the useful features of genetic
animal models is their application to the detection of genetic
correlations; the implication is that they represent activity in
genetically influenced common systems (in this case, 3–7, but
not 1 and 2).

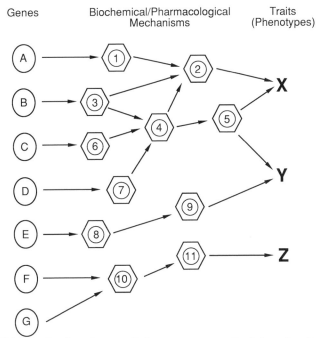

| Genes | Biochemical/Pharmacological Mechanisms | Traits (Phenotypes) |

Fig. 4. Schematic showing partial common genetic determination (genetic correlation) of traits X and Y, and the genetic independence of trait Z. Environmental influences exist, but are not shown. *See* caption to Fig. 1.

Many experiments employing genetic animal models are implicitly seeking evidence for a genetic correlation. The strategy is to use several defined genotypes and then to study the trait of interest, as well as the putative genetic correlate, under carefully controlled environmental conditions. The cooccurrence (better, the formal covariation) of two traits is taken as evidence for their functional relationship. For example, mice selected for genetic differences in ethanol withdrawal seizures also display correlated differences in withdrawal signs following chronic ingestion of benzodiazepines or barbiturates, suggesting that some common mechanisms account for dependence on these three classes of drugs (Phillips and Crabbe, 1991). Several technical issues regarding the appropriate use of genetic animal models to estimate the magnitude of genetic correlations are discussed in a recent review (Crabbe et al., 1990).

1.4. Sources of Genetic Differences

Obviously, there are many ways in which genetic differences could influence drug response phenotypes. An excellent discussion of many of the various sources of influence may be found in Horowitz and Dudek (1983). An important source of genetic variability is allelic. We usually imagine that a genetic difference results because of a change in the nucleotide sequence in a protein-coding region of the gene, thereby altering the amino acid sequence of a critical protein. However, it is important to realize that nucleotide changes in noncoding regions, such as introns or 5'-untranslated sequences, can alter gene expression and RNA splicing. In addition, many genes are regulatory, and changes in these genes will alter the expression of many other genes, thereby providing a mechanism to vary transcriptional regulation between genotypes. For cases such as a drug receptor–effector system, where a complex of proteins is concerned, variations in numbers of subunits, the particular mix of subunits, or their interactions have all been proposed. At this time, the basis for genetic differences in drug responses is not well characterized for any drug, but increasing focus on molecular biological techniques will surely lead to greater understanding.

2. Selected Lines

2.1. Definition

Genetic selection for desired traits has been an important aspect of agriculture and sports for centuries, and has resulted in remarkable milk output for dairy cows and speed in racehorses, to name only two of the multitude of successful selective breeding ventures. The principles worked out by agricultural geneticists have more recently been applied to laboratory animals (primarily mice) for behavioral traits, including drug sensitivity (Falconer, 1989). This is done by testing a genetically heterogeneous stock and selecting the most and least affected animals for breeding. In addition, a tested but unselected control line (or lines) is (are) often maintained. The selection process is repeated generation after generation until an asymptote or maximal and minimal levels of drug sensitivity are achieved. At this point, all genes or alleles relevant to drug sensitivity have theoretically segregated. Thus, the role of a "candidate" gene or gene product can be tested by determining if it is different in the selected lines. It should be noted that this process does not usually select for mutant animals, which are quite rare, but rather segregates the normal genetic variability existing in the starting population. Theoretically, animals are inbred (homozygous) for all trait-relevant genes, but retain the original population polymorphism of all other loci.

2.2. Offspring of Selected Lines

By mating animals from the insensitive line with those from the sensitive line, one may produce a generation that is heterogeneous for trait-relevant genes. These animals are then mated to produce a segregating generation. Each member of this generation will have a combination of genes from the sensitive and insensitive lines, and a valid candidate gene product should covary with drug sensitivity among its members. For example, this approach was used to show that sensitivity to seizures produced by the benzodiazepine inverse agonist methyl 6,7-dimethoxy-4-ethyl-beta-carboline-3-carboxylate (DMCM) is determined by a single autosomal gene and that different genes determine sensitivity to caffeine seizures (Seale et al., 1987). A further refinement of this approach is to inbreed members of the segregating hybrid generation by brother–sister matings to produce recombinant inbred strains (RIs). The RIs are powerful tools for both classical and molecular genetics, as discussed in Section 4.

2.3. Applications

One application of selected lines is to determine if two traits are genetically related. For example, if lines are selected for differences in drug-induced ataxia, we can ask if they also differ in the anticonvulsant action of the same drug. If the drug produces ataxia and anticonvulsant actions by the same mechanism, the lines should differ for both drug effects (*see* Fig. 5). Such behavioral experiments have generated vast literature that was recently reviewed by Phillips and Crabbe (1991). Selected lines have also been used to test neurochemical hypotheses because any neurochemical parameter (e.g., receptor density) that is important for a particular behavioral action of a drug should differ between lines selected for that drug effect. This approach provided key

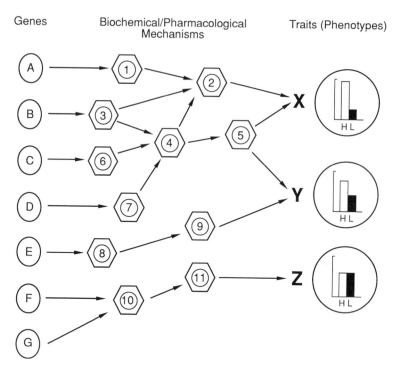

Fig. 5. Schematic representation of the mean phenotypic values of two selected lines. Line H is selected for high response on trait X, whereas line L is selected for low response on trait X. When tested on trait Y, the lines differ somewhat (H > L), owing to the genetic correlation between X and Y. When tested on trait Z, the lines do not differ, owing to the genetic independence of traits X and Z. Environmental influences exist, but are not shown. *See* caption to Fig. 1.

evidence that GABA-activated chloride channels are important in ethanol actions (Harris and Allan, 1989). Neurochemical studies of selected lines are reviewed by Allan and Harris (1991). Another potential application of selected lines is to determine which genes are different between the lines, and thereby identify genes responsible for drug action. This is complicated by the polygenic nature of most behavioral traits and to date has not been accomplished with lines selected for drug neurosensitivity. However, an excellent example of the potential of this approach is the use of Dahl salt-sensitive and salt-resistant rat lines to show that the differential blood pressure response of these animals can be attributed to a single amino acid change in the kidney Na$^+$,K$^+$-ATPase (Herrera and Ruiz-Opazo, 1990). In this way, classical and molecular genetics may be linked by application of such techniques as gene expression, cloning, and sequencing.

Applications of selected lines are obviously limited to traits for which someone has taken the time (and money) to generate and maintain appropriate animals. Because of the large number of animals that must be maintained in a breeding study, and because of their very detailed genomic map, mice are most commonly used. The following mouse lines are currently maintained: LS/SS (Long-Sleep/Short-Sleep; duration of loss of righting reflex after acute injection of ethanol); WSP/WSR (Withdrawal Seizure-Prone and Resistant; severity of withdrawal seizures following chronic ethanol inhalation); HA/LA (High and Low alcohol dependence; severity of withdrawal seizures following chronic ethanol consumption); FAST/SLOW (increase and decrease in motor activity following acute ethanol injection); HOT/COLD (minimal and maximal hypothermia following acute ethanol injection); DS/DR (Diazepam-Sensitive and Resistant; duration of rotarod ataxia following injection of diazepam); HAR/LAR (High and Low Analgesia Response; antinociceptive response to acute injection of levorphanol); and HR/HNR (Haloperidol Responder and NonResponder; catalepsy following acute administration of haloperidol). In addition, several rat lines are maintained: HAS/LAS (High and Low Alcohol Sensitivity; duration of loss or righting reflex after acute injection of ethanol); P/NP (Preferring and NonPreferring; voluntary consumption of ethanol); HAD/LAD (High and Low Alcohol Drinking; voluntary consumption of ethanol); AT/ANT (Alcohol-Tolerant and NonTolerant; insensitive or sensitive to ataxia after acute injection of ethanol); AA/ANA (Alcohol-Avoiding and NonAvoiding; voluntary ethanol consumption); sP/sNP (Sardinian Preferring and NonPreferring; voluntary ethanol consumption); and UChA/UChB (University of Chile A and B lines; low and high voluntary ethanol consumption). Details of selection and application of these animals are presented in Crabbe and Harris (1991).

There are several important considerations for the use of selected lines. The first is that there are replicates available for some lines (e.g., WSP$_1$, WSP$_2$, WSR$_1$ and WSR$_2$) but not for others (e.g., LS/SS). If a correlated response is found for both pairs of lines in a replicated selection, this provides much more convincing evidence of a genetic relationship than if only a single pair of lines can be tested (*see* Section 2.4.). Second, it is clear, at least for ethanol and diazepam, that drug "sensitivity" is dependent on the behavior measured. For example, if one selects for sensitivity or resistance to diazepam-induced ataxia (DS/DR mice), it is not correct to conclude that these animals are sensitive or resistant to all actions of diazepam. DS and DR mice are equally sensitive to the anticonvulsant action of diazepam and the DR mice are more sensitive than DS mice to motor stimulation and anxiolytic actions of diazepam (Gallaher and Crabbe, 1991). Thus, genetic studies have shown that there is no unitary phenomenon of diazepam (or ethanol) "sensitivity" and results obtained with selected lines must be interpreted with this focused light.

2.4. Problems

A potential problem with selective breeding is that some genes unrelated to drug sensitivity will also segregate between selected lines. This is because one does not have access to an infinite population of animals for breeding (nor could one afford the animal care costs required for such a population) and inbreeding is unavoidable. The question of how to determine if a genetic difference between two selected lines is important or accidental can be answered in several ways. One is to test the animals from each generation of selection and determine if the two traits (e.g., neurochemical and behavioral) cosegregate (Falconer, 1989). An example of this approach recently showed that HOT and COLD mice became increasingly divergent in their hypothermic sensitivity to barbiturates as selection progressed from the 6th to the 13th generation (Feller and Crabbe, 1991). Another, more common, approach is to carry out multiple (usually duplicate) identical selections as noted in Section 2.3. Both sensitive lines should differ from both resistant lines for traits related to selection but not for accidentally fixed traits. (This complex issue is discussed by Henderson, 1989 and Crabbe et al., 1990.) If multiple selected lines are not available, genetically segregating generations or recombinant inbred strains (*see* Section 2.2.) may be used to test the importance of a difference between selected lines.

3. Inbred Strains

3.1. Definition

An inbred strain is created by more than 20 generations of mating close genetic relatives, usually brother and sister. Within an inbred strain, each individual is homozygous at virtually every gene, and is genetically identical to all others of that strain (with the exception of genes on the sex chromosomes). Which allele is fixed at each locus is entirely a matter of chance. Thus, these animals have not been selected for any particular trait, but they differ on many traits on a purely random basis. There are more than 100 commercially available inbred strains of rats and mice: The Jackson Laboratory (Bar Harbor, ME) is the historic source of the greatest number of genetically defined mice.

3.2. Applications

Typically, inbred strain analysis is the first indicator of genetic control of a trait. When McClearn and Rodgers (1959) demonstrated that C57BL inbred mice drank nearly 100% of their daily fluid from a bottle containing 10% ethanol (vs tap water), whereas inbred mice of the A strain tested under the same environmental conditions refused alcohol nearly absolutely, they concluded that the genotypic differences between these strains was the primary determinant of the preference difference. Recent reviews of inbred strain studies have demonstrated significant genetic control of many responses to ethanol (Phillips and Crabbe, 1991), benzodiazepines and barbiturates (Gallaher and Crabbe, 1991), opiates (Belknap and O'Toole, 1991), caffeine, amphetamine and cocaine (Seale, 1991), and nicotine (Collins and Marks, 1991).

In Section 2.3., we noted that the genetic control of a drug response is often relatively discrete and may not extend to other drug effects. This is not an inviolable rule, however. When a number of strains are surveyed, one result is that a particular inbred strain may be found to be sensitive or insensitive to multiple drug effects across a range of related traits or even to the same response elicited by multiple drugs. Cunningham et al. (in press) recently tested C57BL/6J and DBA/2J inbred mice for several responses thought to be related to drug reward. DBA/2J mice showed a stronger conditioned place preference for morphine and ethanol than C57BL/6J mice. They also showed greater locomotor stimulation after ethanol. However, C57BL/6J mice showed greater stimulation after morphine than DBA/2J. Both locomotor stimulation and conditioned place preference are thought to index a drug's rewarding effects, and these data derived from the inbred strain method suggest that, at least for ethanol, the DBA/2J strain may be generally sensitive to drug reward. However, these results also point out the fallacy of generalizing from a single behavioral test of inbred strains. As mentioned in Section 1.1., C57BL/6J mice will voluntarily drink ethanol/water solutions whereas DBA mice assiduously avoid these solutions. Thus, it may be incorrect to conclude that this consumption is related to reinforcing properties of the drug.

Inbred strains are also useful for the analysis of the effects of single genes. When a mutation occurs, it can be moved by repeated backcrossing and then maintained in an inbred strain, usually C57BL/6J. The C57 mice bearing the mutation are then *congenic* or *coisogenic* with the standard C57BL/6J mice. If the mutation influences the trait under investigation, congenics will differ from controls. An example of this is the influence of the albino gene on morphine-stimulated activity. Katz and Doyle (1980) found that congenic albino C57BL/6J mice were more stimulated by morphine than controls. Since they differed from control C57BL/6J mice only at the albino locus (and possibly a few other tightly linked genes), this indicates a significant effect of this gene's product on morphine sensitivity.

Another use of inbred strains is to examine several inbreds to discern the presence of genetic correlations. When 20 inbred strains of mice were investigated for several different responses to ethanol, the severity of ethanol withdrawal was found to correlate only with variables related to hypothermia, but not ethanol-stimulated activity, ataxia, or other responses (Crabbe et al., 1983b). This implies that the effects of ethanol to induce hypothermia are mechanistically related to its effects on the syndrome of ethanol withdrawal handling-induced convulsions. The potency of ethanol to inhibit firing of cerebellar Purkinje cells was correlated 0.995 with its in vivo effect to induce loss of righting reflex in eight inbred mouse strains (Spuhler et al., 1982). Yoshimoto and Komura (1987) found a strong negative relationship between alcohol drinking and whole brain levels of serotonin in 10 inbred mouse strains.

An enormous potential utility of inbred strain data derives from the relative stability of the genotypes. Since the rate of mutation is quite low, C57BL/6J mice currently in the Jackson Laboratory population are genetically almost identical to those that were there 15 years ago. Thus, data collected with these mice (and in any other laboratory that buys from Jackson Laboratory) are highly reproducible and may be compared cross-sectionally and longitudinally. The data sets in the most common 10–20 inbred mouse strains are large and growing rapidly; the application of these data to mapping will be treated in a later section.

Whereas many ethanol-related traits have been characterized in a substantial number of inbred mouse strains, the data for other drugs of abuse are more sparse. Are the neural substrates of the ethanol withdrawal syndrome the same as those underlying withdrawal from diazepam and/or barbiturates? Are the neural substrates that adapt during the development of tolerance to the hypothermic effect of a barbiturate largely the same as those seen to change when diazepam is chronically administered? These fundamental pharmacological questions can be addressed by screening a sufficient number of inbred strains, and the data remain available for subsequent correlational analyses. Recently, George (1990) has initiated attempts to characterize a few inbred strains for their propensity to self-administer a number of drugs of abuse, with the intent of finding evidence for such genetic correlations. The field of pharmacogenetics is poised to develop this technique, and increasing numbers of investigators are willing to make the investment in time and money.

3.3. Limitations of This Technique

One limitation is the cost of inbred mice, which averaged approx $13/animal in 1990, including shipping. Another is the number of strains that must be studied to gain a reasonable compromise between sensitivity and amount of work. It is clearly insufficient to study two inbred strains on two traits and conclude from a positive result that a genetic correlation exists. On statistical grounds, this correlation has zero degrees of freedom. On the other hand, if one studies 40 inbred strains, their mean values on two traits will be correlated significantly at $p < .05$ if $r = .30$. However, this correlation accounts for only 9% (r^2) of the variance in either trait, and is unlikely to be biologically meaningful. A reasonable compromise would be to examine 10–15 strains. This will detect a significant correlation with r values between .63 and .51, which accounts for at least one-fourth of trait variance. If the true genetic correlation between the traits is very high, fewer strains may suffice.

Another potential problem with inbred strains is the technical genetic issue of their gene frequencies. Within any inbred strain, all alleles are, by definition, at gene frequency = 1; thus, variation on any trait among members of an inbred strain does not reveal the effects of allelic interactions. For example, Shuster (1975) repeatedly administered morphine to C57BL/6J and A/J inbred mice, and mice of their F_1 hybrid cross. The sensitization of the locomotor stimulant response to morphine was much larger in the F_1 mice than in either parent strain, indicating that allelic interactions (*epistasis*) were important in determining this trait.

3.4. Conclusions

There is a rich historical data base for ethanol-related responses in a number of common inbred strains, and the data for other drugs of abuse are accumulating. In combination with other pharmacogenetic methods, and used as described further in later sections, examination of panels of inbred strains will remain a useful tool for the neuroscientist.

4. Recombinant Inbred Strains

4.1. Definition

Recombinant inbred strains are derived from the segregating cross of two inbred strains. When two parental inbred strains, such as C57BL/6By (C57) and BALB/cBy (BALB), are crossed to form an F_1 population, and the F_1 animals are mated to form the segregating F_2 generation, all alleles present in the F_2 animals are derived from either the C57 or BALB parent. A randomly chosen pair of F_2 mice is then inbred systematically until a new inbred strain (e.g., CXBD) is created. At each gene, all CXBD mice are now homozygous, and the particular allele, which was determined by chance, is either the BALB or the C57 allele. CXBD mice, therefore, represent a random sample of the alleles in the two parental stocks. When a panel of such RIs is created, the power of RI analysis is ready to be employed.

BALB mice carry the recessive *c* allele for albinism at the C locus. C57 mice carry the wild-type C allele, and are pigmented. Of the seven RI strains in this battery, two (CXBG and CXBI) are albino and the other five are pigmented. If a drug-related trait is typed in the seven RIs and two parental strains and is found to display a bimodal pattern (i.e., the RI strains tend to be like either the C57 or the DBA parent, but not intermediate), one may suspect that this strain distribution pattern (SDP) suggests the workings of a single gene. Since SDPs are known in the BXD RI series for genes such as the coat-color C locus described above, as well as for histocompatibility and many other loci, any new pattern may be compared with the existing SDPs for evidence of linkage.

4.2. Applications of RI Analyses

As in any set of inbred strains, some RI strains may turn out to be interesting in isolation. Within the CXB RI series, one strain has emerged of great interest to opiate pharmacology. The CXBK RI strain was found to have a relatively severe deficiency in naloxone binding sites in brain (Baran et al.,1975). In addition, they show deficiencies in analgesia and insensitivity to opiate analgesia. This has been assumed to reflect a mutation in this strain, but its genetic location has yet to be identified.

A more powerful panel of RI strains, called BXD/Ty and derived from the cross of C57BL/6J and DBA/2J, has been typed for several ethanol-related responses. Mice were tested for alcohol preference drinking, and other mice from 18 of the RI strains composing the battery were typed for electrophoretically mobile charge variants for a number of brain proteins. An SDP indicating a weak linkage to a region on Chromosome 1 near the gene Ltw-4 was subsequently confirmed in a panel of 20 inbred strains (Goldman et al., 1987). These studies suggest that a single gene exerts a relatively pronounced influence on ethanol preference drinking. The linkage with Ltw-4 is not, however, close enough to encourage moving to molecular biological methods to identify the gene at this time.

Use of the BXD RI series in particular will be increasingly important. This series of strains is large enough to permit meaningful genetic correlational analysis as described in this section for inbreds. Furthermore, many investigators in a number of disciplines are busy typing this panel of strains for new marker genes. Thus, the list of SDPs is growing almost daily. The most current published list of SDPs (Taylor, 1989) had 163 SDPs identified. Benjamin Taylor, who developed the RI series, recently provided us with a list of over 400 markers (J. K. Belknap, personal communication). This rapid increase is largely owing to the generation of numerous SDPs based on restriction fragment length polymorphisms (RFLPs).

Thus, the advantages of RI strains are virtually all those that describe for the use of inbred strains in general. Several RI panels are available, but the BXD RI panel mentioned above is most commonly used to analyze drug responses. Other applications will be discussed in Section 6. of this chapter.

4.3. Limitations of RI Analyses

The principal limitations of inbred strains also describe their RI variant. In addition, one must remember that the RIs do not represent the whole spectrum of genetic variance in *Mus musculus.*

Each RI battery can display only alleles that were present in one or the other parent inbred strain. One reason that the BXD RI battery has been employed in pharmacogenetic research is that the C57 and DBA parental strains differ markedly in almost any response to any psychoactive drug; they also differ markedly in protein polymorphisms (Taylor, 1972), suggesting marked genetic dissimilarity. This allows a wide range of possible values for the array of BXD RI strains for any response of interest, but does not completely mitigate the fact that they represent a subset of mouse genetic variance. An additional limitation of the RI approach is that it is most suitable for cases where a single gene determines at least 15% of trait variance. It remains to be established whether this is a frequent occurrence in pharmacogenetics.

A final issue is raised by the development of a new set of RI strains by a group at the University of Colorado in Boulder, CO. They developed 26 RI strains from the cross of two selectively bred lines, the LS and SS mice previously discussed (De Fries et al., 1989). The LSXSS RI strains are under active investigation, and will likely produce useful information. They face a large hurdle, however, in that there is currently no available large set of mapped SDPs for these RIs. They will likely overcome this hurdle, however, since Thomas Johnson (University of Colorado, Boulder, CO) is actively generating RFLPs in the strains to provide map locations. However, an investigator without a large supporting group would be ill-advised to consider breeding a new set of RI strains, but rather should concentrate on increasing the density of the genetic map available from one of the established series.

4.4. Conclusions

The RI approach is a nice hybrid genetic approach between classical analysis of panels of inbred strains and the more molecularly based approaches to be discussed later in this chapter. This suggests that the focus on mapping in the RIs will intensify. The BXD RI series, in particular, will become more important to pharmacogenetic research in the coming years.

5. Heterogeneous Stocks

5.1. Definition

Heterogeneous stocks are the opposite of inbred strains in that they are systematically outbred to maximize individual genetic variability and genetic differences. Individual animals from heterogeneous stocks may differ markedly in drug sensitivity and this variability may be used to correlate behavioral or neurochemical parameters with drug sensitivity (*see* Crabbe and Harris, 1991). In this way, a substantial amount of variability in drug sensitivity can be obtained without the time and expense required for selective breeding.

5.2. Applications

One example of the use of HS mice is as a verification of results obtained with LS/SS mice. Because no replicate lines are available for the LS/SS mice, HS mice found to be sensitive (HS–LS) or resistant (HS–SS) to ethanol-induced loss of righting reflex were tested for pentobarbital-induced loss of righting reflex and later killed for preparation of brain membranes for analysis of ethanol and pentobarbital augmentation of GABA-activated $^{36}Cl^-$ influx (Allan and Harris, 1989). Results from the individual HS mice confirmed results from LS/SS mice showing that ethanol and pentobarbital sensitivity did not covary, but behavioral ethanol sensitivity was correlated with ethanol sensitivity of GABA-activated chloride channels. Although comparisons of the extremes of a segregating population do not provide estimates of genotypic correlations (for the reasons discussed in Section 5.2.), taken together with the data from LS/SS mice, they provide further evidence that GABA-activated chloride channels are important for ethanol-induced loss of righting reflex and that the mechanism of action of pentobarbital is distinct from that of ethanol. An important use of heterogeneous stocks is that they represent an excellent starting population for selective breeding because of their genetic diversity. Heterogeneous mice (HS/Ibg) are available from the Institute for Behavioral Genetics in Boulder, CO.

5.3. Problems

There are, however, two major shortcomings of correlations obtained with individual HS animals. First, any correlation obtained with these animals (e.g., sensitivity to drug X vs sensitivity to drug Y) is a phenotypic correlation, not a genotypic correlation (as would be obtained from selected lines or inbred strains). This is because the variability among animals is owing to both genetic and environmental factors. It is difficult to determine the relative contribution of the genetic component, but as Crabbe et al. (1990) noted, studies have shown that 70–80% of the variance in behavioral sensitivity to ethanol is environmental, leaving only 20–30% as genetic. It is likely that a similar relationship will be found for other drugs. However, it has also been shown that environmental and genetic variables may also be correlated. Thus, a phenotypic correlation may be similar to the genetic correlation, but this cannot be assumed *a priori* (Crabbe et al., 1990). This will not be a problem if the goal is to correlate differences in sensitivities regardless of whether they are genetically or environmentally based.

A second problem is that each animal is unique; thus, the drug sensitivity of any animal can be ascertained only after testing, and that test procedure (e.g., drug administration) may itself alter subsequent behavior or neurochemistry of the animal. Another aspect of the genetic uniqueness of each HS animal is that a given result will, theoretically, occur in only one animal, never to be repeated again. This is the opposite of inbred strains, where genetically identical animals can be obtained over a period of years. Nonetheless, when replicated selected lines are unavailable to confirm the presence of a greater correlation, it may be useful to examine segregating populations.

6. Combining Classical and Molecular Genetics

6.1. Expression Systems

A current goal (some might say fantasy) of pharmacogeneticists is to use genetic differences in drug response to identify the genes or gene products responsible for drug actions.

One strategy for accomplishing this is to begin by expressing DNA or mRNA from selected lines (or inbred strains) and determining a difference in the function of the resultant proteins that might reflect differences in drug sensitivity. A first step is to express brain (or brain regional) mRNA in *Xenopus* oocytes, which allows study of many neuronal functions, such as ion channels and neurotransmitter receptors coupled with phospholipase C (Snutch, 1988). As shown in Figs. 6–8, oocytes are isolated from frog ovary, injected with mRNA, and the resultant ion channels are inserted into the membrane. The large size of the oocyte allows for two-electrode voltage-clamping as well as injection of chemicals into the oocyte. Another advantage of oocyte expression is that the mRNA contains information from all of the genes that are actively expressed in a given tissue (e.g., brain region), and, at present, the oocyte is the only system that will express this mixture of RNA. The use of expression systems to study mechanisms of drug action and to discover new drugs was reviewed by Lester (1988).

An example of this approach is the study of Wafford et al. (1990), who expressed total brain mRNA from LS and SS (alcohol-sensitive and -insensitive) mice in *Xenopus* oocytes and measured the function of the ion channels produced from this mRNA electrophysiologically. The GABA-activated chloride channels expressed from LS mRNA were enhanced by ethanol whereas those channels expressed from SS mRNA were inhibited by ethanol. Despite this marked difference, many other channel functions were not different between LS and SS, including GABA action, enhancement of GABA action by pentobarbital or diazepam, NMDA action, and inhibition of NMDA action by ethanol (Wafford et al., 1990). This provides a candidate gene family (the $GABA_A$ receptor complex) that may be a site of action of ethanol and may be responsible for mediating genetic differences in alcohol sensitivity. Identification of the gene or genes responsible for the difference in ethanol action on oocytes expressing LS or SS mRNA is complicated by the complex nature of the $GABA_A$ receptor, which contains four or five different protein subunits; furthermore, these subunits are derived from a number of different genes (Vicini, 1991). However, there is still the

mRNA injections

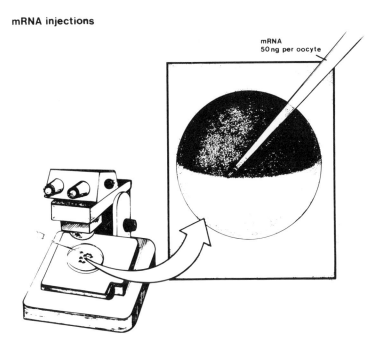

mRNA
50 ng per oocyte

Fig. 6. Oocyte expression: injection of mRNA. Individual oocytes are isolated from ovarian tissue of the South African clawed frog *Xenopus laevis* and the attached follicular cells are removed. They are then individually injected with mRNA. The injection vol is about 50 nL and the amount of brain mRNA injected is 50–100 ng.

possibility of cloning different subunit genes from LS and SS mice and expressing these subunits to define which are critical for ethanol enhancement or inhibition of GABA responses. Cloned genes may be expressed by transfecting cultured cells with DNA, or by making RNA from the cloned DNA, and using this RNA for oocyte expression. Cells transfected with DNA usually express the proteins of interest only transiently, although permanent transfection has been achieved in some cases (Julius et al., 1989). For functional studies (electrophysiology, receptor binding, enzyme activity), mammalian cells are the usual choice for transfection, but if production of a large amount of protein is the goal, then bacterial or insect cells may be superior.

Receptor Expression

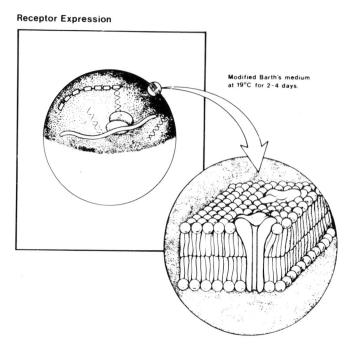

Modified Barth's medium
at 19°C for 2-4 days.

Fig. 7. Oocyte expression: receptor synthesis and assembly. The injected mRNA is translated by the oocyte protein synthesis machinery, modified posttranslationally, and assembled into a functional complex that is inserted into the oocyte membrane. Possible posttranslational modifications include glycosylation and phosphorylation. Expression of receptor function is often detectable 1 or 2 d after injection of mRNA and reaches a maximum after 3–5 d.

6.2. Gene Cloning and Sequencing

The traditional approach to identifying a gene responsible for a genetic difference would be to clone that gene, determine its sequence, and express it to show that it confers the proper functional phenotype. This could be accomplished by identifying a candidate gene that had already been cloned and sequenced. DNA libraries could then be constructed from the selected lines (or inbred strains) of interest and these libraries could be screened with a probe based on the known sequence of the candidate gene. In this way, the gene would be cloned from the selected lines, sequenced, and expressed. This has been successful in the case of the Dahl salt-sensitive and -resistant rats, where the alpha

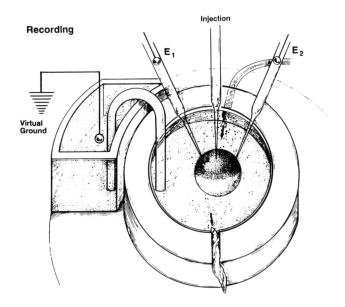

Fig. 8. Oocyte expression: electrophysiological recording. The oocyte is placed in a bath and continually superfused with medium to which drugs may be added. It is impaled with a current passing and a recording electrode to allow clamping of the membrane potential at the desired level (two-electrode voltage clamp). In this mode, effects of drugs are recorded as changes in membrane current. The oocyte may also be impaled with a third pipet to allow intracellular injection of substances during recording. This technique may be used to introduce enzymes (kinases, phosphatases), calcium chelators (EGTA), or membrane-impermeable drugs or peptides. $E_1 =$ current passing electrode; $E_2 =$ recording electrode.

subunit of kidney Na^+,K^+-ATPase was selected as a candidate gene and cloned from the two lines of rats. A sequence comparison indicated that a change in a single nucleotide resulted in an amino acid change that altered the function of the enzyme. This change in function was the same as that observed when kidney total mRNA was expressed in oocytes (Herrera and Ruiz-Opazo, 1990). Such success has yet to be achieved with lines selected for behavioral or drug-related phenotypes. The requirement of accurately selecting a candidate gene, and preferably one for which the complete sequence is available, is particularly difficult for brain genes. A new approach to identifying candidate genes is mapping selected differences to a chromosomal location (*see* Section 6.3.). Although map distances are still too large to allow pinpoint identification of candidate genes, this area is advancing very rapidly. Another attractive, but difficult, approach is subtractive hybridization (*see* Section 6.5.), where one attempts to identify only the genes that are different between animals of two relevant genotypes.

6.3. RFLP/QTL Approaches to Gene Mapping

Gene mapping has been alluded to tangentially in the previous sections, but will now be discussed in more detail. To apply genetic animal models for this purpose, the basic approach is to establish the responses of a number of inbred strains on the trait to be mapped. This new strain distribution pattern is compared with the known SDPs for those strains, which have already been mapped to particular chromosomal locations. As previously discussed, this procedure is much more powerful when using RI strains, because of the greater density of markers. When classical RI analysis is employed and the evidence for gene linkage is taken to be the existence of a bimodal pattern in the RI strains, it is only possible to detect genes with marked influence on the trait. However, it is clear that many genes exert lesser influence, but cumulatively can have an important influence on drug responses. Such genes are called Quantitative Trait Loci (QTL), and a recently proposed analytical scheme can be applied to the BXD RI panel (Gora-Maslak et al., 1991). Comparing BXD values to the 400 known SDPs for these strains can allow the detection of associations for genes with as little as 20% influence on trait variance.

Gora-Maslak et al. (1991) recently applied this analysis to four existing reports employing BXD RI mice. In each case, the older reports had studied responses to ethanol or amphetamine and had detected significant genetic influence. In some cases, they had identified potential single gene effects from the bimodal responses seen, but were unable to establish definitive linkages. That is, the new SDP for the drug response did not exactly match any existing, mapped SDP. Seale et al. (1985) studied *d*-amphetamine effects on body temperature; Crabbe et al. (1983a) reported evidence for single genes influencing ethanol-stimulated, open-field activity, ethanol drinking under thirst motivation, and ethanol withdrawal severity; Goldman et al. (1987), as previously described, had found a gene influencing ethanol preference drinking; and Phillips et al. (1991) examined morphine consumption. Reanalysis of these data using the QTL approach revealed for each response the significant influence of a number of loci that mapped to several chromosomes (Gora-Maslak et al., 1991). These loci were in addition to the major gene effects described by the earlier studies.

6.4. Prospects for the Future

Pharmacogenetic research with genetic animal models has been built on a strong quantitative foundation. One version of this model is elegantly presented by Falconer (1989). This theoretical framework is ideal for the description and analysis of polygenic traits, but it has been little applied by researchers studying drugs of abuse in cases where single genes may exert major influence. Conversely, the great strides made over the last decade in application of molecular biological tools to basic issues in neuroscience have been largely restricted to the manipulation of single genes. Thus, most research at the molecular level has been conducted with candidate genes.

The recent advances in mapping strategies (e.g., QTL analysis, and the several methods that have evolved from the Human Genome Project) are very exciting because they suggest that a fruitful marriage of molecular techniques and use of genetic animal models is beginning to occur. In addition, rapid progress in understanding the molecular basis of such human hereditary diseases as cystic fibrosis and Huntington's Disease (Wexler et al., 1991) will provide techniques and paradigms for animal genetic studies. The application to the study of drugs of abuse seems likely to lead this field because of the long history of research in genetic animal models. For example, the QTL approach to gene mapping is essentially made feasible by the ability to generate new markers from RFLPs. Indeed, one investigator largely responsible for the initial findings in alcohol pharmacogenetics has recently proposed developing selected lines from RFLPs known to be associated with QTLs determining alcohol-related traits (G. E. McClearn, personal communication). The great statistical power of multivariate approaches to gene mapping is highly developed in the context of agricultural genetics (hence the genesis of the QTL approach; *see* Lander and Botstein, 1989).

The methods of molecular biology have significant potential for exploiting existing genetic animal models. For example, there is known to be a substantial genetic variability in the effects of ethanol on the GABA-benzodiazepine receptor complex (Allan and Harris, 1991). Genes for subunits of the $GABA_A$ receptor have been cloned, and these subunits display different cellular and brain regional localization providing for great functional diversity, and may be responsible for genetic differences in sensitivity to some drugs (Luddens and Wisden, 1991; Wafford et al., 1991). However, virtually nothing is known about the basis for strain differences in the GABA system. The approach described in Section 3. of establishing genetic correlation in panels of inbred strains could fruitfully be applied to the analysis of GABA receptor pharmacology.

A second application of molecular biological techniques is the use of subtractive hybridization. Milner and Sutcliffe (1988) have described procedures for comparing the genetic material of two groups directly to establish regions of genetic similarity based on crosshybridization. One could envision the application of such a technique to the analysis of the DNA from two lines selectively bred for differences in sensitivity to a drug. Repeated subtraction could allow one to eliminate large regions of genetic similarity and eventually have only those chromosomal regions that differ. This could be enhanced by using as the starting material a small region of a particular chromosome identified by QTL analyses as the location of a group of genes influencing the drug trait. Such a region could be amplified using polymerase chain reaction to provide sufficient starting material. Whereas current methods suffer from a signal-to-noise ratio making this infeasible at the moment, it seems likely that technical advances in molecular biology could allow this type of analysis in the foreseeable future.

In summary, the use of genetic animal models provides one of the most powerful approaches to determine if there is a relationship between drug sensitivity and another behavioral action or neurochemical action of the drug. In addition, genetic differences in drug sensitivity may prove crucial in determining molecular bases of drug action as molecular and classical genetics are linked by such techniques as gene mapping and expression. An area for future research is the selection of lines based on neurochemistry (e.g., drug receptor density, neurotransmitter levels) and on QTL markers. The approaches could complement transgenic animals that are constructed to over- or underexpress a single gene. These new genetic approaches will lead to revolutionary advances in our understanding of drug action.

References

Allan A. M. and Harris R. A. (1989) Sensitivity to ethanol hypnosis and modulation of chloride channels does not co-segregate with pentobarbital sensitivity in HS mice. *Alcoholism: Clin. Exp. Res.* **13,** 428–434.

Allan A. M. and Harris R. A. (1991) Neurochemical studies of genetic differences in alcohol actions, in *The Genetic Basis for Alcohol and Drug Actions* (Crabbe J. C. and Harris R. A., eds.) Plenum, New York, NY, pp. 105–152.

Baran A., Shuster L., Eleftheriou B. E., and Bailey D. W. (1975) Opiate receptors in mice: Genetic differences. *Life Sci.* **17,** 633–640.

Belknap J. K. and O'Toole L. A. (1991) Studies of genetic differences in response to opioid drugs, in *The Genetic Basis for Alcohol and Drug Actions* (Crabbe J. C. and Harris R. A., eds.) Plenum, New York, NY, pp. 225–252.

Collins A. C. and Marks M. J. (1991) Genetic studies of nicotinic and muscarinic agents, in *The Genetic Basis for Alcohol and Drug Actions* (Crabbe J. C. and Harris R. A., eds.) Plenum, New York, NY, pp. 323–352.

Crabbe J. C. and Harris R. A., eds. (1991) *The Genetic Basis for Alcohol and Drug Actions.* Plenum, New York, NY.

Crabbe J. C., Kosobud A., Young E. R., and Janowsky J. S. (1983a) Polygenic and single-gene determination of response to ethanol in BXD/Ty recombinant inbred mouse strains. *Neurobehav. Toxicol. Teratol.* **5,** 181–187.

Crabbe J. C., Young E. R., and Kosobud A. (1983b) Genetic correlations with ethanol withdrawal severity. *Pharmacol. Biochem. Behav.* **18(Suppl. 1),** 541–547.

Crabbe J. C., Phillips T. J., Kosobud A., and Belknap J. K. (1990) Estimation of genetic correlation: Interpretation of experiments using selectively bred and inbred animals. *Alcoholism: Clin. Exp. Res.* **14,** 141–151.

Cunningham C. L., Niehus D. R., Malott D. H., and Prather L. K. Genetic differences in the rewarding and activating effects of morphine and ethanol. *Psychopharmacol.,* in press.

De Fries J. C., Wilson J. R., Erwin V. G., and Petersen D. R. (1989) LS X SS recombinant inbred strains of mice: Initial characterization. *Alcoholism: Clin. Exp. Res.* **13,** 196–200.

Falconer D. S. (1989) *Introduction to Quantitative Genetics:* 3rd ed. Longman, Essex, U.K.

Feller D. J. and Crabbe J. C. (1991) Effects of alcohols and other hypnotics in mice selected for differential sensitivity to hypothermic actions of ethanol. *J. Pharmacol. Exp. Ther.* **256,** 947–953.

Gallaher E. S. and Crabbe J. C. (1991) Genetics of benzodiazepines, barbiturates and anesthetics, in *The Genetic Basis for Alcohol and Drug Actions* (Crabbe J. C. and Harris R. A., eds.) Plenum, New York, NY, pp. 253–277.

George F. R. (1990) Genetic approaches to studying drug abuse: Correlates of drug self-administration. *Alcohol* **7,** 207–211.

Goldman D., Lister, R. G., and Crabbe J. C. (1987) Mapping of a putative genetic locus determining ethanol intake in the mouse. *Brain Res.* **420,** 220–226.

Gora-Maslak G., McClearn G. E., Crabbe J. C., Phillips T. J., Belknap J. K., and Plomin R. (1991) Use of recombinant inbred strains to identify quantitative trait loci in psychopharmacology. *Psychopharmacology* **104,** 413–424.

Harris R. A. and Allan A. M. (1989) Alcohol intoxication: Ion channels and genetics. *FASEB J.* **3,** 1689–1695.

Henderson N. D. (1989) Interpreting studies that compare high- and low-selected lines on new characters. *Behav. Genet.* **19,** 473–502.

Herrera V. L. M. and Ruiz-Opazo N. (1990) Alteration of alpha1 Na^+,K^+-ATPase $^{86}Rb^+$ influx by a single amino acid substitution. *Science* **249,** 1023–1026.

Horowitz G. P. and Dudek B. C. (1983) Behavioral pharmacogenetics, in *Behavior Genetics: Principles and Applications* (Fuller J. L. and Simmel E. C., eds.), Lawrence Erlbaum Associates, Hillsdale, NJ, pp. 117–154.

Julius D., Livelli T., Jessell T. M., and Axel R. (1989) Ectopic expression of the serotonin 1c receptor and the triggering of malignant transformation. *Science* **244,** 1057–1062.

Kalow W. (1962) *Pharmacogenetics: Heredity and the Response to Drugs.* Saunders, Philadelphia, PA.

Katz R. J. and Doyle R. C. (1980) Enhanced responses to opiates produced by a single gene substitution in the mouse. *Eur. J. Pharmacol.* **68,** 229–232.

Lander E. S. and Botstein D. (1989) Mapping Mendelian factors underlying quantitative traits using RFLP linkage. *Genetics* **121,** 185–189.

Lester H. A. (1988) Heterologous expression of excitability proteins: Route to more specific drugs? *Science* **241,** 1057–1063.

Luddens H. and Wisden W. (1991) Function and pharmacology of multiple GABA$_A$ receptor subunits. *Trends Pharmacol. Sci.* **12,** 49–51.

Mardones J. (1951) On the relationship between deficiency of B vitamins and alcohol intake in rats. *Q. J. Stud. Alcohol* **12,** 563–575.

McClearn G. E. and Rodgers D. A (1959) Differences in alcohol preference among inbred strains of mice. *Q. J. Stud. Alcohol* **20,** 691–695.

Meyer U. A. (1990) Molecular genetics and the future of pharmcogenetics. *Pharmacol. Ther.* **46,** 349–355.

Milner R. J. and Sutcliffe J. G. (1988) Molecular neurobiological strategies applied to the nervous system, in *Discussions in Neurosciences* (Magistretti P., ed.) Elsevier, Amsterdam, vol. **V,** no. 2, pp. 1–63.

Phillips T. J., Belknap J. K., and Crabbe J. C. (1991) Use of recombinant inbred strains to access vulnerability to drug abuse at the genetic level. *J. Addict Disease* **10,** 73–87.

Phillips T. J. and Crabbe J. C. (1991) Behavioral studies of genetic differences in alcohol action, in *The Genetic Basis for Alcohol and Drug Actions* (Crabbe J. C. and Harris R. A., eds.) Plenum, New York, NY, pp. 25–104.

Seale T. (1991) Genetic differences in response to cocaine and stimulant drugs, in *The Genetic Basis for Alcohol and Drug Actions* (Crabbe J. C. and Harris R. A., eds.), Plenum, New York, NY, pp. 279–321.

Seale T. W., Carney J. M., Johnson P., and Rennert O. M. (1985) Inheritance of amphetamine-induced thermoregulatory responses in inbred mice. *Pharmacol. Biochem. Behav.* **23,** 373–377.

Seale T. W., Abla K. A., Roderick T. H., Rennert O. M., and Carney J. M. (1987) Different genes specify hyporesponsiveness to seizures induced by caffeine and the benzodiazepine inverse agonist, DMCM. *Pharmacol. Biochem. Behav.* **27,** 451–456.

Shuster L. (1975) Genetic analysis of morphine effects: Activity, analgesia, tolerance and sensitization, in *Psychopharmacogenetics* (Eleftheriou B. E., ed.) Plenum, New York, NY, pp. 73–97.

Snutch T. P. (1988) The use of *Xenopus* oocytes to probe synaptic communication. *Trends Neurosci.* **11,** 250–256.

Spuhler K. P., Hoffer B., Weiner N., and Palmer M. (1982) Evidence for genetic correlation of hypnotic effects and cerebellar Purkinje neuron depression in response to ethanol in mice. *Pharmacol. Biochem. Behav.* **17,** 569–578.

Taylor B. A. (1972) Genetic relationships between inbred strains of mice. *J. Hered.* **63,** 83–86.

Taylor B. A. (1989) Recombinant inbred strains, in *Genetic Variants and Strains of the Loboratory Mouse.* 2nd ed. (Lyon M. F. and Searle A. G., eds.), Oxford University Press, New York, pp. 773–796.

Vesell E. S. (1973) Advances in pharmacogenetics. *Prog. Med. Genet.* **9,** 291–367.

Vicini S. (1991) Pharmacologic significance of the structural heterogeneity of the GABA$_A$ receptor-chloride ion channel complex. *Neuropsychopharmacology* **4,** 9–15.

Wafford K. A., Burnett D. M., Dunwiddie T. V., and Harris R. A. (1990) Genetic differences in ethanol sensitivity of GABA$_A$ receptors expressed in *Xenopus* oocytes. *Science* **249,** 291–293.

Wafford K. A., Burnett D. M., Leidenheimer N. J., Burt D. R., BeiWang J., Kofuji P., Dunwiddie T. V., Harris R. A. and Sikela J. M. (1991) Ethanol sensitivity of the GABA$_A$ receptor expressed in *Xenopus* oocytes requires eight amino acids contained in the γ_{2L} subunit. *Neuron,* **7,** 27–33.

Wexler N. S., Rose E. A., and Housman D. E. (1991) Molecular approaches to hereditary diseases of the nervous system: Huntington's disease as a paradigm. *Ann. Rev. Neurosci.* **14,** 503–529.

Yoshimoto K. and Komura S. (1987) Reexamination of the relationship between alcohol preference and brain monoamines in inbred strains of mice including senescence-accelerated mice. *Pharmacol. Biochem. Behav.* **27,** 317–322.

Alcohol Tolerance

Methodological and Experimental Issues

Anh Dzung Lê, S. John Mihic, and Peter H. Wu

1. Introduction

Drug tolerance can be broadly defined as a reduction in the effects of a drug as a consequence of repeated exposure to that drug. This is shown as a shift to the right of the dose–response curve (DRC) of a drug, meaning that a higher dose of the drug is required to produce the same degree of effect in a tolerant individual as that seen in a naive subject (Kalant et al., 1971). The field of drug tolerance has been studied by investigators with a variety of interests, ranging from clinical pharmacological problems to tolerance as a form of biological plasticity. It is, however, in the study of drugs of abuse that the phenomenon of tolerance has been most extensively examined. For example, tolerance to opiates and alcohol has been documented in literature as early as the 19th century (Goudie and Emmett-Oglesby, 1989). The main reason for such an interest in tolerance in the field of drug addiction is that it is widely believed to be a critical factor in the acquisition and maintenance of the addictive process (Kalant, 1988).

Traditionally, tolerance has been the domain of pharmacologists or biologists, who viewed it as an adaptive response to the effects of a drug in the body (Kalant et al., 1971; Poulos and Cappell, 1991). Over the last three decades, however, the study of tolerance and its underlying mechanism(s) has become a subject of interest to researchers in a variety of disciplines. A recent review by Goudie and Emmett-Oglesby (1989) of the tol-

From: *Neuromethods, Vol. 24: Animal Models of Drug Addiction*
Eds: A. Boulton, G. Baker, and P. H. Wu ©1992 The Humana Press Inc.

erance literature published between 1966 and 1986 showed that the field has expanded greatly in the last 20 years. Approximately 100 papers on tolerance were published in 1966, increasing in a linear fashion to about 500 papers in 1986. As a direct consequence, many varied perspectives or opinions on what mechanisms underly tolerance have been proposed. There is, however, general agreement that tolerance is a complex phenomenon and its development and expression are regulated by pharmacological, behavioral, and genetic factors (Kalant, 1985, 1988; Kalant et al., 1971; Lê, 1990; Lê and Kalant, 1990).

The main purpose of this chapter is to examine various methodological approaches that have been employed for the production and measurement of functional central nervous system (CNS) tolerance to alcohol. The emphasis is placed on in vivo or whole animal studies, particularly in rodents, because most studies of ethanol tolerance involve the use of these animals. The methods employed to measure or quantify the effects of ethanol are described first. A description of the various types of tolerance (chronic, acute, and rapid) and how they are assessed is then discussed, followed by a detailed description of the experimental approaches that have been employed to study these various forms of tolerance. Problems and issues associated with the study of drug tolerance are also examined.

Although it is not the intention of this chapter to cover fully the huge amount of literature on various behavioral mechanisms or factors that regulate tolerance development, it is, however, quite important to note that the effect of behavioral and environmental processes on tolerance depends on how tolerance is produced, as well as how it is assessed. The behavioral factors shown to affect tolerance development are described and the methods used to produce and assess behaviorally augmented tolerance are discussed.

2. Methods for Measuring the Effects of Ethanol

Any behavioral, physiological, or biochemical process that is affected by alcohol (e.g., motor incoordination, hypothermia) can be measured to assess tolerance. Depending on the dose used, ethanol affects many behavioral and biological processes. The effects of ethanol can be measured in either whole animal preparations (in vivo) or in in vitro preparations, such as the longitudinal muscle myenteric preparation (Mayer et al., 1980), brain tissue slices (Carlen and Corrigall, 1980; Palmer et al., 1985), and neuronal cell cultures (Charness et al., 1983). Ethanol's effects, and the tolerance that develops to them, can also be measured on cell membrane and subcellular preparations using electrophysiological and biochemical techniques (*see* Buck and Harris, 1991). In whole animal studies, one may investigate either the spontaneous effects of ethanol on various physiological systems, such as changes in core body temperature (*see* Kalant and Lê, 1984), or its effects on the performance of certain tasks requiring prior training (behavioral tests). A variety of whole animal physiological and behavioral tests and some of the more common in vitro preparations used to study ethanol's effects and ethanol tolerance are shown in Table 1.

The choice of which ethanol effect to measure is generally governed by the research interests and scientific background of the investigator and the resources available. The effects of ethanol, and the tolerance arising to them, can be measured quantally or quantitatively. Any effect of ethanol, with the exception of lethality, probably can be assessed by quantal as well as quantitative dose–response curves. For example, the loss of righting reflex (LRR) induced by ethanol can be measured quantally by examining the ethanol doses required to produce LRR in 25, 50, and so on to 100% of the subjects (Litchfield and Wilcoxon, 1949; Melchior, 1990). The duration of LRR produced by various doses of ethanol can, in contrast, be used to determine quantitatively the DRC (Tabakoff et al., 1980).

A change in the ED_{50} of an opiate is often used to assess opiate tolerance; an example is the increase in ED_{50} of morphine-induced analgesia associated with tolerance (Fernandes et al., 1977; Kalant, 1977). This quantal approach, however, is not common in studies of the effects of ethanol or the tolerance developing to them. The relatively narrow range of the ethanol DRC and the low degree of tolerance observed (Kalant, 1977, 1989)

Table 1

Physiological and Behavioral Tests and Some Common In Vitro Preparations Used to Study Ethanol's Effects and Ethanol Tolerance

Test procedures	Dose ranges,[a] g/kg	References
Physiological tests		
Lethality	> 4.0	Melchior, 1990
Sleep time	3.0–4.5	Tabakoff et al., Khanna et al., 1989
Hypothermia	1.5–4.0	Ritzmann and Tabakoff, 1976; Lê et al., 1981.
Convulsions	1.0–3.0[b]	McQuarrie and Fingl, 1958; Pinel et al., 1985
Heart rate	1.0–2.0	Pohorecky et al., 1986; Tiffany et al., 1987
Analgesia	0.5–2.0	Pohorecky and Shah, 1987
Spontaneous behavioral tests		
↑ Motor activity	0.25–2.0[c]	Masur and Boerngen, 1980; Masur et al., 1986
↓ Motor activity	> 2.0[c]	Hunt and Overstreet, 1977
Startle threshold	1.0–3.0	Gibbins et al., 1971
Tilting plane test	2.0–3.0[d]	Arvola et al., 1958; Khanna et al., 1991
Dowel test	2.0–3.0[d]	Dunham and Miya, 1957; Gallaher et al., 1982
Jumping test	1.5–2.5	Tullis et al., 1977
Behavioral tests with prior training		
Conflict behavior	0.5–0.8	Koob et al., 1987
Maze test	0.5–1.0	Chen, 1968; LeBlanc et al., 1973
Avoidance task	1.0–2.0	Lê et al., 1986
Operant responding	0.5–1.5	Holloway et al., 1988
Rotarod	1.0–2.0[d]	Jones and Roberts, 1968
Moving belt	1.2–2.2[d]	Gibbins et al., 1968
In vitro preparations[e]		
Brain tissue slices		Carlen and Corrigall, 1980; Palmer et al., 1985
Aplysia abdominal ganglion		Traynor et al., 1979
Neuronal and other cell cultures		Charness et al., 1983; Gordon et al., 1986
Guinea pig illeum		Mayer et al., 1980
Neuromuscular junction		Curran and Seeman, 1977

[a]Since rats are genrally more sensitive to ethanol than mice, the doses required for the rats are generally lower than those required for the mice.

[b]The dose range is dependent on the seizure model employed: electroshock or kindled seizure.

[c]The biphasic effect of ethanol on motor activity is readily seen in mice. In the rat, the increase in activity produced by ethanol is much more difficult to observe.

[d]These tests generally measure the motor-impairing effects of ethanol. The complexity of each test, however, varies greatly.

[e]The concentration of ethanol is not listed because it depends on the parameters that one wishes to measure.

are the main reasons for the limited use of such an approach. This is one of the major disadvantages of all-or-none response studies, i.e., they cannot detect or differentiate among small changes in the effect of ethanol. Because large sample sizes are required, these types of studies are usually performed using mice (Kalant and Khanna, 1990).

Quantitative responses, permitting one to observe small changes in the effects of ethanol, have been more commonly used during the last two decades. These types of measurements permit one to investigate complex phenomena, ranging from the kinetic processes of tolerance development (LeBlanc et al., 1973,1975a) to the mechanisms of tolerance processes (Lê et al., 1981). Tolerance to a number of different behavioral or physiological effects of ethanol can be measured concurrently in a single individual, including correlations between behavioral and biochemical changes. In addition, various manipulations that might affect tolerance can also be examined. The development of sensitive behavioral and biochemical tests, allowing for small changes in the effects of ethanol to be observed, has also encouraged the use of quantitative measures. For example, the motor-impairing effects of ethanol can be measured using doses as low as 1.2 g/kg on the moving belt test (Gibbins et al., 1968), in contrast to a minimal dose of 2 g/kg that might be required for the tilting plane test (Arvola et al., 1958).

3. Types of Tolerance to Ethanol

3.1. Dispositional Tolerance

The intensity and duration of the effects of ethanol are governed by various pharmacokinetic factors: absorption, distribution, and elimination, which determine the availability of ethanol to the brain. Changes in any of these parameters, following the repeated administration of ethanol, lead to a decrease in the amount of ethanol in the brain and a shorter duration of ethanol action. The consequent reduction in the overall effects of ethanol is referred to as dispositional tolerance (*see* Kalant et al., 1971; Lê and Khanna, 1989). Among these three parameters, the increase in the rate of ethanol metabolism induced by repeated exposure (as much as 30–40% above basal) is viewed as the most important component of dispositional tolerance (Kalant et al., 1971; Lê and Khanna, 1989). Changes in the rates of ethanol absorption and distribution following repeated treatment have also been shown to occur and can extensively modify the behavioral effects of ethanol (Melchoir and Tabakoff, 1985; El Ghundi et al., 1989). The magnitude of such changes, however, greatly depends on the paradigm employed for repeated ethanol administration, as discussed in the subsequent section.

3.2. Functional Tolerance

Functional tolerance is usually referred to as a reduction in neuronal sensitivity to a given dose of ethanol, following repeated ethanol administration. Experimentally, it can be measured directly by comparing the behavioral or physiological effects of ethanol with brain ethanol concentration (BrEC) or blood ethanol concentration (BEC). Most often, however, it is defined by means of exclusion: It is a reduction in drug effect that cannot be accounted for by changes in dispositional factors. Based on the time-frame of development, there are three categories of tolerance.

3.2.1. Acute Tolerance

Acute tolerance is seen during a single course of drug exposure. This form of tolerance was first described by Mellanby (1919), who reported that at the same level of BEC, the intoxication displayed was much less severe on the descending than the ascending portion of the time–BEC curve. Acute tolerance develops very rapidly (within a matter of minutes) (Palmer et al., 1985). It is not an all-or-none phenomenon; the degree of acute tolerance becomes greater as the duration of ethanol exposure increases (LeBlanc et al.,1975b).

3.2.2. Rapid Tolerance

This phenomenon was first described by Crabbe et al. (1979), who reported tolerance to the hypothermic effect of a second dose of ethanol given 24 h after injection of the first. The major difference between acute and rapid tolerance is that tolerance to the second injection of ethanol is measured at a time when all of the ethanol from the first dose has been eliminated from the body.

3.2.3. Chronic Tolerance

This form of tolerance is produced by prolonged ethanol treatment, usually over a number of days or weeks. Therefore, it is usually accompanied by dispositional tolerance. Among the three forms of tolerance, chronic tolerance has been studied the most extensively and is probably the most complex because various behavioral processes may affect its development.

Although these forms of tolerance are usually classified according to the time-frame of their development, the amount of ethanol given and the duration of exposure should always be indicated. Acute and rapid tolerance studies are usually performed using a single-dose-administration paradigm; i.e., ethanol is provided as a single ip injection or intubation. For chronic tolerance studies, ethanol may again be given by daily intubation or injection, although exposure can also occur by means of vapor inhalation or silastic implant.

4. Experimental Approaches to Study Acute and Rapid Tolerance to Ethanol

4.1. Acute Tolerance

4.1.1. Experimental Approaches to Demonstrate Acute Tolerance

After the administration of a single dose of ethanol, the BEC and BrEC rise during the ascending phase of the time–BEC curve, peak, and then decline in the descending phase. There are two ways in which one can demonstrate the development of acute tolerance to ethanol: (1) a lower degree of intoxication seen during the descending phase than the ascending phase, at the same BEC or BrEC; or (2) the same degree of intoxication observed but with a higher BEC (or BrEC) measured on the descending phase. The time elapsing after ethanol exposure and the measurement of BEC or BrEC are critical factors in the study of acute tolerance to ethanol. Over the past few years, a number of experimental approaches have been used to demonstrate this phenomenon.

4.1.1.1. RATE OF ETHANOL INFUSION AND/OR ABSORPTION

This type of approach essentially involves the manipulation of the rate of ethanol absorption and the time-dependent decreases in intoxication associated with acute tolerance. The classical study first employing this approach was that of Mirsky et al., 1941. They found that the degree of alcohol intoxication seen in hepatectomized rabbits, maintained at a predetermined blood ethanol concentration, depended on the rate of ethanol infusion or how quickly the final BEC was attained. The slower the rate of infusion, the lower the degree of intoxication observed at the final BEC. By using different routes of ethanol administration, one can manipulate the rate of ethanol absorption, thereby affecting the development of acute tolerance. For example, the degree of hypothermia produced by ethanol is much lower in mice when ethanol is infused slowly by iv administration than when it is given as a bolus ip injection, even though the final BEC achieved after iv administration is at least as high as that seen after ip injection (Gilliam, 1989).

4.1.1.2. DEGREE OF INTOXICATION AT VARIOUS TIMES
AFTER A SINGLE ADMINISTRATION OF ETHANOL

In the paradigm first used by Mellanby (1919), ethanol intoxication and BEC or BrEC are measured concurrently (in the same or different animals) at various times after the administration of a single dose of ethanol. At the same BEC or BrEC, the intoxication seen on the descending phase of the time–BEC curve is much less severe than that seen on the ascending phase (Mellanby, 1919; Tullis et al., 1977; Edwards et al., 1983). Since intoxication is usually measured very shortly after ethanol administration (on the ascending portion of the time–BEC curve), the rate of equilibration of ethanol among the arterial, venous, and brain compartments is critical. Insofar as it is the BrEC that determines the degree of behavioral intoxication measured, brain alcohol measurements at each time-point, from separate groups of animals, are usually required in these studies. The time required for equilibration among venous and arterial BEC and BrEC has been a major issue in the study of acute tolerance and will be discussed later in this chapter.

One common experimental paradigm employed in the study of acute tolerance has been the measurement of BrEC at the onset or disappearance of a physiological or behavioral marker. For example, a higher BrEC at onset than offset of the hypnotic effect of ethanol has been used as a measure of acute tolerance (Tabakoff and Ritzmann, 1979; Tabakoff et al., 1980).

4.1.1.3. DIFFERENT DOSES OF ETHANOL

BEC or BrEC measurement at a fixed end point (physiological or behavioral index) following the administration of different doses of ethanol can also be used to demonstrate acute tolerance. Maynert and Klingman (1960) found that although it takes longer for animals receiving higher doses of ethanol to recover from its effects, they also tend to have a higher BEC upon recovery than those receiving lower doses. The development of acute tolerance to the motor-impairing or hypnotic effects of ethanol in rats (LeBlanc et al., 1975b) and mice (Gallaher et al., 1982; Keir and Deitrich, 1990) has been assessed using this paradigm, with some variations in design.

4.1.1.4. TWO CONSECUTIVE INJECTIONS

Waller et al. (1983) were the first to use a paradigm involving two consecutive injections of ethanol to measure acute tolerance. After the administration of the first dose of ethanol, the BEC is measured when a certain criterion of performance is achieved. The animal is then injected with a second dose of ethanol and BEC is again determined when the criterion level of performance is achieved. The increase in BEC seen after the second injection, compared to the first, is attributable to acute tolerance. Thus far, this type of approach has been used to measure the development of acute tolerance to the motor impairment produced by ethanol. It could be used for various measures ranging from sleep time to hypothermia.

4.1.2. Issues Involved in the Study of Acute Tolerance

4.1.2.1. EQUILIBRIUM AMONG VENOUS, ARTERIAL BEC, AND BrEC

A correlation between the degree of intoxication observed and the BrEC is ideal in defining the development of acute or chronic functional tolerance. Obtaining this correlation, however, is often impractical from an experimental viewpoint because animals have to be sacrificed immediately after each physiological or behavioral measurement. Measurement of BEC is therefore commonly used as an index of BrEC in many experimental studies, because of the convenience of collecting blood samples and because BEC, unlike BrEC measurements, can be made repeatedly at various times after ethanol administration.

Since intoxication is usually assessed soon after ethanol administration, in the study of acute tolerance, the time required for arterial and venous BEC and BrEC to equilibrate is an important factor. It was once suggested that acute tolerance, as first reported by Mellanby (1919), was an experimental artifact owing to the inequilibrium between venous and arterial BEC shortly after ethanol administration (Kalant et al., 1971).

Equilibration of ethanol between the venous and arterial blood and the brain in mice and rats has been reported to occur within approx 10–30 min after ethanol administration (LeBlanc et al., 1975b, Sunahara et al.,1978; Sunahara,1979; Lumeng et al., 1982; Smolen and Smolen, 1989). The limiting factor is the time required for the BEC of the tail vein, the most common site of blood sample collection, to attain equilibrium with BrEC. Peak BEC of tail vein blood samples has been reported to occur within 10 min (Czaja and Kalant,1961; Tullis et al., 1977; Lumeng et al., 1982) after ethanol administration in the rat. However, the time required for tail vein BEC to equilibrate with BrEC occurs 15–60 min (Sunahara, 1979; Lumeng et al., 1982) after an ip administration of ethanol. In mice, the venous BECs from orbital sinus and tail veins have been shown to peak within 5 and 15 min, respectively, and to reach equilibrium 30 min after ethanol administration (Gentry et al., 1983). Goldstein (1983), on the other hand, found that tail vein BEC equilibrates with BrEC only 60 min after ethanol administration. Before equilibrium has been achieved, venous BEC, regardless of where the blood sample was collected (Lumeng et al., 1982; Goldstein, 1983; Smolen and Smolen, 1989), tends to underestimate BrEC. For this reason, less intoxication observed at a later time might simply be a reflection of the decline in BrEC owing to redistribution and not acute tolerance.

4.1.2.2. PARADIGMS AND TESTS EMPLOYED
TO MEASURE ACUTE TOLERANCE

As previously mentioned, different experimental approaches have been used to demonstrate the development of acute tolerance. Although acute tolerance has not yet been extensively studied, the available data can be used to illustrate some of the problems in interpreting results from these studies. For example, the classical approach of assessing acute tolerance as a difference in the degree of intoxication observed between the ascending and descending limbs of the time–BEC curve results in a number of problems in data interpretation. First, as previously stated, the BrEC and BEC do not reach equilibrium until about 15 min after ethanol administration. Second, during the initial absorption and redistribution period, the blood alcohol level is rising so rapidly that it is difficult to obtain reproducible results.

The ability to detect acute tolerance has also been found to vary depending on which paradigm and test are used to measure it. For example, no evidence for acute tolerance to the hypnotic effect of ethanol was found in Short Sleep (SS) mice, selected for their resistance to the hypnotic effects of alcohol, when acute tolerance was defined as a higher BrEC at offset than onset of sleep (Tabakoff and Ritzman, 1979; Tabakoff et al., 1980). However, acute tolerance to the hypnotic effect of ethanol was observed in these mice when tolerance was defined as a higher BEC and BrEC at offset of sleep from higher doses of ethanol, compared to lower doses (Keir and Deitrich, 1990).

4.2. Rapid Tolerance

Grieve and Littleton (1979) first reported the rapid onset of tolerance within hours of exposure of mice to ethanol vapor. Gallaher et al. (1982) found that the second in a series of ip injections of ethanol also resulted in tolerance to its ataxic effects, as indicated by a higher BEC at the same level of motor impairment. Although the reduction in sensitivity to ethanol observed in these studies was referred to as rapid tolerance, it is difficult, in this case, to differentiate it from acute tolerance since a sustained BEC was present between the treatment and test or supplement doses. The phenomenon of rapid tolerance was experimentally first differentiated from acute tolerance by Crabbe et al. (1979). In their studies, the hypothermic response to an ip injection of ethanol (2–4 g/kg) given to mice was attenuated by a single equivalent dose of ethanol given 24 h earlier. Tolerance was not seen if the second injection of ethanol was given 48 or 72 h after the first (Crabbe et al., 1979). It is difficult to conclude that the observed tolerance to the second test dose is a manifestation of acute tolerance since all the alcohol from the first injection should be eliminated from the body 12–16 h before the second injection. The phenomenon of rapid tolerance, assayed by two injections of ethanol given 24 h apart, has been shown to occur to a variety of effects of ethanol, ranging from motor impairment to hypnosis in mice and rats (Lê and Kiianmaa, 1988; Buck et al., 1991; Khanna et al., 1991).

The characteristics, as well as the possible mechanisms, of the rapid tolerance phenomenon have not been fully explored. Given the minimal amount of ethanol exposure required as well as the time-frame restriction, i.e., demonstrable within 24 h but not 48 h after the first injection, it is possible that rapid tolerance reflects the retention of acute tolerance.

5. Experimental Approaches to Study Chronic Tolerance to Ethanol

5.1. Overview

Chronic functional tolerance to ethanol has been the focus of most investigations on ethanol tolerance. Chronic tolerance studies can be generally divided into four different categories, based on the purpose of investigation. The first category is descriptive, involving studies on the characteristics of tolerance, such as the kinetics of its acquisition and loss, and so forth. The second area involves an examination of various behavioral and environmental factors underlying tolerance, such as classical and instrumental learning. The third category is concerned with investigations of the relationship between tolerance and other processes of addiction, such as physical dependence and alcohol intake. The last area, which has been studied most extensively, involves studies of the neurochemical and biochemical mechanisms of tolerance. These diverse interests have resulted in a variety of methods of varying degrees of complexity being employed to produce tolerance. In the last 20 years, it has become apparent that the intensity, manifestation, and rate of tolerance development can be readily modified by the manner in which ethanol is chronically administered, as well as how tolerance is monitored (Kalant, 1985,1988; Lê and Kalant, 1990).

5.2. Methods for Administering Ethanol Chronically

There are two important pharmacological/chemical features of ethanol that one has to consider when planning to administer it chronically to animals. First, ethanol has a low relative potency and at high concentrations can cause tissue damage. For this reason, ethanol must be given in a large volume; intracranial, im, or sc routes of administration are generally not practicable. Second, the oxidation of ethanol represents a substantial proportion of caloric intake in the chronically treated animals, necessitating the use of appropriate controls for the caloric value of ethanol. In addition, one must ensure that the animals' nutritional balance is maintained. Despite some of these limitations, several methods are available for the chronic administration of ethanol, to produce tolerance. The choice of method rests on the ease of drug administration, the nature and/or complexity of the study in question, as well as the resources available to the experimenter. Essentially, ethanol can be administered by either voluntary consumption or forced administration.

5.2.1. Voluntary Consumption

Some of the earliest methods employed for administering alcohol chronically involved mixing it in the drinking water and giving it as the sole fluid, or allowing animals to choose between ethanol and water solutions. To increase the amount of alcohol consumed, the ethanol solution is sometimes sweetened with saccharin or some other palatable substance (*see* Kalant and Khanna, 1990). These techniques, however, fail to control for the nutritional and caloric values of alcohol. More importantly, the amount of alcohol consumed by animals tends to show large variations within and among strains.

To address the problems of nutritional and caloric balance, as well as to enhance the amount of alcohol consumed by animals, Lieber et al. (1965) developed an alcohol-containing liquid diet. In this diet, an ethanol solution is mixed together with a nutritionally balanced diet that constitutes the sole source of both food and fluid for the animals. Control animals receive an identical experimental diet except that a caloric substitute for ethanol (e.g., sucrose) is added. Rats and mice have been shown to consume an average of 8–12 g/kg of ethanol daily on this diet and BECs exceeding 100 mg/dL have been observed in these animals. The variation in the amount of alcohol consumed among the animals can be controlled by a pair-feeding technique in which each ethanol-treated animal has its own individual pair-fed control.

Apart from interindividual variation in alcohol consumption, the other major disadvantage inherent in any method of voluntary alcohol consumption is that one cannot be certain about the exact amount of alcohol consumed and the various pharmacokinetic parameters of the BEC in relation to the estimated dose of consumption. For example, although animals might consume 8 g/kg of ethanol/d, one cannot readily predict the nature of the BEC, including the magnitude of the peak and daily fluctuation in BEC, because the pattern of alcohol consumption varies throughout the day. For these reasons, the usefulness of voluntary consumption methods to address questions on the nature of tolerance to ethanol is limited.

5.2.2. Forced Administration

Gastric intubation and ip injection are two common routes for the acute or chronic administration of ethanol. Since the dose(s) and time of administration are known to the investigator, the pharmacokinetic properties of ethanol can be estimated reliably. However, it should be noted that the large volume in which ethanol is often administrated, as well as the stress of intubation and injection, are problems associated with the use of these techniques. Repeated ip injections can also produce intestinal damage, particularly when a high concentration of ethanol is used, altering the rate of ethanol absorption (Hawkins et al.,1966). The daily routine associated with gastric intubation and ip injection, such as daily weighing, removal of the animal from its cage, inserting of the needle, and so on, can serve as a predictive cue of the effects of ethanol for the animals (Siegel, 1983). This associative learning can profoundly affect tolerance, as will be discussed later.

The chronic delivery of ethanol through a tube implanted sc under the neck of the rat was first used by Erickson et al. (1978). Ethanol diffuses out of the silastic tube and is absorbed in the circulation. This technique has had limited use, however, because the BEC achieved is relatively low, and sc administration offers no advantages over other available routes. Ethanol can also be given through an (iv) cannula. Because the rate of ethanol metabolism is well defined, it is possible to maintain a constant desired BEC in the animal by varying the concentration or the rate of infusion of ethanol (Kalant et al., 1986). This route of administration, however, has also had limited use in chronic studies since implantation of indwelling cannulae requires considerable skill as well as being labor-intensive. Further, the ease of use of the vapor inhalation model also detracts from the use of this method.

From a pharmacokinetic and experimental viewpoint, the best route available for the chronic delivery of ethanol is by the vapor inhalation method (Goldstein and Pal, 1971). Ethanol, because of its volatile nature, can be given in the inspired air at a predetermined concentration. The major advantages of this technique are that the BEC can be readily estimated through the animal's expired air and that, as a result of the continuous exposure to ethanol vapor, fluctuations in BEC are minimal compared to other methods of administration. Another advantage of this technique is that the inhalation chamber can be built to house a large number of mice or rats (6–10), facilitating the simultaneous treatment of large numbers of animals. However, since profoundly drunk animals may not eat enough, one must monitor them closely to ensure that a nutritional deficiency does not become an unwanted experimental variable.

5.3. Behavioral Processes Associated with Ethanol Treatment and Tolerance Development

It has still not been conclusively determined, from a theoretical viewpoint, if behavioral factors affecting tolerance represent a separate process or just simply a modulation of tolerance development. It is, however, well documented that tolerance development and manifestation can be affected by various behavioral processes (Kalant et al., 1971; Kalant, 1988). The theoretical issues concerning the role of behavioral factors in tolerance are not considered here, but their effects on tolerance from an experimental viewpoint are discussed.

5.3.1. Pavlovian Conditioning

Rituals and environmental cues associated with the repeated administration of ethanol have been shown to be critical determinants of tolerance (Lê et al., 1979; Melchior and Tabakoff, 1981). These cues, serving as conditioned stimuli, evoke a conditioned compensatory response to counteract the drug's effects, resulting in tolerance development (Siegel, 1983). When chronic treatment was carried out using a paradigm of daily or bidaily injections of ethanol, tolerance was seen only when animals were tested with ethanol in the presence of environmental or ritual cues associated with chronic treatment (Lê et al., 1979,1987b; Melchior and Tabakoff, 1981). The influence of such conditioning processes on tolerance depends on the treatment dose of ethanol used; when the daily treatment dose of ethanol was 4 g/kg ip or 5 g/kg orally, tolerance could be demonstrated in the presence or absence of environmental cues (Lê et al., 1987b). However, using a daily treatment dose of 2 g/kg ip, no tolerance was seen when animals were tested in the absence of such cues, although partial tolerance was found when a dose of 2.5 g/kg was used (Lê et al., 1979,1987b).

Regardless of the route or method employed for the chronic administration of ethanol, tolerance is usually assessed using an ip injection of ethanol because of the well defined pharmacokinetic properties of ethanol absorption, redistribution, and metabolism following ip injection. The specific procedure involved in administering ethanol, as well as the pharmacokinetic properties of ethanol resulting from the use of such a route, can serve in part as conditioned stimuli for eliciting the compensatory response. The interoceptive cues resulting from a low BEC can serve as an effective conditioned stimulus for the development of tolerance to ethanol itself (Greeley et al., 1984). Similarly, tolerance to morphine has been shown to be affected by changing the route of administration during tolerance testing, compared to that used during chronic treatment (Mucha and Kalant, 1981). Consideration should therefore be given to these factors when evaluating tolerance.

When animals are chronically treated with ethanol through vapor inhalation, liquid diet, or voluntary consumption in a two-bottle choice paradigm, the influence of conditioning on the observed tolerance (tested using the ip route of administration) is minimal. The main reasons for the minimal influence of conditioning are (1) the absence of a discrete association between the drug's effects and environmental cues during chronic administration; and (2) the difference in procedure and environmental cues associated with chronic drug treatment (animals not handled in ethanol vapor chamber) and those associated with tolerance testing (animals handled during ip injection outside ethanol chamber).

5.3.2. Intoxicated Practice

Practice of a task while intoxicated was the first behavioral factor found to affect tolerance development (Chen, 1968). Animals that are forced to practice a task while intoxicated acquire tolerance more rapidly or to a greater extent than those that are not (Chen, 1968; LeBlanc et al., 1973,1975a; Wenger et al., 1980; Lê et al., 1987a,1989). The tolerance acquired using the intoxicated practice paradigm is also retained for a longer time after the cessation of ethanol treatment, compared to tolerance produced without the benefit of intoxicated practice (Lê et al., 1989).

The influence of intoxicated practice on tolerance development varies with the treatment dose employed (Lê et al., 1987a,1989). When the daily treatment dose is low, intoxicated practice can differentiate between the presence or absence of tolerance. However, when high treatment doses are used, intoxicated practice might instead modify the rate of tolerance development rather than its extent. The test dose employed to measure tolerance is also an important factor because it can affect the extent of tolerance observed in the intoxicated and nonintoxicated practice groups (Lê et al.,1989). Experimentally, one can minimize the influence of intoxicated practice by using different groups of animals, with each group being tested only once. This type of design, however, requires large numbers of animals and is more costly and labor-intensive.

6. Experimental and Interpretational Issues in the Study of Tolerance to Ethanol

6.1. Interrelationships Among Test, Treatment Dose, and Test System

A shift of the DRC and/or LDRC to the right is often used to define tolerance (Kalant et al., 1971). In vivo assessment of ethanol tolerance, however, can only be carried out using the DRC because of the narrow concentration range of the effects of ethanol. For example, when using the moving belt test to measure motor impairment, the dose range used is 1.2–2 g/kg (Gibbins et al., 1968). To test for the hypnotic effect of ethanol, 2.5–4 g/kg are used (Tabakoff and Ritzmann, 1979; Tabakoff et al., 1980). Ideally, the DRC should be used to evaluate tolerance particularly when a new task or new test system is employed or when any theoretical aspects of tolerance are studied. If only a single test dose is employed to measure tolerance and it falls on the upper plateau of the DRC, then the existence or degree of tolerance might not be seen or detected accurately. The relative importance of intoxicated practice on the development of tolerance, for example, also depends on the test dose employed (Lê et al., 1989).

An important issue often neglected in the study of tolerance is the effect of varying the intensity (e.g., higher doses) of chronic treatment on ethanol tolerance. There is considerable debate in the literature on whether tolerance is a consequence of learning (intoxicated practice, Pavlovian conditioning) or a pharmacological adaptation, or both. The importance of learning to the development of tolerance can only be assessed if one uses a variety of doses of ethanol combined with varying durations of treatment. For example, the argument that tolerance is akin to learning either in the form of intoxicated practice (Chen, 1968; Wenger et al., 1980) or conditioning (Siegel, 1975) was primarily attributable to the use of single treatment and test doses. The relative contribution of each of these factors to tolerance was clarified when different chronic treatment regimens varying in their intensities and duration were employed (LeBlanc et al., 1973,1975a; Lê et al., 1979,1987a,b,1989; Baker and Tiffany, 1985).

An equally important isssue is the relationship between treatment dose, test dose, and test system employed to evaluate tolerance. A treatment regimen, regardless of whether learning plays a role, might produce tolerance to some effects of the drug but not necessarily to all (Lê et al., 1989). The presence or absence of tolerance to various effects of a drug, using a particular treatment regimen, may also relate to the dose that is required for an effect on that test. For example, the ethanol dose required to produce hypnosis is higher than that required to produce hypothermia and motor impairment. Within a single effect of ethanol (e.g., motor impairment), the test dose required to produce an effect in different test systems also varies. A dose of 1–1.2 g/kg produces close to maximal impairment on the Rotarod test whereas a dose of 2.5–3 g/kg might be required to achieve the same level of impairment on the tilting plane test. These factors should be considered when deciding the treatment regimen, test system, and test dose to use in a tolerance study.

6.2. Necessity for the Measurement of BEC or BrEC

Ideally, BEC and/or BrEC measurements should be made to determine if the observed tolerance is attributable to functional or dispositional changes after chronic treatment. Most often, when the BEC or BrEC is determined, it is at a single time-point and usually at the end of intoxication testing (e.g., offset of

LRR), in order that behavioral measures are not affected. If blood samples are to be taken during the period of behavioral testing, care must be taken to choose an appropriate behavioral or physiological index for measuring tolerance, so as not to interfere with subsequent behavioral measures. Also important is the time required for equilibration between BEC (from the area of measurement) and BrEC. For example, the use of BEC measured from rodent tail vein at onset and offset of LRR as an index of acute tolerance is meaningless because equilibrium between tail vein BEC and BrEC is not reached within 3–4 min, the time required for LRR to occur after ip ethanol administration. In this instance, the use of BrEC is more appropriate. On the other hand, measurement of BEC at the offset of LRR in control and ethanol-treated animals is appropriate for the interpretation of functional tolerance, if such samples are taken after BEC and BrEC have equilibrated (>30 min after ip administration). The hypothermic effect of ethanol, a common index for the measurement of tolerance, can also be correlated with BEC and, by inference, BrEC. The hypothermia induced by ethanol is usually measured at 30-min intervals for a period of 120 min after ethanol administration. Since the peak hypothermic response usually occurs 30–60 min after ethanol injection, the BEC taken at the end of testing (120 min or after) might support the contention that there is no difference in the ethanol elimination rate between control and ethanol-treated animals, but it can not rule out the possibility that differences in the degree of hypothermia produced might be attributable to differences in the rate of absorption.

A number of arguments have been put forth against the necessity to make BEC or BrEC determinations. The time of behavioral or physiological measurement is often so soon after ethanol administration that metabolic factors play a negligible role, even though there might be differences in metabolic tolerance between control and experimental groups. In some experiments, two groups of rats may receive exactly the same amount of ethanol, but only one group will show tolerance. This can be seen in a study in which one treatment group receives ethanol *before* practicing a task, and another receives the same dose of ethanol *after* the task, or where a group shows tolerance in the home room but not in a different test room. It could be argued that in these cases, there is no need for BEC or BrEC measurements because all the animals are receiving the same dose of ethanol. However, some behavioral manipulations of tolerance, such as conditioning, have been shown to affect the rate of absorption and the volume of distribution of ethanol (Melchior and Tabakoff, 1985; El-Ghundi et al., 1989). Further, in instances where repeated testing is involved, the stress of testing can alter the rate of ethanol absorption (Wallgren and Tirri, 1963; El-Ghundi et al., 1989). Differences in the rate of absorption can markedly affect the intensity of a drug's effect, confounding the interpretation of results. For these reasons, a complete evaluation of the pharmacokinetic properties of ethanol in control and ethanol-treated animals is necessary in some experiments, particularly when theoretical issues in tolerance are the main theme of the study.

6.3. Other Factors

The development of acute and chronic tolerance is influenced by genetic factors that can affect tolerance indirectly through control of initial sensitivity; the acquisition of tolerance is inversely related to initial sensitivity (Crabbe et al., 1982; Khanna et al., 1985,1989). For example, C57 mice, which are initially more sensitive to the hypothermic effect of ethanol than DBA mice, acquire better tolerance to this effect of ethanol than DBA mice (Moore and Kakihana, 1978; Crabbe et al., 1982). Genetics can also directly influence the development of tolerance (Lê and Kiianmaa, 1988,1990). Alcohol tolerant (AT) rats develop better tolerance to ethanol than alcohol nontolerant (ANT) rats regardless of whether tolerance is measured to an effect of ethanol in which both lines display the same (hypothermia) or different (motor impairment) initial sensitivity (Lê and Kiianmaa, 1990).

Depending on the test employed, the ambient temperature (Ta) in the treatment and test rooms can be an important variable in the development of tolerance. For example, the degree of hypothermia produced by a dose of ethanol is related to the Ta, with more pronounced hypothermia seen at low Ta. The body temperature during intoxication can, in turn, influence the animal's sensitivity to the behavioral effects of ethanol after acute or chronic treatment (Malcolm and Alkana, 1981; Pohorecky and Rizek, 1981; Alkana et al, 1987).

The biological responses to and disposition of ethanol are influenced by circadian rhythms (Deimling and Schnell, 1980; Brick et al., 1984). This relationship between sensitivity to ethanol and circadian rhythm varies depending on the nature of the testing procedure used. For example, the greatest sensitivity to the depressant effects of ethanol, on startle response and motor activity tests, occurs during the light phase, but the greatest sensitivity to the stimulation of locomotor activity by low-dose ethanol is observed at the beginning of the dark phase (Brick et al., 1984). Since the degree of intoxication produced by ethanol, regardless of the dose used, plays a critical role in the development of tolerance, it is very likely that the time at which ethanol is administered during the light/dark cycle affects tolerance development.

In addition to the possible effects on the disposition of ethanol, stress has also been shown to affect the development of tolerance directly (Maier and Pohorecky, 1986; Peris and Cunningham, 1987). Stress can result from the handling of animals, injection and testing procedures, and from the animals undergoing ethanol-withdrawal reactions (Peris and Cunningham, 1987; Maier and Pohorecky, 1986,1987). For example, the stress resulting from handling or foot shock accelerates the development of tolerance to the hypothermic and motor-impairing effects of ethanol (Peris and Cunningham,1987; Maier and Pohorecky, 1986). When the amount of ethanol administered and total duration of treatment are controlled for, tolerance to the hypothermic and motor-impairing effects of ethanol develops more rapidly in animals repeatedly experiencing withdrawal (Maier and Pohorecky, 1987). This is an important consideration given the number of different techniques used for chronically administering ethanol and the variations in the degree of alcohol withdrawal associated with each. For example, using the vapor inhalation technique, animals are not subject to ethanol withdrawal until tested for tolerance. In contrast, single daily intubations or ip administrations of ethanol subject animals to daily withdrawal symptoms owing to the large daily fluctuations in BEC. How this might affect tolerance development has not been examined systematically.

7. Conclusions

The continuing yearly increase in the number of publications related to the study of tolerance indicates a great deal of interest in this phenomenon. All this work, at the same time, also indicates the complexity of this field. Many important issues remain to be investigated, including the relationships among the various forms of tolerance (acute, rapid, and chronic). Various methods for the treatment of animals with ethanol and paradigms to assess the resulting tolerance are now available. Each method has its own special features with regard to the pharmacokinetic properties of administered ethanol and the behavioral features associated with it. How tolerance produced by one method of treatment differs from that produced by another and the mechanisms underlying such differences are poorly understood. A systematic approach to these questions is obviously required.

Of particular importance and interest are the physiological, biochemical, and molecular mechanisms of tolerance to ethanol. One of the major problems associated with these studies is that investigators often pay little attention to the various behavioral characteristics of tolerance and how tolerance is produced. For example, tolerance has been shown to develop at different rates to different drug effects (Lê et al., 1979; Melchior and Tabakoff, 1981) and that, depending on how chronic ethanol treatment is carried out, various factors, such as intoxicated practice and conditioning, can play a role. The recent advances in the field of molecular biology, when used in conjunction with various electrophysiological and neurochemical techniques, will soon permit investigators to identify the biochemical loci associated with tolerance. In conducting such studies, however, the various features of tolerance, as well as variables associated with the methods employed to produce and measure tolerance, must be carefully considered.

Acknowledgment

This work was supported in part by a grant from NIAAA (Ro AA 08254). The authors thank J. Sheppard for her secretarial assistance.

References

Alkana R. L., Bejanian M., Syapin P. J., and Finn D. A. (1987) Chronic functional ethanol tolerance in mice influenced by body temperature during acquisition. *Life Sci.* **41,** 413–420.

Arvola A., Sammalisto L., and Wallgren H. (1958) A test for level of alcohol intoxication in the rat. *J. Stud. Alcohol* **19,** 563–572.

Baker T. B. and Tiffany S. T. (1985) Morphine tolerance as habituation. *Psychol. Rev.* **92,** 78–108.

Brick J., Pohorecky L. A., Faulkner W., and Adams M. N. (1984) Circadian variations in behavioral and biological sensitivity to ethanol. *Alcoholism: Clin. Exp. Res.* **8,** 204–211.

Buck K. J. and Harris R. A. (1991) Neuroadaptive responses to chronic ethanol. *Alcoholism: Clin. Exp. Res.* **15,** 460–470.

Buck K. J., Heim H., and Harris R. A. (1991) Reversal of alcohol dependence and tolerance by a single administration of flumazenil. *J. Pharmacol. Exp. Ther.* **257,** 984–989.

Carlen P. L. and Corrigall W. A. (1980) Ethanol tolerance measured electrophysiologically in hippocampal slices and not in neuromuscular junctions from chronically ethanol-fed rats. *Neurosci. Lett.* **17,** 95–100.

Charness M. E., Gordon A. S., and Diamond I. (1983) Ethanol modulation of opiate receptors in cultured neural cells. *Science* **222,** 1246–1248.

Chen C. S. (1968) A study on alcohol tolerance effect and an introduction of a new behavioural technique. *Psychopharm-acologia (Berl.)* **12,** 433–440.

Crabbe J. C., Rigter H., Uijlen J., and Strijbos C. (1979) Rapid development of tolerance to the hypothermic effect of ethanol in mice. *J. Pharmacol. Exp. Ther.* **208,** 128–133.

Crabbe J. C., Janowsky J. S., Emmett R. Y., Kosobud A., Stack J., and Rigter H. (1982) Tolerance to ethanol hypothermia in inbred mice: Genotypic correlations with behavioural responses. *Alcoholism: Clin. Exp. Res.* **6,** 446–458.

Curran M. and Seeman P. (1977) Alcohol tolerance in a cholinergic nerve terminal: relation to the membrane expansion-fluidization theory of ethanol action. *Science* **197,** 910,911.

Czaja C. and Kalant H. (1961) The effect of acute alcoholic intoxication on adrenal ascorbic acid and cholesterone in the rat. *Can. J. Biochem. Physiol.* **39,** 327–334.

Deimling M. J. and Schnell R. C. (1980) Circadian rhythms in the biological response and disposition of ethanol in the mouse. *J. Pharmacol. Exp. Ther.* **213,** 1–8.

Dunham N. W. and Miya T. S. (1957) A note on a simple apparatus for detecting neurological deficit in rats and mice. *J. Am. Pharm. Assoc.* **46,** 208,209.

Edwards F., Schabinsky V. V., Jackson D. M., Starmer G. A. and Jenkins O. (1983) Involvement of catecholamines in acute tolerance to ethanol in mice. *Psychopharmacology* **79,** 246–250.

El-Ghundi M., Kalant H., Lê A. D., and Khanna J. M. (1989) The contribution of environmental cues to cross-tolerance between ethanol and pentobarbital. *Psychopharmacology* **97,** 194–201.

Erickson C. K., Koch K. I., Mehta C. S., and McGinity J. W. (1978) Sustained release of alcohol: Subcutaneous silastic implants in mice. *Science* **199,** 1457–1459.

Fernandes M., Kluwe S., and Coper H. (1977) The development of tolerance to morphine in the rat. *Psychopharmacology* **54,** 197–201.

Gallaher E. J., Parsons L. M., and Goldstein D. B. (1982) The rapid onset of tolerance to the ataxic effects of ethanol in mice. *Psychopharmacology* **78,** 67–70.

Gentry R. T., Rappaport M. S., and Dole V. P. (1983) Serial determination of plasma ethanol concentration in mice. *Physiol. Behav.* **31,** 529–532.

Gibbins R. J., Kalant H., and LeBlanc A. E. (1968) A technique for accurate measurement of alcohol intoxication in small animals. *J. Pharmacol. Exp. Ther.* **159,** 236–242.

Gibbins R. J., Kalant H., LeBlanc A. E., and Clark J. W. (1971) The effects of chronic administration of ethanol on startle threshold in rats. *Psychopharmacologia (Berl.)* **19,** 95–104.

Gilliam D. M. (1989) Alcohol absorption rate affects hypothermic response in mice: evidence for rapid tolerance. *Alcohol* **6,** 357–362.

Goldstein D. B. (1983) *Pharmacology of Alcohol.* Oxford University Press, New York, NY.

Goldstein D. B. and Pal N. (1971) Alcohol dependence produced in mice by inhalation of ethanol: Grading the withdrawal reaction. *Science* **172,** 288–290.

Gordon A. S., Collier K., and Diamond I. (1986) Ethanol regulation of adenosine receptor-stimulated cAMP levels in clonal neural cell line: An in vitro model of cellular tolerance to ethanol. *Proc. Natl. Acad. Sci. USA* **83,** 2105–2108.

Goudie A. J. and Emmett-Oglesby M. W. (1989) Tolerance and sensitization: Overview, in *Tolerance and Sensitization to Psychoactive Drugs* (Goudie A. J. and Emmett-Oglesby M. W., eds.) Humana Press, Clifton, NJ, pp 12–16.

Greeley J. D., Lê A. D., Poulos C. X., and Cappell H. (1984) Ethanol is an effective stimulus in the Pavlovian control of tolerance to ethanol. *Psychopharmacology* **83,** 159–162.

Grieve S. J. and Littleton J. M. (1979) The rapid development of functional tolerance to ethanol by mice. *J. Pharm. Pharmacol.* **31,** 605–610.

Hawkins R. D., Kalant H., and Khanna J. M. (1966) Effects of chronic intake of ethanol on rate of ethanol metabolism. *Can. J. Physiol. Pharmacol.* **44,** 241–257.

Holloway F. A., Bird D. C., Holloway J. A., and Michaelis R. C. (1988) Behavioural factors in the development of tolerance to ethanol's effects. *Pharmacol. Biochem. Behav.* **29,** 105–113.

Hunt G. P. and Overstreet D. H. (1977) Evidence for parallel development of tolerance to the hyperactivating and discoordinating effects of ethanol. *Psychopharmacology* **55,** 75–81

Jones B. J. and Roberts D. J. (1968) The quantitative measurement of motor incoordination in naive mice using an accelerating rotarod. *J. Pharm. Pharmacol.* **20,** 302–304.

Kalant H. (1977) Comparative aspects of tolerance to, and dependence on, alcohol, barbiturates and opiates, in *Alcohol Intoxication and Withdrawal, vol. 3B, Advances in Experimental Medicine and Biology.* (Gross A. J., ed.) Plenum, New York, NY, pp. 169–186.

Kalant H. (1985) The 1985 Upjohn award lecture: Tolerance, learning, and neurochemical adaptation. *Can. J. Physiol. Pharmacol.* **63,** 1485–1494.

Kalant H. (1988) Alcohol tolerance, dependence and withdrawal: an overview of current issues. *Austr. Drug Alcohol Rev.* **7,** 27–34.

Kalant H. (1989) Drug tolerance and sensitization: A pharmacological overview, in *Tolerance and Sensitization to Psychoactive Drugs.* (Goudie A. J. and Emmett-Oglesby M. W., eds.) Humana Press, Clifton, NJ, pp. 549–577.

Kalant H. and Lê A. D. (1984) Effects of ethanol on thermoregulation. *Pharmacol. Ther.* **23,** 313–364.

Kalant H. and Khanna J. M. (1990) Methods for study of tolerance, in *Modern Methods in Pharmacology, vol. 6: Testing and Evaluation of Drugs of Abuse* (Adler M. W. and Cowan A., eds.) Wiley-Liss Inc., New York, NY, pp. 43–66.

Kalant H., LeBlanc A. E. and Gibbins R. J. (1971) Tolerance to, and dependence on, some non-opiate psychotropic drugs. *Pharmacol. Rev.* **23,** 135–191.

Kalant H., Khanna J. M., Llewellyn M., Rinkel G. J., and Woodworth I. (1986) Functional tolerance to chlormethiazole and cross-tolerance to ethanol in the rat: importance of test and mode of drug administration. *Acta Psych. Scand.* **Suppl. 329,** 54–65.

Keir W. J. and Deitrich R. A. (1990) Development of central nervous system sensitivity to ethanol and pentobarbital in Short- and Long-Sleep mice. *J. Pharmacol Exp. Ther.* **254,** 831–835.

Khanna J. M., Lê A. D., LeBlanc A. E., and Shah G. (1985) Initial sensitivity versus acquired tolerance to ethanol in rats selectively bred for ethanol sensitivity. *Psychopharmacology* **86,** 302–306.

Khanna J. M., San-Marina A., Kalant H., and Lê A. D. (1989) Relationship between initial sensitivity and chronic tolerance to ethanol and morphine in a heterogeneous population of mice and rats, in *Genetic Aspects of Alcoholism* (Kiianmaa K., Tabakoff B., and Saito T., eds.) The Finnish Foundation for Alcohol Studies, vol. 37, Helsinki, Finland, pp. 207–217.

Khanna J. M., Kalant H., Shah G., and Weiner J. (1991) Rapid tolerance as an index of chronic tolerance. *Pharmacol. Biochem. Behav.* **38,** 427–432.

Koob G. F., Wall T. L., and Schafer J. (1987) Rapid induction of tolerance to the antipunishment effects of ethanol. *Alcohol* **4,** 481–484.

Lê A. D. (1990) Factors regulating ethanol tolerance. *Ann. Med.* **22,** 265–268.

Lê A. D. and Kalant H. (1990) Learning as a factor in ethanol tolerance, in *Neurobiology of Drug Abuse: Learning and Memory* (Erinoff L., ed.) NIDA Res. Monogr. 97, DHHS Publ. # (ADM) 90-1677 pp. 193–208.

Lê A. D. and Kiianmaa K. (1988) Characteristics of ethanol tolerance in alcohol drinking (AA) and alcohol avoiding (ANA) rats. *Psychopharmacology* **94,** 479–483.

Lê A. D. and Kiianmaa K. (1990) Role of initial sensitivity and genetic factors in the development of tolerance to ethanol in AT and ANT rats. *Psychopharmacology* **102,** 11–16.

Lê A. D. and Khanna J. M. (1989) Dispositional mechanisms in drug tolerance, in *Tolerance and Sensitization to Psychoactive Drugs.* (Goudie A. J. and Emmett -Oglesby M. W., eds.) Humana Press, Clifton, NJ, pp. 281–351.

Lê A. D., Poulos C. X., and Cappell H. (1979) Conditioned tolerance to the hypothermic effect of ethyl alcohol. *Science* **206,** 1109,1110.

Lê A. D., Khanna J. M., Kalant H., and LeBlanc A. E. (1981) The effect of lesions in the dorsal, median and magnus raphe nuclei on the development of tolerance to ethanol. *J. Pharmacol. Exp. Ther.* **218,** 525–529.

Lê A. D., Kalant H., Khanna J. M., and Grossi F. (1986) Tolerance and crosstolerance among ethanol, pentobarbital and chlordiazepoxide. *Pharmacol. Biochem. Behav.* **24,** 93–98.

Lê A. D., Kalant H., and Khanna J. M. (1987a) Effect of treatment dose and intoxicated practice on the development of tolerance to ethanol-induced motor impairment. *Alcohol Alcohol.* **Suppl. 1,** 435–439.

Lê A. D., Khanna J. M., and Kalant H. (1987b) Role of Pavlovian conditioning in the development of tolerance and cross-tolerance to the hypothermic effect of ethanol and hydralazine. *Psychopharmacology* **92,** 210–214.

Lê A. D., Kalant H., and Khanna J. M. (1989) Roles of intoxicated practice in the development of ethanol tolerance. *Psychopharmacology* **99,** 366–370.

LeBlanc A. E., Gibbins R. J., and Kalant H. (1973) Behavioural augmentation of tolerance to ethanol in the rat. *Psychopharmacologia (Berl.)* **30,** 117–122.

LeBlanc A. E., Gibbins R. J., and Kalant H. (1975a) Generalization of behaviourally augmented tolerance to ethanol, and its relation to physical dependence. *Psychopharmacologia (Berl.)* **44,** 241–246.

LeBlanc A. E., Kalant H., and Gibbins R. J. (1975b) Acute tolerance to ethanol in the rat. *Psychopharmacologia (Berl.)* **41,** 43–46.

Litchfield J. T and Wilcoxon F. (1949) A simplified method of evaluating dose-effects experiments. *J. Pharmacol. Exp. Ther.* **96,** 99–113.

Lieber C. S., Jones D. P., and DeCarli L. M. (1965) Effects of prolonged ethanol intake: Production of fatty liver despite adequate diet. *J. Clin. Invest.* **44,** 1109–1121.

Lumeng L., Waller M. B., McBride W.J., and Li T. K. (1982) Different sensitivities to ethanol in alcohol-preferring and -non preferring rats. *Pharmacol. Biochem. Behav.* **16,** 125–130.

Maier D. M. and Pohorecky L. A. (1986) The effect of stress on tolerance to ethanol in rats. *Alcohol Drug Res.* **6,** 387–401.

Maier D. M. and Pohorecky L. A. (1987) The effect of repeated withdrawal episodes on the acquisition and loss of tolerance to ethanol in ethanol-treated rats. *Physiol. Behav.* **40,** 411–424.

Malcolm R. D. and Alkana R. L. (1981) Temperature dependence of ethanol depression in mice. *J. Pharmacol. Exp. Ther.* **217,** 770–775.

Masur J. and Boerngen R. (1980) The excitatory component of ethanol in mice: a chronic study. *Pharmacol. Biochem. Behav.* **13,** 777–780.

Masur J., Olivera de Souza M. L., and Zwicker A. P. (1986) The excitatory effect of ethanol: Absence in rats, no tolerance and increased sensitivity in mice. *Pharmacol. Biochem. Behav.* **24,** 1225–1228.

Mayer J. M., Khanna J. M., Kalant H., and Spero L. (1980) Cross-tolerance between ethanol and morphine in the isolated guinea pig ileum myenteric plexus preparation. *Eur. J. Pharmacol.* **63,** 223–227.

Maynert E. W. and Klingman G. I. (1960) Acute tolerance to intravenous anesthetic in dogs. *J. Pharmacol. Exp. Ther.* **128,** 192–200.

McQuarrie D. G. and Fingl E. (1958) Effect of single doses and chronic administration of ethanol on experimental seizures in mice. *J. Pharmacol. Exp. Ther.* **124,** 264–271.

Melchior C. L. (1990) Conditioned tolerance provides protection against ethanol lethality. *Pharmacol. Biochem. Behav.* **37,** 205,206.

Melchior C. L. and Tabakoff B. (1981) Modification of environmentally cued tolerance to ethanol in mice. *J. Pharmacol. Exp. Ther.* **219,** 175–180.

Melchior C. L. and Tabakoff B. (1985) Features of environmental-dependent tolerance to ethanol. *Psychopharmacology* **87,** 94–100.

Mellanby E. (1919) Alcohol: its absorption into and disappearance from the blood under different conditions. *Special Report Series No. 31,* Medical Research Committee, London, UK.

Mirsky I. A., Piker P., Rosenbaum M., and Lederer M. (1941) "Adaptation" of the central nervous system to varying concentrations of alcohol in the blood. *Q. J. Stud. Alcohol* **2,** 35–45.

Moore J. A. and Kakihana R. (1978) Ethanol-induced hypothermia in mice: influence of genotype on development of tolerance. *Life Sci.* **23,** 2331–2337.

Mucha R. F. and Kalant H. (1981) Naloxone prevention of morphine LDR curve flattening associated with high-dose tolerance. *Psychopharmacology* **75**, 132,133.

Palmer M. R., Basile A. S., Proctor W. R., Baker R. C., and Dunwiddie T. V. (1985) Ethanol tolerance of cerebellar Purkinje neurons from selectively outbred mouse lines: In vivo and in vitro electrophysiological investigations. *Alcoholism: Clin. Exp. Res.* **9**, 291–296.

Peris J. and Cunningham C. L. (1986) Handling-induced enhancement of alcohol's acute physiological effects. *Life Sci.* **38**, 273–279.

Peris J. and Cunningham C. L. (1987) Stress enhances the development of tolerance to the hypothermic effect of ethanol. *Alcohol Drug Res.* **7**, 187–193.

Pinel J. P. J., Mana M. J., and Renfrey G. (1985) Contingent tolerance to the anticonvulsant effects of alcohol. *Alcohol* **2**, 495–499.

Pohorecky L. A. and Shah P. (1987) Ethanol-induced analgesia. *Life Sci.* **41**, 1289–1295.

Pohorecky L. A. and Rizek A. E. (1981) Biochemical and behavioural effects of acute ethanol in rats at different environmental temperatures. *Psychopharmacology* **72**, 205–209.

Pohorecky L. A., Peterson J. T., and Carpenter J. A. (1986) Development of tolerance to ethanol-induced tachycardia in rats. *Alcohol Drug Res.* **6**, 431–439.

Poulos C. X. and Cappell H. (1991) Homeostatic theory of drug tolerance: a general model of physiological adaptation. *Psychol. Rev.* **98**, 390–408.

Ritzmann R. F. and Tabakoff B. (1976) Body temperature in mice: A quantitative measure of alcohol tolerance and physical dependence. *J. Pharmacol. Exp. Ther.* **199**, 158–170.

Siegel S. (1975) Evidence from rats that morphine tolerance is a learned response. *Can. J. Comp. Physiol. Psych.* **89**, 498–506.

Siegel S. (1983) Classical conditioning, drug tolerance, and drug dependence, in: *Research Advances in Alcohol and Drugs Problems, vol. 7* (Smart R. J., Glaser F. B., Israel Y., Kalant H., Popham R. E., and Schmidt W., eds.) Plenum, New York, NY, pp. 207–246.

Smolen T. N. and Smolen A. (1989) Blood and brain ethanol concentrations during absorption and distribution in Long-Sleep and Short-Sleep mice. *Alcohol* **6**, 33–38.

Sunahara G. I. (1979) Effect of ethanol on stimulated acetylcholine release *in vitro* from rat cortical and hippocampal tissue. M.Sc. thesis, Department of Pharmacology, University of Toronto.

Sunahara G. I., Kalant H., Schofield M., and Grupp L. (1978) Regional distribution of ethanol in the rat brain. *Can. J. Physiol. Pharmacol.* **56**, 988–992.

Smolen T. N. and Smolen A. (1989) Blood and brain ethanol concentrations during absorption and distribution in Long-Sleep and Short-Sleep mice. *Alcohol* **6**, 33–38.

Tabakoff B. and Ritzmann R. F. (1979) Acute tolerance in inbred and selected lines of mice. *Drug Alcohol Depend.* **4,** 87–90.

Tabakoff B., Ritzmann R. F., Raju T. S., and Deitrich R. A. (1980) Character-ization of acute and chronic tolerance in mice selected for inherent differences in sensitivity to ethanol. *Alcoholism: Clin. Exp. Res.* **4,** 70–73.

Tiffany S. T., McCal K. J., and Maude-Griffin P. M. (1987) The contribution of classical conditioning to the antinoceptive effects of ethanol. *Psy-chopharmacology* **92,** 524–528.

Traynor M. E., Schlapper W. T., Woodson, P. B. J., and Barondes S. H. (1979) Crosstolerance to effect of ethanol and temperature in aplysia: pre-liminary observation. *Alcoholism: Clin. Exp. Res.* **3,** 57–59.

Tullis K. V., Sargent W. Q., Simpson J. R., and Beard J. D. (1977) An animal model for the measurement of acute tolerance to ethanol. *Life Sci.* **20,** 875–880.

Waller M. B., McBride W. J., Lumeng L., and Li T. K. (1983) Initial sensitivity and acute tolerance to ethanol in the P and NP lines of rats. *Pharmacol. Biochem. Behav.* **19,** 683–686.

Wallgren H. and Tirri R. (1963) Studies on the mechanisms of stress-induced reduction of alcohol intoxication in rats. *Acta Pharmacol. Toxicol.* **20,** 27–38.

Wenger J. R., Berlin V., and Woods S. C. (1980) Learned tolerance to the behaviourally disruptive effects of ethanol. *Behav. Neural Biol.* **28,** 418–430.

Animal Models of Drug Addiction

Barbiturates

Michiko Okamoto

1. Pharmacologic Considerations for Developing a Barbiturate Addiction Model

1.1. Physical-Chemical Properties of Barbiturates

The barbiturates consist of a pyrimidine nucleus derived from the condensation of malonic acid and urea. Pharmacologically, barbituric acid *per se* does not have central nervous system (CNS) depressant action. The CNS action is produced only after the appropriate alkyl or aryl groups are substituted on the carbon 5,5' positions. Accordingly, barbiturates used in therapy possess various substitutional groups at the carbon 5,5' positions. The following requirements when substituted appear to be important for production of the pharmacological activity (cf Sharpless, 1970; Andrews et al., 1979).

1. The total number of carbon atoms in positions 5,5' must be between 4–8 for production of the optimal CNS depressant action.
2. Only one of the substitution groups at positions 5,5' can be a closed ring.
3. Introduction of the double bonds in the alkyl substitution groups at positions 5,5' renders barbiturates more vulnerable for oxidation in the liver. Those barbiturates are short-acting.

From: *Neuromethods, Vol. 24: Animal Models of Drug Addiction*
Eds: A. Boulton, G. Baker, and P. H. Wu ©1992 The Humana Press Inc.

4. The compounds with short chains in positions 5,5' resist oxidation and, therefore, are long-acting. In contrast, the compounds with long chains are readily oxidized and are, therefore, short-acting.

5. Certain alkyl groups at positions 5,5' render molecules to be convulsant. Examples are 1,3-dimethylbutyl ethyl barbituric acid and crotyl n-butyl thiobarbituric acid.

6. The replacement of the carbonyl oxygen atom at position 2 by a sulfur atom produces thiobarbiturates. This replacement renders the molecule highly lipid-soluble and short-acting.

7. The replacement of the carbonyl oxygen atom at position 2 by an —NH group produces a guanidine derivative. This will destroy the hypnotic activity of the molecule.

8. The stereoisomers of barbiturate produce opposing pharmacological actions. In general, the excitatory actions are prominent with the R($^+$) isomer, whereas the depressant actions are characteristic with the S($^-$) isomer.

Barbituric acid and its 5,5' disubstituted derivatives are weak organic acids. They are capable of combining with fixed alkalines and of forming soluble salts. The salts dissociate in water; their solutions are strongly alkaline.

1.2. The Pharmacokinetic Characteristics of Commonly Used Barbiturates

The barbiturates recognized by the US Pharmacopeia are: phenobarbital, amobarbital, pentobarbital sodium, secobarbital, and thiopental sodium. In addition, aprobarbital, butabarbital sodium, mephobarbital, metharbital, and tabutal are available commercially. Sodium barbital has been used extensively in animal studies. These barbiturates represent compounds of a wide range of time-action properties, namely long-acting (e.g., barbital, phenobarbital), intermediate and/or short-acting (e.g., amobarbital, pentobarbital, secobarbital), and ultrashort-acting (e.g., thiopental). Pharmacodynamically, however, they are all similar. The chemical structure of these is shown in Table 1. In this chapter, only barbiturates that are used commonly in animal drug addiction studies are discussed.

Table 1
Structures and Physicochemical Properties of Barbiturates

Barbiturate	R_1	R_2	X
Thiopental	Ethyl	1-Methylbutyl	S
Secobarbital	Allyl	1-Methylbutyl	O
Pentobarbital	Ethyl	1-Methylbutyl	O
Phenobarbital	Ethyl	Phenyl	O
Barbital	Ethyl	Ethyl	O

Barbiturate	pK_a[a]	Partition coefficient[b]	Protein binding Plasma[c]	Brain[d]	Degradation by liver[e]	Excretion by kidney[f]	LD_{50}
Thiopental	7.4	580	0.65	0.50	0.53	–	120(68)
Secobarbital	7.9	52	0.44	0.39	0.30	–	110(35)
Pentobarbital	8.0	39	0.35	0.29	0.21	–	75(50)
Phenobarbital	7.3	3	0.20	0.19	0.00	30	190
Barbital	7.8	1	0.05	0.06	–	65–90	300

[a]pK_a: Ionization exponent at 25°C (Bush, 1963).
[b]Partition coefficient: (Concentration in methylene chloride)/(concentration in aqueous phase) of the nonionized form at around 25°C.
[c]Plasma protein binding: Fraction of 0.001M barbituric acids bound by 1% bovine albumin in 1/15 M phosphate buffer at pH 7.4 (Goldbaum and Smith, 1954).
[d]Brain protein binding: Fraction of barbiturate bound by rabbit brain homogenates (Goldbaum and Smith, 1954).
[e]Degradation by liver: Fraction degraded by liver slices in 3 h (Dorfman and Goldbaum, 1947).
[f]Excretion by kidney: Percentage of total dose excreted unchanged in urine in humans (Sharpless, 1970).
[g]LD_{50}: LD_{50} by intraperitoneal (intraveneously) administered barbiturate in mg/kg in rats (Barnes and Eltherington, 1966).

1.2.1. Absorption and Routes of Administration

An excellent overview can be obtained in the pharmacology textbook, in the chapter on "hypnotics and sedatives," by Sharpless (1970). Short-acting, lipid-soluble barbiturates are, in principle, absorbed more rapidly than the long-acting barbiturates. The rate-limiting step in absorption of barbiturate from the oral route is, however, usually not owing to the penetration of barbiturate through the gastrointestinal mucosa, but the dissolution and dispersal of the drug in the gastrointestinal contents. Thus, the rate of absorption of various intermediate/short-acting barbiturates was not significantly different when they were given encapsulated free-acid form (Sjorgen et al., 1965). The sodium salts are, therefore, more readily absorbed than are the free acids because of their rapid dissolution. Absorption takes place primarily from the intestine, despite the favorable protonated/unprotonated pH partition of barbiturate in the stomach resulting from vascularity of the intestine. Food in the stomach decreases the rate, but not the bioavailability, of absorption.

Subcutaneous injection may result in necrosis and sloughing of skin because of the extreme alkalinity of the salt solution. The sodium salts of barbiturates have been injected intramuscularly in <10% solution; however, the drug should be delivered into a deep depot. Subcutaneous and im routes should be avoided in small animals, especially for chronic experiments; because of their relatively small muscle mass accessible for the im drug injection, pain caused at the injection sites might bias experimental results.

Generally, a <5% aqueous solution of a sodium salt of barbiturate is used for iv injection. The injection must be made slowly, especially with the long-acting barbiturates, because a considerable time may be required for the drug to penetrate through the blood–brain barrier and to produce the drug action. This time lag could be >10 min to produce maximum CNS depression after an iv bolus injection of long-acting barbiturates, for example, phenobarbital and barbital. Fall in blood pressure and apnea may occur after a rapid injection. The iv injection of barbiturates may produce laryngospasm, coughing, and other respiratory irregularities under hypnotic and light anesthetic states in animals.

1.2.2. Distribution

The important physicochemical factors that determine the passage of barbiturates across cytostructural barriers and the manner in which they influence drug distribution and excretion have been discussed extensively in many pharmacology textbooks (cf Sharpless, 1970; Harvey, 1985). This information is particularly complete for barbiturates because of careful studies on large numbers of barbiturate congeners in order to understand fundamental principles that govern the processes (Table 1).

The three most important factors affecting the distribution and fate of the barbiturates are lipid solubility, protein binding, and extent of ionization. The lipid solubility of barbiturates correlates well with the pharmacological activity of the compounds; generally, the more highly lipid-soluble barbiturates are of rapid onset and have a short duration of action. They are almost completely reabsorbed by the renal tubule. Highly lipid-soluble barbiturates, e.g., thiopental, attain their maximum concentration in the brain within two or three circulation times. These drugs are found in highest concentration in those brain areas that have greatest blood flow (e.g., cortex, geniculate bodies, and colliculi) during the first few minutes after injection. Barbital and phenobarbital, on the other hand, with low lipid/water partition coefficients, penetrate the blood–brain barrier much more slowly. If sufficient time is allowed after administration, however, both the short- and long-acting barbiturates will achieve uniform distribution throughout the brain. Regional differences in blood flow do not affect the distribution of the slowly penetrating barbiturates.

No impenetrable barrier to the diffusion of barbiturates exists in the body; consequently, if the drug persists in plasma for a sufficiently long time, it will be distributed to all tissues and fluids (Raventos, 1954; Bush, 1963). Accordingly, barbiturates readily cross the placental barrier and distribute to the fetus. Considerable amounts of barbiturates may also appear in the milk, especially after the repeated administration of the drug. Coupled with the fact that newborns generally do not have the full ability to metabolize and eliminate barbiturates, this accessibility may result in drug cumulation in the fetus after repeated administration.

A fraction of the barbiturate in blood is reversibly bound to plasma protein, chiefly to albumin (Goldbaum and Smith, 1954). Protein binding appears to require the same structural features that determine affinity for nonpolar solvents, since those barbiturates with the highest methylene chloride:water partition coefficients bind to plasma protein to the greatest extent (Table 1).

The cerebrospinal fluid is virtually protein-free. Accordingly, the maximum concentration of barbiturates attained in the cerebrospinal fluid is less than the plasma concentration. In organ tissues, the barbiturate concentration is generally as high as or slightly higher than in plasma. The capacity of organ tissues to concentrate barbiturates depends largely on protein-binding properties of the drug; various barbiturates show the same relative affinities to tissue protein as to plasma protein. Somewhat higher concentrations are achieved in liver and kidney than in other tissues. Fat depots may contain a very high concentration of the ultrashort-acting barbiturates because of their high lipid solubility. Body fat depots are generally not accessible to intermediate- and long-acting barbiturates by single administration because of their poor blood circulation, but may become drug reservoirs after long-term chronic treatment. Accordingly, accumulation of barbiturates in fat depots may change their overall elimination kinetics after chronic treatment (Waters and Okamoto, 1972).

Changes in pH in a particular tissue environment and, hence, ionization of a barbiturate often affect its distribution in body compartments. If the plasma pH is increased by hyperventilation or by infusion of sodium bicarbonate, the plasma concentration of the undissociated, lipid-soluble form of the drug falls. This results in an increase in the outward flux of the drug from tissue to plasma. Waddell and Butler (1957) observed a total plasma concentration of phenobarbital and lightening of anesthesia in experimental animals resulting from a decrease in the concentration of the drug in the brain, following hyperventilation or the infusion of sodium bicarbonate solution. An increase in plasma pH also makes the urine more alkaline, decreases the tubular reabsorption, and, thus, increases excretion of the barbiturates. As a rule, an increase in plasma pH produces a greater change in the ionization of barbiturates that have dissociation constants close to body pH (e.g., phenobarbital) (Table 1). Administration of large amounts of the sodium form of a barbiturate that is required to override dispositional tolerance may elevate plasma pH considerably during chronic treatment.

1.2.3. Termination of Action

Three processes are responsible for the termination of the CNS depressant action of the barbiturates: physical redistribution, metabolic degradation, and renal excretion. All of these processes reduce the plasma concentration of the barbiturate, which in turn removes the drug from the CNS. The lipid solubility of each individual barbiturate governs its route of elimination. The long-acting barbiturates, e.g., barbital, are excreted largely unchanged by the kidney. Physical redistribution plays an important role in the ultrashort-acting, highly lipid-soluble barbiturates with regard to their ultra-short duration of action. Except for those excreted unchanged by the kidney, all barbiturates are biotransformed, primarily by the liver. The biotransformed products are generally pharmacologically inactive.

1.2.3.1. REDISTRIBUTION

The role of redistribution of the drug from brain to other tissues has been studied most thoroughly for thiopental. Thiopental is almost completely metabolized, and only a very small fraction of the dose is excreted unchanged in the urine. However, the rate of metabolism is quite slow, much too slow to account for the short duration of action of this compound (Brodie et al., 1950). The recovery from anesthesia after a single anesthetic dose of thiopental depends on a mobilization of the drug from the brain to other tissue compartments according to the concentration gradient. The rates at which the various tissue compartments take up thiopental from the blood are related to their blood flow. Because of their high perfusion rates, the brain and certain visceral organs (liver, kidney, heart, and so on) that together receive 70% of the total cardiac output, exhibit maximum concentrations of thiopental within 30 s after its iv administration. On the other hand, >15 min are required for muscle and skin to become saturated with the drug. As the muscle and fat take up thiopental, the plasma concentration falls, and the drug diffuses out of the brain along its concentration gradient. By 30 min, the brain and viscera may have given up as much as 90% of their initial peak concentrations to the nonvisceral lean tissues and fat depots. It is for this reason that an animal may awake within 15 min after a single iv injection of thiopental, even though the amount metabolized in this time is very small. The fact that thiopental remains in the body, but not in the CNS may lead to the drug cumulation when the doses are repeated, and when the nonvisceral lean tissues and fat depots become saturated.

1.2.3.2. RENAL EXCRETION

Barbiturates not destroyed in the body are excreted unchanged in the urine. Only barbital depends mainly on renal excretion for the termination of its pharmacological action; 90% of the total dose appears in the urine unaltered. As much as 50% of the total dose of phenobarbital is also eliminated unchanged.

The renal clearance rates of the barbiturates depend on a number of factors. Glomerular filtration of the drug is reduced by binding to plasma proteins. The barbiturates in the glomerular filtrate are reabsorbed by passive back diffusion. Since the tubular epithelium serves as a diffusion barrier to the dissociated form of barbiturates, the relatively polar compounds (barbital, phenobarbital, and so on) escape tubular reabsorption to a significant degree.

1.2.3.3. METABOLIC DEGRADATION

The liver is the primary organ in terminating the central depressant action of intermediate/short-acting barbiturates. It transforms lipid-soluble agents into more polar derivatives that can then be excreted by the kidney. Numerous articles on processes of barbiturate biotransformation are available for review (Maynert and Van Dyke, 1949; Mark, 1963; Williams and Parke, 1964; Sharpless, 1970; Harvey, 1985). When they are converted into a form that can be excreted, most (but not all) barbiturates lose their pharmacological activity. Barbiturates are transformed primarily by the following four routes (Dorfman and Goldbaum, 1947; Raventos, 1954):

1. Oxidation of radicals at C_5;
2. N-dealkylation;
3. Desulfuration of thiobarbiturates; and
4. Destruction of the barbituric acid ring.

1.2.4. Stimulation and Inhibition of Barbiturate-Metabolizing Enzymes

The microsomal enzymes responsible for the oxidative metabolism of barbiturate are susceptible to a variety of influences. Thus, in experimental animals, nutritional deficiencies, presence of other drugs, and so on, may cause a sufficient degree of enzyme inhibition to increase the duration of hypnotic action of barbiturates. Stress itself on animals may prolong the action of short-acting barbiturates, possibly through the effect of adrenocorticosteroids on drug-metabolizing enzymes. Such drugs as ethanol also interfere with the metabolism of barbiturates in experimental animals. Frequently, these drugs have a biphasic action, the initial inhibition of barbiturate-metabolizing enzymes being followed by a period of increased enzyme activity.

Of great importance is the fact that barbiturates stimulate the enzymes responsible for their own metabolism (Conney and Burns, 1962; Remmer, 1964; Conney, 1967). This fact accounts in part for the tolerance (drug-dispositional tolerance) seen with barbiturates, especially with the intermediate/short-acting ones. It has been shown in a wide variety of animal species that chronic administration of barbiturates is followed by a reduction in the duration of action, which correlates with an increased rate of barbiturate metabolism by the liver (Kato et al., 1964; Stevenson and Turnbull, 1968). It is noteworthy that barbital, which is not itself metabolized by hepatic enzymes, markedly enhances the enzyme activity responsible for metabolizing other barbiturates (Conney and Burns, 1962; Conney, 1967).

1.2.5. Sterospecificity

Clinically used intermediate/short-acting barbiturates (pentobarbital and secobarbital) are most frequently employed as racemic mixtures. When the individual isomers were studied separately, the S-(–) isomers of both pentobarbital and secobarbital have generally been shown to be more potent than the R-(+) isomers (Wahlstrom, 1966; Haley and Gidley, 1970; Huang and Barker, 1980; Ho and Harris, 1981). It has been reported that S-(–) isomers are from 1.5 to 4 times more potent than the R-(+) isomers in studies of lethality and anesthetic potency (Buch et al., 1969; Christensen and Lee, 1973; Waddell and Baggett, 1973), responding under various schedules of reinforcement (Rastogi et al., 1985; Wenger, 1986; Wenger et al., 1986), discriminative stimulus properties (Young et al., 1984; Wessinger and Wenger, 1987), and spontaneous motor activity (Wenger, 1986).

The stereospecificity of tolerance development to a single isomer of a barbiturate with optically active centers (Wenger, 1988) was tested in rats and pigeons trained with self-administration paradigms. After responding had stabilized, dose–response curves were determined for R-(+)-pentobarbital, S-(–)-pentobarbital, R-(+)-secobarbital, and S-(–)-secobarbital in both species. Upon the completion of the acute dose–response curves, the rats and pigeons were given 10 mg/kg/d of S-(–)-pentobarbital for 30 consecutive days prior to the redetermination of all four dose–response curves. The study showed that, with repeated administration of S-(–)-pentobarbital in rats, the development of tolerance was observed to the effects of S-(–)-pentobarbital. Furthermore, a crosstolerance was observed to R-(+)-pentobarbital and to both isomers of secobarbital. A similar self-administration protocol in pigeons, however, failed to show a clear tolerance development (Wenger, 1986). Although the issues of whether the observed tolerance and crosstolerance in rats are pharmacokinetic and/or pharmacodynamic in nature have not been answered directly in those studies, the observations are interesting and should be pursued. No signs of physical dependence were observed in either species with this treatment method.

2. Selection of Animal Species

Lester and Freed (1972) and Cicero (1979) have listed the criteria that are essential for establishing relevant animal models in ethanol addiction studies. Similar criteria should also apply to the animal models in barbiturate addiction studies. These are:

1. The animal should self-administer the drug primarily by the oral route in pharmacologically significant amounts. The drug also should be consumed in preference to other drugs on the basis of its drug properties. Furthermore, self-administration should result in pharmacologically relevant blood barbiturate concentrations;
2. Pharmacodynamic tolerance should be produced after a period of chronic self-administration, and this should be objectively demonstrated; and
3. Withdrawal signs and symptoms should be objectively assessed.

Table 2 lists the animal species that have been utilized for chronic barbiturate studies; not all of these have met the criteria for the ideal model described above. Furthermore, many research objectives may not require fulfillment of all the criteria listed above; however, in the author's opinion, any new findings related to the issues of barbiturate abuse and dependence problems should be reported with reference to chronic blood barbiturate concentrations and/or physical dependence state produced, and should not simply relate to the chronic drug dosage. In this way, the pharmacological relevance of the reported results would be greatly enhanced with regard to historic perspectives.

One of the great advantages of studies of barbiturates is the relatively uniform CNS responsiveness to the drugs in the animal kingdom, which facilitates selection of animal species for studies of barbiturates that simulate effects on humans. Aside from investigations of psychological aspects of drug dependence that require a high degree of behavioral complexity in order to simulate human drug addiction behavior in animal models (in this case, the primate model is perhaps the most appropriate), the basic pharmacological CNS responses to barbiturates seen in all animal species are qualitatively similar. Accordingly, the greatest restriction in selecting an animal model for the chronic

Table 2
Barbiturate Chronic Administration Procedures

Route of administr.	Type of barbiturates	Species	Dose, mg/kg body wt	Duration of treatment	Reference
Forced drug administration					
po	Phenobarbital	Mouse	150–350	7 d	Belknap et al., 1973
po	Barbital	Rat	100–400	32 d	Stevenson and Turnbull, 1968
po	Barbital	Rat	313–396	111–159 d	Essig, 1966
po	Barbital	Cat	61–279	3–217 d	Essig and Flanary, 1959
po	Barbital	Cat	190–335	6–267 d	Essig and Flanary, 1961
po	Amobarbital	Dog	40	2 mo	Swanson et al., 1937
po	Amobarbital	Dog	55	180–195 d	Fraser and Isbell, 1954
po	Barbital	Dog	106–168	216–339 d	Fraser and Isbell, 1954
po	Pentobarbital	Dog	60–104	180 d	Fraser and Isbell, 1954
po	Secobarbital	Dog	35–42	180–195 d	Fraser and Isbell, 1954
po	Barbital	Monkey	143	7–52 d	Mott et al., 1926
		Rabbit	190–650		
ig	Amobarbital	Cat	400	5 wk	Boisse and Okamoto, 1978c
ig	Barbital	Cat	30–50	35 d	Okamoto et al., 1975
ig	Pentobarbital	Dog	61.6–100	4.5–38 mo	Seevers and Tatum, 1931
ig	Barbital	Monkey	75	3 mo	Yanagita and Takahashi, 1973
sc	Barbital	Mouse	Osmotic mini pump 50 mg/mouse	95–101 h	Siew and Goldstein, 1978
sc	Barbital	Mouse	16 mg pellet/mouse	3 d	Turnbull and Watkins, 1976

Route	Drug	Species	Dose	Duration	Reference
sc	Phenobarbital	Mouse	75 mg pellet/mouse	3 d	Ho et al., 1975
sc	Pentobarbital	Rat	6,18,36	7 wk	Stanton, 1936
sc	Phenobarbital	Rat	8–23	7 wk	Stanton, 1936
im	Pentobarbital	Monkey	30–45	6 wk	Yanagita and Takahashi, 1970
im	Phenobarbital	Monkey	50–100	6 wk	Yanagita and Takahashi, 1970
ip	Thiopental	Mouse	50	13 d	Hubbard and Goldbaum, 1949
ip	Barbital	Rat	200	25 d	Wahlstrom, 1968
ip	Barbital	Rat	150	5–15 d	McGee and Bourn, 1978
ip	Pentobarbital	Rat	35–40	14–32 d	Moir, 1937
ip	Pentobarbital	Rat	29	10 d	Gruber and Keyser, 1946
ip	Pentobarbital	Rat	20–30.5	3–96 d	Aston, 1965
ip	Pentobarbital	Rat	30–40	4 d	Wahlstrom, 1968
ip	Pentobarbital	Rat	85	10 h	Turnbull and Watkins, 1976
ip	Secobarbital	Rat	69	10 d	Gruber and Keyser, 1946
ip	Thiopental	Rat	25–50	0–72 h	Singh, 1970
ip	Pentobarbital	Guinea pig	7.5–20	4–6 wk	Carmichael and Posey, 1933
ip	Phenobarbital	Dog	60–100	25 d	Butler et al., 1954
iv	Hexobarbital	Mouse	70	1–4 d	Rumke et al., 1963
iv	Amobarbital	Rabbit	35–40	1–10 d	Masuda et al., 1938
iv	Amobarbital	Rabbit	37.5–40	10 d	Gruber and Keyser, 1946
iv	Pentobarbital	Rabbit	29	10 d	Gruber and Keyser, 1946
iv	Secobarbital	Rabbit	69	10 d	Gruber and Keyser, 1946
iv	Pentobarbital	Cat	88–114	1–21 d	Jaffe and Sharpless, 1965
iv	Pentobarbital	Dog	28–42	27–83 d	Ettinger, 1938

(continued)

Table 2 (continued)

Route	Drug	Species	Dose	Duration	Reference
iv	Thiopental	Dog	20	10 d	Green and Koppanyi, 1944
iv	Pentobarbital	Monkey	50–400	2 wk	Yanagita and Takahashi, 1970
icv	Barbital	Rat	2.4 mg/rat	12 h	Stolman and L. H., 1975
icv	Phenobarbital	Rat	800 µg/mouse	4–5 d	Mycek and B. H., 1976
Self drug administration					
po	Phenobarbital	Rat	12.6	122–142 d	Tang et al., 1980
po	Methohexital	Monkey	20.4–93.8		Caroll et al., 1984
po	Pentobarbital	Monkey	1–4		DeNoble et al., 1982
po	Pentobarbital	Monkey	12	15 d	Meisch et al., 1982
ig	Barbital	Monkey	75	3 mo	Yanagita and Takahashi, 1973
ig	Pentobarbital	Monkey	150–220	32 d	Altschuler et al., 1975
ig	Pentobarbital	Monkey	0.5–1.0	2–20 wk	Woolverton and Schuster, 1983
im	Pentobarbital	Pigeon	1,2,4,5.6	2 wk	Dews, 1955
iv	Amobarbital	Rat	1.5	5 d	Davis, 1963
iv	Pentobarbital	Monkey	0.5, 0.25, 4.0	10 d	Winger et al., 1975
iv	Pentobarbital	Monkey	0.25 or 0.5	13 d	Johanson, 1987
iv	Pentobarbital	Monkey	5	2 wk	Yanagita and Takahashi, 1970
iv	Pentobarbital	Monkey	.25–2.0	6 wk	Goldberg et al., 1971
iv	Pentobarbital	Monkey	0.5–3.0	14 d	Johanson, 1982

studies of barbiturates, is not so much concern about the appropriateness of direct effects of the barbiturate on the nervous systems *per se*, but pharmacokinetic considerations. Some considerations of advantages and disadvantages of commonly used animal species for studies of barbiturate addiction are presented here.

2.1. Pharmacokinetic Considerations

Biotransformation and elimination of barbiturates are considerably faster in rodents, especially in mice, than in larger animals, such as cats, dogs, monkeys, and humans (Table 3). Additionally, for barbiturates that are primarily biotransformed in liver, Quinn and colleagues (1958) have shown that there is an inverse relationship between the activity of the hepatic drug-metabolizing enzyme systems and the duration of hexobarbital "sleeping time" in various animal species (Table 4). It was found that microsomal enzyme activity in the mouse was nearly 17 times that in the dog. Consequently, the kinetic exposure of barbiturates to the CNS should be considerably shorter in rodents than those in larger animals, including humans, if the drugs are administered repeatedly at the same time intervals in those animals. The experiments require the persistent presence of the drug in the CNS, pharmacokinetic influences should be taken into consideration with respect to the peak and the persistence of the drug concentration in the body during chronic treatment. Furthermore, induction of hepatic enzymes, and hence, pharmacokinetic tolerance production in rodents are also extensive; these factors will contribute to decreasing the overall exposure time of metabolizable barbiturate to the CNS.

An additional consideration that is required for the rodent animal model is its nocturnal nature of feeding behavior. If barbiturates are to be given through the oral route in drinking water/food, most of the drug will be consumed during the night, the consumption peaking at around midnight. A considerable amount of the drug will be gone by morning, and the amount of the drug that persists during the day will be very low. Accordingly, drug concentration monitoring is essential for establishing treatment criteria, especially when the animals are exposed

Table 3
Relationships Between Elimination Half-Lives of Barbiturates
in Different Animal Species and the Incidence
of Spontaneous Withdrawal Convulsions

Species	Half-Lives of Barbiturates, %				Incidence convulsion, References
	Barbital		Pentobarbital		
Human	3–4 d	(0%)	40 h	(78%)	Wulff, 1959
					Fraser et al., 1954
Cat	30 h	(56%)	8 h	(100%)	Essig and Flanary, 1959
					Okamoto et al., 1976
Rat	10 h	(100%)	1 h	(0%)	Waters, 1973
Mouse	3 h	(10%)	0.5 h	(0%)	Waters and Okamoto, 1972

Table 4
Species Differences in Hexobarbital Metabolism[a]

Species	Duration of action	Hexobarbital half-life, min	Enzyme activity $\mu g/g/h$
Mouse	12 +/– 8	19 +/– 7	598 +/– 184
Rabbit	49 +/– 12	60 +/– 11	196 +/– 28
Rat	90 +/– 15	140 +/– 54	134 +/– 51
Dog	315 +/– 105	260 +/– 20	35 +/– 30

[a]All animals received hexobarbital 100 mg/kg ip, except dog (50 mg/kg iv) (Quinn et al., 1958).

to low barbiturate concentrations with this treatment method. Furthermore, the feeding behavior pattern might be altered with chronic barbiturate treatment (Groh et al., 1988).

2.2. Genetic Strains

Several strains of mice have been reported to display genotypically variable neurosensitivity to CNS depressants. Accordingly, one of feasible genetic maneuvers is to identify a genotypically pure strain that shows high sensitivity to the drugs and/or displays a high liability of drug dependence characteristics and investigate the underlying mechanisms that make the responses of the particular strain of animals unique to the drugs. For example, C57BL mice have been reported to have lower "phenotypic buffering," i.e., susceptibility to environmental effects, than DBA mice and yet are most susceptible to the effect of phenobarbital administration on behavior testings (Ginsburg, 1958). It has also been established that C57BL mice have a shorter barbiturate-induced "sleep time" than DBA (Jay, 1955; Vesell, 1968; Siemens and Chan, 1976). However, the low sensitivity to barbiturates may be owing to the capacity of C57BL mice to metabolize barbiturates faster than DBA mice; since, in most studies, DBA mice showed a higher brain barbiturate concentration at awakening than did C57BL mice, this indicated that DBA mice had a lower CNS sensitivity to barbiturate than did C57BL mice (Siemens and Chan, 1976). This situation was, however, reversed by chronic barbiturate treatment; DBA mice exhibited a greater sensitivity to barbiturate lethality testing and were less able to develop tolerance to barbiturates than were C57BL mice (Tabashima and Ho, 1981).

Genotype (strain)–environment (drug) interaction is evident from early ages, and has been demonstrated for the effect of barbiturate administration on behavior of neonates (Diaz and Schain, 1978). Pups of the inbred strains DBA/1 and C57BL/10 were fostered by HS/Ibg dams and were given daily sc injections of 50 mg/kg Na phenobarbital during postnatal days 2–21. At 28–30 d, the animals were tested in an open field and in a barrier cage. After the treatment, DBA/1 mice were greatly affected by the drug, whereas the behavior of C57BL mice was little affected by the phenobarbital administration (Yanai et al., 1989). Untreated C57BL mice are usually more active in the open field than are DBA mice (Guttman et al., 1969).

Selective bleeding techniques can also be utilized to produce highly inbred strains that are selected on the basis of a preset physiological and/or pharmacological testing response criterion. It has been shown that highly inbred strains of mice treated with ethanol chronically show many-fold differences among strains in severity of withdrawal convulsions produced by handling (Crabbe et al., 1985a,b). Differences among strains have also been reported following chronic pentobarbital treatment (Yamamoto and Ho, 1978). The clear evidence that genotype can account for a large proportion of individual differences in susceptibility to withdrawal from CNS depressants (Crabbe et al., 1985a,b) led to investigation of whether the genes determining susceptibility to withdrawal from barbiturates were the same as those determining ethanol withdrawal susceptibility (Yamamoto and Ho, 1978).

Lines of mice were genetically tooled to display large differences in ethanol withdrawal severity (Crabbe et al., 1985a,b). Ethanol Withdrawal Seizure Prone (WSP) and Withdrawal Seizure Resistant (WSR) mice have been bred selectively to display severe and minimal handling-induced convulsions after a 3-d regimen of chronic ethanol vapor inhalation. After five generations of selective breeding, WSP mice had approximately three times more severe withdrawal from ethanol than WSR mice following identical ethanol exposure. When phenobarbital rather than ethanol was used as the dependence-producing drug, the WSP vs WSR differences in handling-induced convulsions were similar to those seen with ethanol. These results implied that ethanol and the barbiturate possessed underlying mechanisms that were shared between these two drugs, acting as determinants in producing withdrawal convulsions. Both WSP and WSR mice showed equivalent brain phenobarbital concentrations and equal functional tolerance development as measured by several behavioral impairment testings.

3. Considerations Regarding Chronic Barbiturate Administration

Barbiturates are pharmacologically classified as general CNS depressants (cf Sharpless, 1970; Smith, 1977; Harvey, 1985). They produce a wide spectrum of effects in the CNS ranging from calming and sedation at relatively low doses, hypnosis, anesthesia, coma and respiratory paralysis, and finally death with a progressively increasing dose. Although in recent years specific molecular mechanisms of action of barbiturates in the CNS through GABA-chloride-coupled receptor system have been described, for the most part, the clinical effects exhibited by barbiturates are nonspecific and nonselective in the CNS. Any selectivity, in terms of the clinical indications for which an individual barbiturate is used, is related to:

1. The dose and schedule of administration;
2. The slope of the dose–response curve along the continuum of CNS depression; and
3. The pharmacokinetic profile of the compound.

Accordingly, a clear understanding of these pharmacological principles is essential to the appropriate selection of a barbiturate for addiction studies. Chronic dosing schedule of a barbiturate should adhere to strict dose–response principles.

The difficulty in establishing a dosing regimen for a barbiturate and/or any of CNS depressant drugs is compounded by the development of tolerance during the chronic treatment. Masuda and his associates (1938) first noted the acquired tolerance to barbiturates. Gruber and Keyser (1946), however, were the first investigators to describe in detail the phenomenon of tolerance and crosstolerance to barbiturates in a series of chronic animal studies utilizing loss of righting reflex as a response index. Another important conclusion of their studies was that induction of tolerance was directly related to the time intervals between each consecutive barbiturate administration. They also showed that the same magnitude of tolerance could be produced even when the time intervals between doses were lengthened if they used a long-acting barbiturate. They concluded that the discrepancies in the literature could be attributed to the following problems:

1. The imprecise use of the terms "drug tolerance" and "drug addiction;"
2. Differences in the time intervals between repeated injection of drugs;
3. Differences in the quantity of drug injected per dose;
4. Differences in criteria used to determine acquired tolerance; and
5. Different animal species used by the various investigators.

Those issues still remain as fundamental pharmacological problems in designing experiments and interpreting results in the current literature.

3.1. Barbiturate Tolerance and Its Characteristics

The term "tolerance" has two different connotations; one is "initial tolerance" and the other is "acquired tolerance." Initial tolerance implies that there is a basic difference in individual existence that makes certain individuals less susceptible to a drug and requires the higher than average dose to produce a set drug effect. The extent of tolerance could be variable owing to differences in species, sex, age, congenital factors, environment, and so forth. The acquired tolerance reflects acquired changes in sensitivity to the drug by repeated administration of the drug in the same individual. This chapter deals primarily with "acquired tolerance" to barbiturates.

Barbiturate acquired tolerance is defined as a decreasing response to repeated administration of the same dosage of a barbiturate or as an increase of dosage needed to obtain the initial response. This definition of tolerance applies primarily to a single parameter of drug-response measurement. The development of tolerance varies in its intensity and the speed of acquisition for each independent CNS function that is utilized in experimental response measurement. Furthermore, each of those CNS functions can be different in their underlying neuronal make-up(s) and hence, the sensitivity to the drug. Therefore, if a response measurement is a composite type and constructed by many independent response entities (pharmacological actions) in whole animals, the tolerance characteristic may present different features with different chronic treatment procedures. The indices used to assess the tolerance vary considerably depending on the species utilized. Methods of these measurements are discussed in Section 3.1.3.

Characteristics of barbiturate tolerance development have been summarized in several review articles (cf Kalanat, 1977; Kalant et al., 1971; Ho and Harris, 1981; Okamoto, 1985; Le and Khanna, 1989). There are generally two distinct rates by which acquired tolerance develops in an organism. One is relatively fast and called "acute;" the other is slow and called "chronic" because it generally requires prolonged chronic treatment.

3.1.1. Acute Tolerance

The term "acute" applies to at least two different types of tolerance. One is the type that is seen after one administration of the drug. The other applies to the changes in sensitivity that are detected within a time–action duration after a single-drug administration.

Evidence of acute tolerance to thiopental and pentobarbital was first demonstrated by Brodie et al. (1951) and by Maynert and Klingman (1960). Plasma concentrations of the drug were determined along with the measurement of barbiturate effects in dogs. The results showed that, as the administered dose of barbiturate was increased, the plasma concentration of the drug at the time when the ataxia disappeared was also significantly increased. The most direct proof of acute tolerance, however, can be provided by correlating the pharmacological effect with actual concentration of the drug in the brain. Such experiments have clearly demonstrated acute tolerance development; an example is the study of ethanol with the treadmill test (Gibbins et al., 1968).

3.1.2. Chronic Tolerance

Chronic tolerance to barbiturates can be detected in laboratory animals within a matter of days. The extent of the tolerance is relatively small and ranges from a 25 to 60% decrease in the barbiturate effect. This contrasts to the degree of tolerance seen with opioids, the magnitude of which is often >10 times the initial dose. The underlying cause(s) for this difference has not been addressed. However, one of the reasons may be the relative insensitivity of respiratory neurons in acquiring tolerance, thereby limiting the necessary increase of chronic barbiturate dose for producing the maximal degree of tolerance in other neuronal systems. Additionally, when a chronic treatment procedure does not produce steady blood drug concentrations throughout the day, but produces a peak-trough concentration fluctuation between each consecutive drug administration (i.e., iv infusion vs repeated oral administration), total drug exposure time on the drug-resistant neuronal function will be less than on the drug-sensitive functions in the CNS. Accordingly, the greater the resistance a particular response system is to the drug, the less time the system is exposed to develop tolerance during intermittent-type treatment.

3.1.2.1. DISPOSITIONAL AND FUNCTIONAL TOLERANCE

Biotransformation of barbiturates (Maynert and VanDyke, 1949; Williams and Parke, 1964; Stevenson and Turnbull, 1968) and induction of drug-metabolizing enzymes by barbiturates and other drugs have been extensively reviewed (Conney and Burns, 1962; Remmer, 1964; Conney, 1967). It has been shown in a wide variety of animal species that chronic administration of barbiturates known to be metabolized in liver is followed by a reduction in the duration of action (i.e., the period of righting reflex loss); this accompanies an increased rate of metabolism of the compounds by the liver (Kato et al., 1964; Stevenson and Turnbull, 1968). Accordingly, a part of tolerance seen with barbiturates can be explained by changes in the rate of barbiturate metabolism. This type of tolerance, which occurs primarily because of changes in drug disposition and, hence, because of the reduction in the amount of the drug reaching the CNS, is called "dispositional tolerance" or "pharmacokinetic tolerance" (cf Le and Khanna, 1989). Dispositional tolerance, by definition, also includes underlying causes owing to changes in drug absorption, distribution, and renal elimination. For barbiturates, an increase in drug metabolism is the major underlying cause of dispositional tolerance. However, drug disposition alone cannot explain the whole scope of barbiturate tolerance. For example, it cannot explain the tolerance phenomena when the drug-treated animals show higher plasma and/or brain barbiturate concentrations than the concentration seen with the drug-naive animals at the time when the recovery of response takes place (Kalant et al., 1971). In addition, tolerance seen with chronic Na barbital administration cannot be explained by the drug disposition, since barbital is metabolized only to a limited extent in liver in most animal species. Siew and Goldstein (1978) showed a significant decrease in "sleeping time" produced by a challenge dose of Na barbital before and after (administered 24 h after the chronic treatment termination) the chronic barbiturate treatment. This type of tolerance is described as "functional tolerance" or "pharmacodynamic tolerance." The neurons in the CNS are capable of adapting themselves and improve their function even in the presence of the same concentration of the drug in the brain. The development of functional tolerance during pentobarbital chronic treatment was further supported by the experiments (Ho, 1976; Flint and Ho, 1980) that used the pellet implantation procedure in the mouse. The intracerebral LD_{50} of Na pentobarbital in the pentobarbital pellet-implanted group of mice was significantly higher than with the control group. The hypothermia induced by intracerebral administration of Na pentobarbital was also less; furthermore, body temperature recovered to normal much faster in mice implanted with barbiturate pellets than in those implanted with placebo pellets. It is important to recognize that, with most of barbiturates that are known to be biotransformed in liver, both types of tolerance would develop during chronic drug administration.

To identify and characterize the two components of the tolerance, i.e., dispositional and functional tolerance, in overall tolerance, one can combine measurement of pharmacological drug responses and drug pharmacokinetic studies. This was illustrated, as an example, by the "maximally tolerable dose" technique developed by Okamoto and her associates (1975; Boisse and Okamoto, 1978a) in cats. The dispositional tolerance to pentobarbital developed to its maximal level within 1 wk, indicated by an increase in pentobarbital elimination half-life values in blood, and was maintained at that level throughout the remainder of the treatment period (Boisse and Okamoto, 1978b). Functional tolerance, measured by predetermined CNS functional response against the level of blood drug concentration, however, developed more gradually and progressed with continued treatment (Boisse and Okamoto, 1978b). Importantly, however, the maximal level of both the dispositional and the functional tolerance, which was achieved by the chronic treatment, was chronic dose-dependent (Okamoto et al., 1978).

Behavioral tolerance has been demonstrated for some of the abused drugs, including barbiturates (Tang et al., 1981; Harris and Snell, 1980). Behavioral tolerance is a subvariety of functional tolerance that can be distinguished by the demonstration that behavioral factors are important determinants of the rate and/or degree of tolerance development to a drug's behavioral effects (Dews, 1978). The standard method used to demonstrate that tolerance has a behavioral component is to compare the effects of repeated pretest drug administrations to those of an equal number of posttest administrations. The logic of such a comparison is that pretest drug administration allows the subject to engage in the preset behavioral task while drugged, during which time behavioral compensation can be developed. Posttest administrations, by contrast, result in the same amount of exposure to the drug, but do not allow the subject to engage in the task while under the drug's influence. Squirrel monkeys were trained to press a lever under a multiple schedule of food presentation. Tolerance developed to rate-increasing and reinforcement-frequency-decreasing effects of pentobarbital during the interresponse time schedule when the drug was administered immediately before the daily session, but did not develop when it was given immediately after the session (Branch, 1983).

3.1.3. Measurement of Tolerance

Acquisition of tolerance can be considered as a change in the relationship between the dose of a drug and the effect that it produces. Therefore, any method suitable for measuring the acute drug effect should, in principle, be suitable for measuring the acquisition of tolerance. It is important to recognize, however, that functional tolerance does not necessarily develop uniformly to all response functions; therefore, magnitude of functional tolerance development depends on the level of the drug exposure on each responding function and the sensitivity of each responding function to the drug. Accordingly, not only is presetting the criteria for choosing appropriate response function(s) important, but also the description of acquired tolerance should pertain to that particular function(s) that is (are) preselected for the measurement. Generally, the most useful methods for measurement of acquired tolerance are those methods that:

1. Permit clear identification and analysis of the response affected by the drug;
2. Are continuously variable rather than quantal or discrete;
3. Are sensitive to the changes produced by small dose increments; and
4. Produce reproducible and stable responses during observation periods.

Only a few studies have investigated relative degrees of tolerance development utilizing multiple-response entities simultaneously (Richter et al., 1982).

Physiological measures that have been used ranged from gross indices, such as change in minimum lethal dose or LD_{50} (Masuda et al., 1938; Holck et al., 1950; Isbell and Fraser, 1950), to the very fine and sensitive changes in critical flicker-fusion frequency (Idenstrom, 1954), spontaneous electroencephalographic (EEG) frequencies, and response to photic stimulation (DeSalva, 1956; Essig and Fraser, 1958; Frey and Kampman, 1965; Jaffe and Sharpless, 1965), as well as the threshold and duration of electrical silence in the EEG (Stump and Chiari, 1965; Wahlstrom, 1968). Tolerance has also been measured by the change in the seizure thresholds induced by electroshock or pentylenetetrazole (DeSalva, 1956; Frey and Kampman, 1965; Jaffe and Sharpless, 1965). Other commonly used indices for tolerance to barbiturates are the time of induction and the duration of loss of righting reflex ("sleeping time") in animals (Isbell and Fraser, 1950; Rumke et al., 1963; Aston, 1966). Tolerance has also been determined by testing various motor functions, such as the ability of an animal to climb a pole (Appel and Freedman, 1968), to remain on a tilted surface (Kato, 1967) or on a rotating narrow rod (Dunham and Miya, 1957; Jones and Roberts, 1968), or to walk along a narrow elevated board (Remmer and Siegert, 1960; Molinengo, 1964) without falling off.

With such a wide variety of techniques for measurement of tolerance available, the selection of methods for a particular study of drug tolerance appears somewhat arbitrary. However, choices can be narrowed by exercising a few general principles.

The first is the relevance of the dose range within which a particular test is performed in relation to other pharmacological responses of the particular drug. Measurement of LD_{50}, for example, involves doses that are far above the dose that is thera-peutically relevant for the study of tolerance to barbiturates. Furthermore, the maximal degree of functional tolerance to lethality that can be attained by barbiturates is rather small. Administration of repeated near-lethal doses of barbiturates may cause irreversible neurological damage and/or accidental death (Krop and Gold, 1946). Tolerance to opiates, on the other hand, is of such magnitude that an escalating scale of dosage used to develop tolerance in behavioral effects does indeed go far beyond the normal LD_{50} (Isbell and Fraser, 1950), so that an increase in LD_{50} is also an appropriate index of tolerance. Chronic experiments with barbiturates, therefore, involve smaller doses. This may be one reason why very little increase in the minimum lethal dose of barbiturates (Okamoto et al., 1978) or the LD_{50} of barbiturates (Gruber and Keyser, 1946; Ho, 1976) has been observed.

Specificity of the phenomenon is another important con-sideration. Change in food or water intake during chronic administration and withdrawal of barbiturates (Essig, 1966; Isbell, 1950) is a nonspecific indication of the subject's general state of well-being. There may be a depression of food intake during the initial stage of drug action in the course of chronic treatment and withdrawal of barbiturates. In contrast, food intake is specifically reduced by amphetamine and related com-pounds (Kosman and Unna, 1968), and hypopraxia occurs dur-ing withdrawal (Kramer et al., 1967; Smith, 1969). Therefore, food intake would be a rather nonspecific index for studying toler-ance to CNS depressants, but is quite appropriate for amphet-amine (Harrison et al., 1952; Kosman and Unna, 1968) and its related compounds.

Another consideration in the choice of methods is the goal of the intended study. Relatively crude, but simple methods, such as change in LD_{50} or subjective scoring of ataxia, may be quite adequate for qualitative answers as to whether or not tolerance occurred. In contrast, quantitative kinetic studies on tolerance development, especially during treatment with mod-erate doses, require a technique that can objectively quantify and discriminate the effects produced by small dose increments.

Quantal response measurement for tolerance evaluation is based on those tests in which the responses are quantitated in an all-or-none fashion. The rotarod test of motor performance in rats or mice (Dunham and Miya, 1957), for example, involves placing the animal on a rod that rotates along its longitudinal axis at constant speed. The test runs for a set time, and the animal is scored as either staying on the rod for the full time or falling off. To establish a dose–response curve, therefore, it is necessary to have a considerably large number of animals at each dose to permit a statistically valid estimate. A simpler version of the test consists of placing a rat at the middle of a fixed rod 1 mm long and 1 cm thick, and observing whether it can successfully reach either end (Molinengo, 1964). The chimney-climbing test (Boissier et al., 1960) and lethality test are other examples of the quantal approach.

In contrast, a continuous graded-type response measurement generally permits more precise, sensitive, and economical measures for construction of dose–response curves and an evaluation of tolerance development. This is illustrated by a modification of the rotarod test (Jones and Roberts, 1968) that consists of applying an increasing speed of rotarod rotation, so that each animal will eventually fall off the rod. This point can be measured as a time or speed of rotation. An analogous transformation can be made in the inclined plane test. As a test of motor incoordination, tilting of the board is done at constant angular velocity, and the angle at which the animal begins to slide off is measured (Kato, 1967).

One of the oldest and most widely used techniques that employs continuous measurement scales is loss of righting reflex ("sleeping time;" Ebert et al., 1964; Aston, 1965; Milner, 1968; Wahlstrom, 1968; Svensson and Thieme, 1969). The duration of "sleeping time" is considered to be the time between loss and recovery of righting reflex. The "sleeping time" index has been used in the dog, monkey, guinea pig, cat, rat, and mouse. Although this method is simple to execute, one limitation could be that it measures the time function rather than the intensity of the drug effect. Accordingly, it is more sensitive to reflect the pharmacokinetic tolerance than pharmacodynamic tolerance, unless the blood barbiturate concentration at the time of regaining righting reflex is determined and correlated against the measured response. In contrast, other barbiturate-induced pharmacological responses, such as hypothermia (Ho, 1976), measure intensity of pharmacological effect, i.e., degree of temperature change. Other graded response measurements that have been utilized effectively for evaluation of other general CNS depressants and could be used for evaluation of barbiturates include the inclined plane test (Arvola et al., 1958), and tests of spontaneous and forced locomotor activity (Boyd, 1960; Phillip et al., 1962; Goldberg et al., 1967).

Progressive behavioral impairment of CNS function by CNS depressants has been used to measure the acute effect of, as well as tolerance to, ethanol-utilizing behavioral rating systems and forcing the resulting ratings to a continuously variable linear scale (Newman and Abramson, 1941). Different degrees of motor impairment in dogs, ranging from slight ataxia to complete inability to stand, were ranked in order of increasing impairment and assigned numerical ratings from 1 to 10; the numerals were used as scores to be plotted against the drug dose or drug concentration in the blood. This type of composite behavioral measurement technique was used to evaluate tolerance to barbiturates in rhesus monkeys (Yanagita and Takahashi, 1970) and in cats (Okamoto et al., 1975,1976). The behavioral measurements included degree of motor coordination, posture, degree of ataxia, standing, walking, self-righting, impairment in corneal reflex, nictitating membrane tone, muscle tone, and pain withdrawal. These procedures assume that each interval in the behavioral rating scale represents an equal increment in impairment of CNS function; this assumption is also made for composite behavioral response scores, which are derived from the sum of all behavioral ratings entered into the total scoring. This assumption should be tested before the scoring system is used, but it has not been examined routinely. It has been shown that such composite CNS depression rating scales to barbiturates, which have utilized common behavioral response measurements that reflected progressive CNS depression in cats, were not equal in intervals when they were plotted against barbiturate dose and blood concentration (Boisse and Okamoto, 1978a).

3.1.4. Effect of Chronic Drug Dose
and Duration of the Treatment

Knowledge of the pharmacological determinants of barbiturate tolerance production is essential for designing chronic dosing regimens. The parameters, which must be quantitatively fulfilled in order to describe the drug load, are:

1. Chronic dose (Wikler et al., 1955; Fraser et al., 1958; Essig and Flanary, 1959; Wikler and Essig, 1970; Yanagita and Takahashi, 1970,1973; Waters, 1973);
2. Dosing frequency (Fraser and Isbell, 1954; Essig, 1972; Yanagita and Takahashi, 1970,1973); and
3. Duration of treatment (Goldstein, 1972; Waters, 1973).

The above pharmacological determinants control both the speed and the degree of tolerance production. Very few studies on tolerance have been designed to establish quantitative relationships between these pharmacological determinants and the tolerance produced; most studies in the literature have dealt primarily with demonstration of either the presense or absence of tolerance.

One of the common, but important, variables that affect the degree of tolerance production is the route of the drug administration. The blood or brain concentrations of the drug by a given dose vary considerably with different routes (Crawford, 1966; Kalant et al., 1971). With the possible exception of an iv infusion-type drug administration by which a steady drug concentration in blood and brain can be achieved, an intermittent-type chronic drug administration, although clinically relevant, has an inherant problem with setting ideal experimental conditions for the study of tolerance. Intermittent drug administration produces not only the peak and trough drug concentration fluctuation, but also produces a condition in which the duration of drug action at high concentration is shorter than at low concentration in the CNS. If the extent of tolerance development is governed not only by the drug concentration, but also by the duration of drug exposure, the neuronal system more sensitive to the drug (therefore, it is affected to a greater extent at the low drug concentration range) will be exposed to the drug for a longer time.

Furthermore, a clear-cut relationship between the dose and the degree of tolerance development is also difficult to establish by the intermittent dosing design, since a larger dose almost inevitably increases both the maximum concentration and the effective duration of drug exposure. Nevertheless, it has been indicated that tolerance to drugs such as ethanol, develops more rapidly when the treatment is started with higher initial doses in chronic treatment (LeBlanc et al., 1969).

Aston (1965) has examined the relationship between dose of pentobarbital and magnitude of tolerance development. Groups of rats were injected with different doses of the drug ranging from 25 to 45 mg/kg, and 24 h later were given a second dose of pentobarbital 40 mg/kg. All drug-treated groups showed a significantly shorter mean "sleeping time" on the second day than that of a group given 40 mg/kg without pretreatment. Moreover, the degree of tolerance was found to be directly proportional to the size of the first dose. The "sleeping time" measurement, however, might be misleading if functional tolerance is to be characterized, and the above result could well be the reflection of dispositional tolerance development that resulted from hepatic enzyme induction by pentobarbital. It should be noted that induction of hepatic enzymes by barbiturates also occurs in a dose-dependent manner.

If the development of functional tolerance involves compensatory changes in the affected neurons that render them less sensitive to the drug, then the extent to which compensation develops in a particular neuronal system may well depend on how much it is challenged, i.e., on the level of chronic dosing. This principle might explain the observation of Maynart and Klingman (1960) that the degree of tolerance to barbiturates is less at greater levels of neurological impairment (e.g., inability to walk vs mild ataxia). Moreover, the common belief that chronic administration of barbiturates in doses sufficient to induce tolerance to the sedative and hypnotic effects produces little or no increase in the lethal dose (Gruber and Keyser, 1946; Isbell and Fraser, 1950; Seevers and Deneau, 1963; Jaffe, 1985) might also be explained by postulating that the level of neuronal adaptation depends on the level of chronic dosing. The quantitative

CNS depression rating scales developed for producing physically dependent animals by the "maximally tolerable dosing" technique (Okamoto et al., 1975) have been designed to assess a wide range of CNS functional impairment produced by barbiturates. Utilizing this model system, the effect of "low" and "high" chronic barbiturate doses was studied and compared for the magnitude of functional tolerance development. The results showed that the magnitude of functional tolerance was greatest for those functions that were more sensitively and persistently depressed, whereas functional tolerance was absent for those functions that were not obviously affected during chronic treatment. It was found that functional tolerance did not develop uniformly for all levels of neuronal function and that the magnitude of the challenge to neuronal function was an important determinant for chronic functional tolerance development to barbiturates.

In comparing rate of functional tolerance development produced by different barbiturates, potency and pharmacokinetic differences inherent with pentobarbital and barbital were corrected for by administering equipotent doses and adjusting the intervals of the drug administrations. The "chronically equivalent dosing method" (Boisse and Okamoto, 1978a,b) was designed not only to make the daily drug peak effects constant throughout the chronic treatment by monitoring drug peak effect and adjusting the daily dose to produce equal peak effect, but also to correct for the pharmacokinetic differences between pentobarbital and barbital by making the residual effects at the time of succeeding doses equivalent, i.e., the total CNS depression is based on area under time–action curves made equal between the two drugs. The results showed that functional tolerance to pentobarbital and barbital developed at the same slow rate for this "chronically equivalent" treatment. This suggests that functional tolerance development is independent of the particular barbiturate, reflecting the adaptability of the CNS to chronic depression.

3.1.5. Persistence of the Change and Carryover

In order to characterize tolerance to barbiturates fully, it is important to know not only the rate of acquisition and the magnitude of change to the drug, but also the duration and the rate of decay of that change. Although the rate of disappearance of tolerance has been studied (Aston, 1965; LeBlanc et al., 1969; Okamoto et al., 1985), the factors that could influence its decline have not been explored systematically. The rate of disappearance of tolerance should be systematically examined as a function of speed of acquisition, level of tolerance attained, duration of maintenance, and history of previous exposure to the drug. Such studies should also take into account the possible effect of the test doses themselves upon this decline.

In view of profound differences in the rate of tolerance acquisition between dispositional and functional tolerances, the recovery processes of these produced by the "maximally tolerable dosing" technique in cat were investigated (Okamoto et al.,1985). These workers found that the rates of recovery processes were the mirror images of their acquisitions, being fast with dispositional tolerance recovery and slow with functional tolerance. Furthermore, the neuronal functions most sensitive to barbiturates (i.e., sedation and loss of fine motor coordination), and therefore exposed longer to the drug by intragastric administration, exhibited a greater degree of functional tolerance than did other functions; the recovery of these neuronal functions took a longer period of time. However, the rates of recovery of functional tolerance at all levels of CNS function seemed relatively constant, indicating that there are uniform readaptation mechanisms for all the CNS functions.

Another important aspect of tolerance is its characteristic of being carried over (Kalant et al., 1971). An increase in the rate of subsequent reacquisition of tolerance and/or a more rapid development of maximal tolerance in successive cycles of chronic exposure to the drug has been demonstrated with ethanol. The rate of development of acute tolerance in rats was found to increase from session to session until maximal adaptation was achieved. In addition, tolerance was reacquired more rapidly on each of three successive cycles of chronic treatment and recovery in rats. This carryover has also been demonstrated with ethanol on physical dependence production; when the ethanol exposures were repeated, but separated by alcohol-free periods, the withdrawal signs intensified as the animals were reexposed to ethanol (Majchrowicz, 1975; Baker and Cannon, 1979; Clemmesen and Hemmingsen, 1984; Maier and Pohorecky, 1989). Similar results have been reported with other sedative-hypnotics, including barbiturates, in rodents (Nakamura and Shimizu, 1983).

3.2. Characteristics of Barbiturate Dependence

3.2.1. Definition of Dependence

General CNS depressants, including barbiturates, produce CNS dependence and withdrawal syndrome, both in people and in experimental animals. Traditionally, dependence is classified as "physical" and "psychological" dependence. This distinction is based on the definition of drug dependence by the World Health Organization (Eddy et al., 1965).

> Dependence is defined as a state of discomfort produced by withdrawal of a drug from a subject who has been repeatedly exposed to it, and alleviated by readministration of that drug or another with similar pharmacological actions. The discomfort may consist of non-specific and ill-defined dissatisfaction giving rise to a desire (ranging from a mild wish to intense craving) for the perceived effect of drug; this is called psychological dependence, and may persist for a long time after drug termination. The discomfort may also extend to a more specific set of physiological disturbances with varying intensity and related in a fairly characteristic way to the dosage and pharmacological action of the drug; this is called physiological dependence, and is usually confined to the first few days or weeks after drug withdrawal.

The distinction between these classifications of dependence, however, may become unsettled as the molecular, biochemical, and physiological mechanisms of pyschological functions are identified in the future.

The above definition of physical dependence conceptually equates physical dependence to withdrawal syndrome displayed after abrupt termination of drug administration. Accordingly, the severity of physical dependence to barbiturates is conventionally measured by the intensity of withdrawal syndrome produced after abrupt termination of the drug administration. Since there are no methods currently available to assess directly the degree of dependence without withdrawing the drug, the expression of "physical dependence" can be conceptually misleading when it is equated to the phenomenon of a withdrawal syndrome. The experimental results of Boisse and Okamoto (1978d) have clearly indicated that the rate of barbiturate elimination was the key factor in exposing underlying physical dependence on barbiturates, and that the phenomena of physical dependence should be dissociated from the actual expression of physical dependence as withdrawal phenomena in evaluating barbiturates and other CNS depressants. Cats were treated with "chronically equivalent" doses (Boisse and Okamoto, 1978a) of short- and long-acting barbiturates (Na pentobarbital and Na barbital) for 5 wk and abruptly withdrawn. The withdrawal signs for barbital (long-acting) appeared later, developed more slowly, and persisted longer than those for pentobarbital (short-acting) (Boisse and Okamoto, 1978c). However, when barbital was eliminated as quickly as pentobarbital by hemodialysis, the withdrawal signs became more severe and appeared sooner, resembling those produced by pentobarbital (Boisse and Okamoto, 1978d). If there were specific receptor antagonists to barbiturates available, such as is the case with opioids and benzodiazepines, the intensity of withdrawal signs could at least directly reflect the degree of physical dependence on barbiturates quantitatively without being influenced by the pharmacokinetics, since those antagonists could precipitate withdrawal almost instantaneously regardless of their pharmacokinetic profiles.

3.2.2. Barbiturate Withdrawal Signs and Syndrome

Excellent descriptions of the withdrawal syndrome are available in human experimental studies (Isbell et al., 1955; Mendelson, 1964; Victor and Adams, 1953) on alcohol; perhaps these descriptions depict an overall clinical picture of withdrawal syndrome for all sedative-hypnotics and antianxiety drugs (cf Martin et al., 1979; Mackinnon and Parker, 1982; Jaffe, 1985), including barbiturates (Isbell, 1950; Wikler et al., 1955; Fraser,

1957; Wulff, 1959; Essig, 1964; Epstein, 1980). A series of clinical symptoms increases in severity from tremulousness, insomnia and irritability, and hallucinatory states to delirium tremens and convulsions. These symptoms are characterized by varying degrees of hyperexcitability and hyperreactivity of all parts of the nervous system—central, peripheral, somatic, and autonomic. In general, the signs and symptoms displayed during the withdrawal are the opposite of those characterizing the picture of acute barbiturate intoxication (cf Jaffe, 1985).

Virtually identical reactions have been described for withdrawal signs produced by barbiturates in animals. Furthermore, the picture appears to be essentially the same in all species tested. Thus, it is important to recognize the fundamental similarity in responsiveness of human CNS to barbiturates and the beneficial value of animal models for the study of barbiturate dependency. Barbiturate withdrawal reactions have been reported in dogs (Seevers and Tatum, 1931; Essig, 1963; Roth-Schechter and Mandel, 1976), monkeys (Yanagita and Takahashi, 1970,1973), cats (Essig and Fraser, 1959; Essig and Flanary, 1959,1961; Jaffe and Sharpless, 1965; Okamoto et al., 1976; Boisse and Okamoto, 1978c), rats (Crossland and Leonard, 1963; Essig, 1966; Crossland and Turnbull, 1972; Morgan, 1976; Morgan et al., 1977,1978; Wahlstrom, 1978; Nordberg and Wahlstrom 1977,1979), and mice (Freund, 1971; Waters and Okamoto, 1972; Gates and Chen, 1974; Ho et al., 1975; Belknap et al., 1973,1978,1988; Tabakoff et al., 1978). Although Na barbital is the agent most frequently used to produce physical dependence, especially in small animals, Na phenobarbital and Na pentobarbital are also used. Tremor, convulsions, and behavior suggesting hallucinations have been observed in dogs (Fraser et al., 1954), cats (Essig and Flanary, 1959; Jaffe and Sharpless, 1965), and monkeys (Yanagita and Takahashi, 1970). In mice, similar signs, such as gross tremor, rigidity, and convulsions, have been reported (Freund, 1971). The threshold for electroshock-, handling-, and pentylenetetrazole-induced seizures was found to be lower than normal in mice (Waters and Okamoto, 1972; Ho et al., 1973; Ho, 1976; Siew and Goldstein, 1978; Belknap et al., 1988), rat (Crossland and Leonard, 1963), and cat (Jaffe and Sharpless, 1965) during the withdrawal.

3.2.3. Measurement of Barbiturate Withdrawal

Barbiturate withdrawal intensity has been quantitated primarily by visual observation of overt withdrawal signs and assessment of their severity by experienced observers (Lim et al., 1956; Lim, 1967; Irwin 1976; Gogerty, 1976). For example, comprehensive assessments of withdrawal in rats have been reported (Nozaki et al., 1981; Martin et al., 1982;) in which the degree of physical dependence to pentobarbital has been estimated by subjectively grading the withdrawal reaction utilizing the "Rats SHAAD BRS (Sedative Hypnotic Anti-anxiety Drug Behavioral Rating Scale)" and Withdrawal Observation Form. The incidence of occurrence of a variety of abstinence signs was counted during each sequential 1-h observation period. Those abstinence signs included twitches and jerks, explosive awakenings or violent turning (both of which are episodic movements in which the rats are propelled forcibly against the side and the roof of the cage), poker tail, hot foot behavior, vocalization, retropulsion, chewing, pawing in the air, stretching, rearing, rigid walking, head bobbing, ear flipping, focal or clonic convulsions, tonic-clonic convulsions, scratching, and shivering. A hostility scale was also built that had five weighted items: scratching of handler (1), clinging to the wire mesh (1) and vocalization (1), and struggling (1) and biting (2) on being handled. Agitation was assessed using a three-grade ordinal scale. Respiratory rate was counted once each hour. Body weight, rectal temperature, and water consumption were also measured.

Since physical dependence can only be measured from the intensity of the withdrawal syndrome, development of quantitative methods of evaluating the withdrawal reaction is essential for further gaining insight from experimental studies, as is selection of quantitative criteria for chronic treatment procedures. The most difficult task for the subjective withdrawal evaluation system, in this respect, is the establishment of the linearity between the intensity of the responses with the subjectively measured scales for constructing dose–response curves. This is particularly important when one attempts to compare the physical dependency liability among different CNS depressants. Since the rating system is based primarily on arbitrarily assigned numbers, the observed responses do not necessarily assume a straight linear regression, nor does the line necessarily cross the 0 origin. Accordingly, various methods have been developed in applying correction factors for normalizing the responses (Martin et al., 1982). Yanagita and Takahashi (1973) have defined in rhesus monkeys three stages of withdrawal severity based on predominating overt withdrawal signs appearing during the withdrawal from pentobarbital, phenobarbital, and barbital. Abstinence signs predominating in each stage of withdrawal intensities were:

1. Mild: apprehension, hyperirritability, mild tremors, anorexia, and piloerection;
2. Intermediate: severe tremors, muscle rigidity, impaired motor activity, retching and vomiting, and weight losses of over 10%; and
3. Severe: grand mal convulsions, delirium, and hyperthermia.

The above information indicates and assumes that the CNS functions that contribute to production of certain withdrawal signs, such as tremors and twitches, are more vulnerable to chronic barbiturate dosing; those signs appear with low-intensity dependence/withdrawal, and continue to intensify with high-intensity dependence/withdrawal.

Boisse and Okamoto (1978c) assessed the severity of pentobarbital and barbital withdrawal in cats by counting the number of grand mal convulsions from the recordings of motor activities (Essig and Flanary, 1959) and also by subjectively rating 20 additional motor, autonomic, and behavioral signs, including tremors, twitches, myoclonic jerks, postural disturbances, and motor incoordination (Table 5). Analyses of abstinence evaluations obtained from each animal included:

1. The average total number of signs observed per animal;
2. The incidence of each sign for a group of animals; and
3. The average peak intensity of each graded sign.

The mean intensity of some of the graded signs was also computed as a function of time from the last dose for a plot of time-course. In order to obtain more concise measures of with-

Table 5

Barbiturate Abstinence Evaluation in Cat (Okamoto et al., 1976)

Motor	
Ear and facial twitches	0,1,2,3
Resting tremors	
Head	0,1,2,3
Body and limbs	0,1,2,3
Intention tremors	
Forepaw	0,1,2,3
Weakness	0,1,2,3
Motor functions	
Muscle tone	N = normal, W = weak, SR = spastic and rigid
Posture	N = normal standing posture
	HP = high posture; stands with limbs fully extended and arched back
	LP = low posture stands with limbs partially flexed or extended
	CP = crouched posture; unable to stand
	SE = spread eagle posture; unable to stand; lies flat on chest with limbs fully extended and spread apart
Motor coordination quantitative	0 = stands, walks, and lands; normal
	1 = stands and walks; cannot land
	2 = stands; cannot walk or land
	3 = cannot stand, walk, or land
Movement qualitative	N = normal: Ak = akinetic: At = ataxic
	Hs = high-stepping gait
Patella tendon reflex	N, increase, or decrease

Overall motor activity	N, increase, or decrease
Myoclonic jerks	0,1,2,3
Seizures	
Numbers recorded	
Characteristic type	C = clonic, TC = tonic-clonic, BC = brief clonic, CC = continuous clonic
Body shake (wet dog)	0,1.5,3
Autonomic	
Piloereation	0,1,2,3
Body temperature	°C
Respiratory pattern	N = normal, P = panting
Pupillary dilatation	0 = normal, 1.5 = dilated, 3 = near maximally dilated
Pupillary light reflex	0 = normal, 1.5 = diminished, 3 = absent
Eating	0 = none, 3 = well
Behavioral	
General behavior	N = normal, Ap = apprehensive, Ag = aggressive, P = passive, Bz = bizarre
Startle response	
Auditory	0,1,2,3
Tactile	0,1,2,3
Drowsing	+ or −
Miscellaneous	
Body wt	kg
Death	+ or −

drawal severity, "total intensity scores" were developed by multiplying each grade number assigned for peak intensity by the corresponding incidence and summing all grades for each sign. Therefore, the highest total intensity score achievable for any particular graded sign would correspond to a 100% incidence of the highest grade possible. In general, the use of the incidences of individual withdrawal signs alone is a relatively coarse indicator for evaluating withdrawal severity. Indeed, statistically significant differences in incidence between barbital and pentobarbital were not obtained for six graded signs that started to be seen with the "mild withdrawal" when the sample number was relatively limited. However, when the peak intensities were calculated on the bases of "total intensity scores" and compared, statistically significant differences were obtained for all graded signs. This points to the value of quantitation of withdrawal intensity to obtain meaningful comparisons in the most efficient manner. This finding was confirmed by Boisse and Okamoto (1978c) when comparisons of withdrawal intensity were expressed as "total intensity scores;" it was more sensitive and reliable than comparisons of incidence for detecting differences in withdrawal severity produced by different barbiturates (Boisse and Okamoto, 1978c,d) and has been used in studies of crossdependence and/or treatment of barbiturate withdrawal (Okamoto et al., 1983) with smaller sample sizes. Subjective behavioral response measures, such as those mentioned above, for evaluation of withdrawal intensity require a certain amount of training of observers to evaluate behavioral responses of animals. Furthermore, if a collective withdrawal sign measurement is performed independently by several observers on each animal at each observation period, it should enhance the reliability of the evaluation. Interobserver reliability of evaluating withdrawal responses should also be established. The animals should be coded so the observers are not aware of the treatment.

In humans as well as in most large-size animals, i.e., monkey, dog, and cat, long-acting barbiturates (Lutz, 1929; Meerloo, 1930; Kalinowsky, 1942) have produced less severe withdrawal signs than have intermediate/short-acting barbiturates (Broder, 1936; Palmer and Braceland, 1937; Alexander, 1965). A long-acting barbiturate, such as Na barbital or Na phenobarbital, however, has been the drug of choice for the induction of barbiturate physical dependence in small laboratory animals (Fraser and Isbell, 1954; Essig, 1963,1966,1968,1976; Essig and Flanary, 1959,1961; Crossland and Leonard, 1963; Essig and Fraser, 1966; Leonard, 1965,1966a,b,1967,1968; Mcbride and Turnbull, 1970; Norton, 1971; Stevenson and Turnbull, 1969,1970; Crossland and Turnbull, 1972; Gay et al., 1983). Only a few studies have utilized Na pentobarbital in rodents (Stanton, 1936; Moir, 1937; Gruber and Keyser, 1946; Aston, 1965; Wahlstrom, 1968; Turnbull and Watkins, 1976; Yutrzenka et al., 1985). The severe physical dependence manifested by spontaneous withdrawal convulsions has often not been shown with intermediate/short-acting barbiturates (e.g., pentobarbital) in rats and mice (Yutrzenka and Kosse, 1989). This apparent discrepancy has been thought to be owing to differences in drug elimination rates among different animal species. Table 2 shows the approximate elimination half-lives of barbital and pentobarbital in different animal species and the incidences of severe withdrawal convulsions reported. It is clear from the Table that rodents, particularly mice, eliminate barbital and pentobarbital at a rate >30 times faster than in humans. It has been generally agreed that the persistent presence of the barbiturate and hence, the continued depression of the CNS by the drug (the response) are responsible for the development of the drug tolerance and dependence (Okamoto et al., 1986). It is reasoned, therefore, that pentobarbital is eliminated so quickly in rodents that a sufficient amount of CNS depression cannot be maintained to produce severe physical dependence; on the other hand, the slightly slower elimination of barbital favors prolonged time–action and drug accumulation for production of dependence compared to pentobarbital. In the experiment of Waters and Okamoto (1972), in which mice were treated with 400, 600, and 1000 mg/kg/d for 8 wk in drinking water and withdrawn, only the animals receiving 1000 mg/kg/d had spontaneous clonic-tonic withdrawal convulsions, but this occurred only in 10% of animals. As shown in Table 2, the elimination half-time of barbital in mice was estimated to be 3 h (Waters and Okamoto, 1972) compared to 30 h in cat (Boisse and

Okamoto, 1978b). On the other hand, once the CNS has become dependent on the barbiturate, severe withdrawal is likely to be produced with the drug, which is quickly eliminated, i.e., pentobarbital, which renders the drug-adapted CNS less able to normalize when the drug administration is abruptly terminated. The Na barbital elimination rate in humans and also in large-size animals is slow enough for the CNS to readapt to the absence of the drug without causing severe withdrawal. Crossland and Leonard (1963) were the first to show that a barbital withdrawal convulsion can be induced in rats. This phenomenon has been substantiated by other investigators (Essig, 1966; Crossland and Turnbull, 1972; Morgan, 1976; Morgan and Bryant, 1977; Morgan et al., 1978; Wahlstrom, 1978; Nordberg and Wahlstrom, 1977,1979); since then phenobarbital-induced withdrawal convulsions have also been reported (Freund, 1971; Belknap et al., 1973,1978).

Other methods for detecting relatively low-grade physical dependence and withdrawal (which does not produce spontaneous withdrawal convulsions) produced by barbiturates have been developed. These methods utilize the characteristics of increased susceptibility to seizures that are noted during barbiturate withdrawal. Jaffe and Sharpless (1965) and others (Waters and Okamoto, 1972) used the threshold dose of pentylenetetrazole to produce seizures as the index for the degree of pentobarbital physical dependence. This method was also used in mice that were continuously administered pentobarbital by pellet implantation (Ho et al., 1975). Pentylenetetrazole challenge has been used to plot the time-course of barbital withdrawal in mice; this method is sensitive enough to detect the susceptibility changes resulting from nocturnal variation in normal animals (Waters and Okamoto, 1972). The incidences of convulsion increased in response to the threshold dose of pentylenetetrazole (75 mg/kg ip) during the barbital withdrawal (peak, 16 h, lasting >56 h), even though physical dependence produced was very low-grade and the animals did not display any overt withdrawal signs; normal animals were found to be more susceptible to the pentylenetetrazole during the night.

Other testing conditions that can be applied are the measurement of thresholds for electroshock (Essig, 1966), audiogenic stimulation, and stimulation of CNS by chemicals (e.g., bemegride and picrotoxin; Crossland and Turnbull, 1972) in mice, as well as testing for the threshold for startle response by electric shock to the feet (Gibbins et al., 1971). Relatively few investigators have applied objective quantitative measures, such as weight loss, body temperature change, food and water consumption, locomotion, open-field, and number of struggle responses, for measurement of withdrawal intensity (Stanton, 1936; Schmidt and Kleinman, 1964; Ho, 1976; Palermo-Neto and DeLima, 1982; Kulig, 1986).

3.2.4. Relationship Between the Drug Load and the Intensity of Physical Dependence

The schedule of drug intake or administration used to produce physical dependence on barbiturates has been quite variable. A careful study of human volunteers taking pentobarbital and secobarbital (Fraser et al., 1958) showed that no withdrawal symptoms occurred in subjects receiving <0.4 g daily (approx 5 to 6 mg/kg); with an increasing dosage up to a minimum of 2.2 g daily, the frequency of occurrence and the severity of withdrawal symptoms were proportional to the chronic dose administered. It is, therefore, reasonable to predict that the speed and intensity of development of physical dependence on barbiturates vary with the dosage. When rats were permitted to drink water for only a 30-min period every 24 h, chronic treatment with pentobarbital produced a dose-dependent increase in water intake. When the drug administration was terminated, a sharp decrease in drinking occurred; the magnitude and duration of these effects were related to the preceding drug dosage (Schmidt and Kleinman, 1964). The fall in electroconvulsive threshold during barbiturate withdrawal are also related to the duration of the preceding period of drug intoxication.

If the induction of physical dependence not only depends on the magnitude of CNS depression produced by the barbiturate, but also requires continued suppression of the CNS by the drug, the chronic drug load has to be defined on the basis not only of the drug dose, but also of drug persistence, i.e., the residual drug amount at the time of succeeding dose administration, when an intermittent drug treatment method is used. Therefore, in order to investigate the effect of dose (i.e., peak CNS depression) levels on the degree of physical dependence production, the residual CNS depression levels at the time of the next dose have to be made equivalent to each other (i.e., adjusting the interval between the dose administrations). This was designed in experiments on cats (Okamoto et al., 1978), and it was found that intensity of physical dependence expressed as severity of withdrawal was indeed dependent on the peak CNS depression produced. Furthermore, in order to study the influence of persistence of drug effects on the CNS in production of physical dependence in isolation, the animals were treated to produce consistently a preset level of peak CNS depression; in addition, the frequency of the drug administration was adjusted such that the effect of residual drug effect on dependency production was varied. Interestingly, the drug after-effect, which persisted and contributed to the physical dependence production, was found to be considerable. The production of physical dependence was apparent when the frequency of drug administration was shorter than intervals that were four to five times the half-life of the drug (Okamoto et al., 1986).

3.2.5. Relationship Between Physical Dependence and Pharmacodynamic Tolerance

Experimental studies in humans (Isbell et al., 1955; Mendelson, 1964) showed that, during sustained high intake of ethanol, the behavioral impairment gradually became less marked, unless the dose was raised. At the same time, withdrawal signs began to appear between doses, although major signs did not occur in most cases until ethanol intake was curtailed or stopped. Similarly, in animal studies with ethanol (Ellis and Pick, 1970; Essig and Lam, 1968), signs of withdrawal began to persist for longer periods between doses, suggesting that the duration of the drug effect after each dose was diminishing. These observations indicated that tolerance and physical dependence were developing in a roughly parallel fashion. This relationship was clearly demonstrated quantitatively (Gibbins et al., 1971) in rats treated chronically with ethanol. The startle response thresholds were measured either 30 min after the preceding dose of ethanol while the blood ethanol concentration was high or 23 h after, when the level was negligible. Tolerance and dependence, measured by the treadmill test and the startle response, respectively (LeBlanc et al., 1969), were shown to develop in parallel over the same time-course. Less complete, but essentially similar observations have been reported in relation to concurrent development of tolerance to and physical dependence on pentobarbital (Jaffe and Sharpless, 1965). Moreover, the recommended method for withdrawal of barbiturates without precipitation of a serious abstinence reaction (Blachly, 1964; Essig, 1970a) consists essentially of a titration of the patient's tolerance by giving just enough drug to produce minimal intoxication, and then gradually reducing the dose over many days. This demonstrates that a dose that is required to overcome the acquired tolerance also meets the requirement to mask the appearance of the physical dependence/withdrawal.

Parallel occurrence of tolerance and physical dependence is also suggested by observations (Essig, 1962,1964) of barbital withdrawal convulsions. The number of withdrawal convulsions was greater and more severe in animals exposed the second time to barbital treatment after recovery from the first. Furthermore, rats that have previously experienced physical dependence and withdrawal from barbital will develop physical dependence on other CNS depressants, e.g., meprobamate, at a lower dosage than barbiturate-dependent naive animals (Norton, 1970).

Development of physical dependence and tolerance may not, however, necessarily continue to develop in parallel at all stages. When no further development of tolerance was observed during continued administration of barbiturates to rats (Remmer, 1962; Aston, 1966), further prolonged administration of barbiturates was required to produce a maximal withdrawal reaction when production of spontaneous withdrawal convulsions was used as the measurement (Crossland and Leonard, 1963; Essig, 1966). However, the dispositional component of tolerance has very seldom been identified and distinguished in measurement of overall tolerance. Since kinetically the rate of dispositional tolerance is much faster than that of functional tolerance development, the kinetic picture of overall tolerance is predominated by the dispositional tolerance component in the initial phase; the pharmacodynamic tolerance component is not apparent. This might give an impression of overall tolerance reaching its plateau. Furthermore, it is relatively difficult in rodents to establish ideal drug treatment procedures with which to maintain steady and high barbiturate concentrations in the body (maintenance of high concentrations is needed to produce spontaneous withdrawal convulsion) for establishing ideal conditions for production of physical dependence. This is because of rodents' efficient drug elimination mechanisms and also nocturnal drinking/eating (if the drug is loaded in drinking water/food) behaviors.

3.2.6. Crosstolerance
and Crossdependence to Barbiturates

Crosstolerance and -dependence occur among drugs that produce general CNS depression, regardless of their chemical structure. For example, ethanol has been found to crosstolerate barbiturates (Gruber and Keyser, 1946; Kalant, 1977; Le et al., 1986) and to be effective in the prevention of barbiturate withdrawal (Fraser, 1957; Norton, 1970). Severe ethanol withdrawal reactions, such as grand mal convulsions, are effectively prevented by a variety of CNS depressant drugs, including barbiturates (Essig et al., 1969; Okamoto et al., 1983; Ulrichsen et al., 1986), benzodiazepines (Sereny and Kalant, 1965; Okamoto et al., 1983), paralydehyde, and chloral hydrate (Fraser, 1957; Kaim et al., 1969). Barbital withdrawal convulsions in rats were prevented by chlorodiazepoxide or meprobamate, but not by morphine or chlorpromazine (Norton, 1970). The requirement for certain chemical structures for CNS depressants to prevent withdrawal convulsions is illustrated by anticonvulsants. The barbiturate withdrawal convulsions are prevented by a barbiturate anticonvulsant agent, phenobarbital, and by primidone (Norton, 1970; Okamoto et al., 1977), but not by diphenylhydantoin (77; Essig and Carter, 1962; Okamoto et al., 1977) nor trimethadione

(Okamoto et al., 1977). On the other hand, Essig (1963) reported the effectiveness of aminooxyacetic acid on barbiturate withdrawal in dogs; this was assumed to demonstrate that the GABA system was important for production of barbiturate dependence.

The chemical nonspecificity among "general CNS depressant" drugs, and their capability for substituting for each other, can be utilized for drug screening procedures to test dependence liability to CNS depressant-type drugs (Jones et al., 1976). Dogs are made dependent on Na barbital and new drugs are tested for their ability to postpone and/or prevent a withdrawal reaction when substituted for barbital. This procedure is the same approach used for testing opiate-type dependence liability in morphine-dependent monkeys.

Although crossdependence among chemically dissimilar general CNS depressants has been established, the question of whether this crossdependence is complete, has not been addressed routinely. Certain clinical differences between barbiturate and ethanol withdrawal have been noted (Wikler et al., 1955). "Crossdependence" studies also suggest that physical dependence on barbiturates and dependence on ethanol are not identical (Fraser et al., 1957; Victor, 1966; Norton, 1970; Yanagita and Takahashi, 1973; Gougos et al., 1986; Kaneto et al., 1986; Khanna et al., 1988; Curren et al., 1988). However, not all of these studies have been done by controlling and matching conditions of chronic administration of drugs. When these experimental conditions were set and completeness of crossreactivities of three representative CNS depressants, i.e., ethanol, pentobarbital, and diazepam, was studied in ethanol- and pentobarbital-dependent animals by quantitatively measuring 21 different motor and behavioral withdrawal signs, diazepam was found to crossreact better with pentobarbital withdrawal than with ethanol withdrawal. Certain withdrawal signs, such as tremors and bizarre behaviors indicative of hallucinations produced during ethanol withdrawal, were not suppressed by diazepam (Okamoto et al., 1983). Similar incomplete crossdependence has been reported among barbiturates, benzodiazepines (Martin et al., 1982; Tagashira et al., 1983), and methaqualone (Suzuki et al., 1988).

4. Methods of Chronic Barbiturate Administration

4.1. Forced-Administration Methods

In contrast to the narcotic analgesics where physical dependence can be induced in experimental animals in short periods with intensive drug exposure, the production of physical dependence on barbiturates requires relatively prolonged periods of drug exposure at a high dosage. Therefore, the primary problem to be solved before initiating a barbiturate-dependence study is to establish a reliable system for the delivery of the drug in sufficient amount and consistently over prolonged periods, such that adequate levels of physical dependence can be induced in the experimental animals.

Many treatment procedures for production of barbiturate dependence have utilized the so-called forced-administration methods. The major use of these animal models is to study the pharmacological effects of long-term barbiturate treatment and/or the underlying mechanisms of tolerance to and physical dependence on barbiturates. The objectives are, therefore, to produce a high barbiturate concentration in the body throughout the chronic treatment period, and to produce experimentally reproducible tolerance and dependence in those animals. Virtually all methods involve, therefore, intoxication of animals for sufficiently long periods.

4.1.1. Forced Oral Administration

4.1.1.1. In Drinking Water

A number of experiments have attempted to produce a high oral intake of barbiturates in drinking water, particularly by rodents, by restricting animals to drink barbiturate solution as their only source of fluid for a long period of time. These studies have been termed "forced" treatment, since the animals must choose between the consumption of drug solution or fluid deprivation. Unless the drug concentration is extremely high, the animal chooses to drink the drug solution. The general procedure is to confine the animals to take the highest possible concentration they will tolerate, ideally, without compromising their volume of fluid intake. However, the volume of daily fluid intake is likely drug-concentration-dependent. Higher drug concentration does not necessarily yield higher dose (amount of the drug/kg/d) intake. Accordingly, the monitoring of fluid intake in comparison to that in control is important during the treatment.

Among barbiturates, Na barbital has been exclusively the choice in rodents, because this is the only barbiturate that has been reported to produce a high level of physical dependence by the oral route in drinking water in this species. In mouse, production of a high level of physical dependence on barbiturates, i.e., judged from an induction of spontaneous convulsions upon drug withdrawal, by this oral method is almost nil (only 10% incidence has been reported with 1 g/kg/d for 5-wk treatment; Waters and Okamoto, 1972). Rodents eliminate barbiturates quickly; even long-acting barbital (a half-life of 4–5 d in humans) is eliminated in a matter of 10 h in rats and 3 h in mice (half-life values; Waters and Okamoto, 1972). The reported dose of Na barbital in drinking water is approx 100 mg/kg/d initially and, in most of the methods, elevated weekly to reach the desired dose levels (200 for mild dependence to 500 for severe in the rat; 600 for very mild to 1000 for moderate in mice). However, this induction period could be shortened by elevating the dose daily by increments of 10% for a period of 10 d (Waters, 1973). The concentration of the drug in drinking water is calculated from the average body wt of the animal and the average volume of water consumed during 3–4 d of predrug treatment. When the bitter taste of the Na barbital is masked with Na saccharin, 0.02–0.04 w/v%, the animals generally consume a larger volume of the drug solution and consume it more consistently. Many investigations have maintained mice and rats on barbiturate solutions for periods of 4–5 wk to even 5–6 mo. Minimal duration of the treatment is 4–5 wk in most of the literature studies (4 wk for buildup and 1 wk maintenance at 400 mg/kg/d). However, when induction of the treatment was shortened to 10 d and then maintained at various chronic dosage levels (300, 400, and 500 mg/kg/d), an additional 10 d maintenance was sufficient to produce maximally attainable dependence for each dosage level (incidences of spontaneous withdrawal convulsion were: 300 mg/kg/d, 0.0% by as long as 21 wk treatment; 400

mg/kg/d, 75–80% and unchanged by 3 wk treatment and longer; 500 mg/kg/d, 100%, by 3 wk treatment). The incidence of spontaneous withdrawal convulsions was, therefore, strictly dependent on chronic dose level and not on the duration of the treatment after 3 wk (Waters, 1973). The blood barbital concentrations were found to be in the range of 60 μg/mL (300 mg/kg/d) to 125 μg/mL (500 mg/kg/d) in rats and 2.5 mg/mL (600 mg/kg/d) to 6.0 mg/mL (1000 mg/kg/d) in mice during the peak daily barbital. The barbital concentration fluctuates considerably during a 24-h period owing to the rodents' nocturnal feeding rhythm and especially to the short half-life of barbital in mice.

Several variations of the "forced" method have been attempted, i.e., administering Na phenobarbital (0.3 mg/mL, optimal concentration) in diluted Metrecal as a complete nutritional source of diet for mice (Freund, 1971), adding acid form of phenobarbital (0.225%) in a milled rodent chow (Letz and Belknap, 1975; Belknap et al., 1988) instead of in drinking water in mice, and also phenobarbital (0.05%) in NIH-07 diet in rats for 6 mo (Glauert et al., 1986). The above procedures are obviously simple to utilize, require a minimal amount of daily effort, are relatively inexpensive, and large numbers of animals can be treated simultaneously. Hence, for many experiments, particularly neurobiological or neurochemical studies that require rather substantial numbers of animals, the above treatment procedures have great appeal. The disadvantages are that relatively low drug concentrations can be attained in the range that does not affect normal nutritional intake, and therefore, only low-intensity tolerance and withdrawal are produced. Consequently, the ability to detect barbiturate-induced changes may be somewhat more difficult with this procedure than with others. In addition, and perhaps of greater importance, nutritional variables cannot be well controlled in these treatment procedures.

4.1.1.2. INTUBATION

This technique involves simply the introduction doses of barbiturate into the stomach, to produce high enough blood drug concentrations for periods sufficient to induce tolerance to and dependence on barbiturates. There have been reports demonstrating the efficacy of intubated barbiturates in producing tolerance and dependence. The differences between such investigations are, however, substantial in three aspects: the total daily dose of the drug administered, the frequency of the drug administration each day, and the length of time daily administrations are continued before tolerance and dependence are assessed. Intubation techniques were used to administer large doses of Na barbital to dogs (Seevers and Tatum, 1931; Isbell, 1950; Isbell et al., 1950). These techniques require restraint of the animal several times during the day to administer the drug forcibly. The procedure is rather difficult for both the experimental animal and the investigator. It was found that rodents were much less sensitive to the taste of barbiturates in their drinking water and could be force-fed a sufficient amount of the barbiturate solution over extended periods, whereas larger animals, i.e., cat and dog, are more sensitive to the bitter taste of the drugs. Accordingly, barbiturates are almost exclusively administered in drinking water and not by intubation methods to rodents.

In primates, the intubation method has been used commonly. For example, barbital was suspended in 0.25 methylcellulose and intubated by means of nasal tube into the stomach at doses ranging from 50 to 70 mg/kg, twice a day (initial dose, 40 mg/kg/d) for 3 mo in rhesus monkey (Stockhaus, 1986). In some cases, the dose is gradually increased; in others it is kept constant. Animals have been intubated with the drug anywhere from 1 to 30 d or more. Using these intubation procedures, very high blood drug concentrations can be continuously maintained over a 24-h period without concern for the animals' willingness to accept the drug voluntarily.

The advantages of the intubation method are numerous. Very high blood drug concentrations can be maintained over a 24-h period and, of particular pharmacological importance, it is possible to determine rather precisely dose–response relationships. The technique of nasogastric intubation is also relatively simple, and requires no special equipment; thus, large numbers of animals can be treated simultaneously.

Aside from producing stress conditions for the animals by this treatment procedure, the other disadvantage is gastrointestinal irritation, combined with the grossly intoxicated state of the animal, which could result in a marked reduction of food intake. Moreover, if the animals are kept on solid food, a pair-feeding design is virtually impossible to implement. Hillbom (1975) and Noble et al. (1976) have developed methods for etha-nol that appear to overcome these difficulties to a certain extent. These investigators also intubated ethanol in their animals, but provided them with a liquid diet rather than the standard lab chow and water. Similar methods could be applied for adminis-tration of barbiturates.

Na barbital has been given in capsule form for oral admin-istration in dogs and cats. Capsules containing 95 mg of Na barbital, or multiples thereof, were given orally once a day for periods ranging from 23 to 217 d in cats. The dose was increased according to each animal's sensitivity to the drug so as to main-tain a state of chronic intoxication (somnolence and staggering gait) (Essig and Flanary, 1959). The same approach was used in dogs (Essig, 1962); the duration of the treatment ranged from 94 d to 593 d. Unfortunately, there have been very few explana-tions regarding requisites for duration of treatment.

4.1.2. Intragastric Administration

Several intragastric administration methods by which bar-biturates were delivered directly into the stomach have been developed and utilized in several animal species. Although sterile surgical preparation is required prior to establishing the route of administration, each of these methods has some advantages over the direct oral route, especially in large animals. The opera-tions are relatively minor. The animals resume eating and func-tion normally within a day, and require minimal nursing care. The methods greatly facilitate the handling of large numbers of animals, and simplify and assure proper dosing of the drug. The drug can be delivered without restraining animals, making them less stressful, which often becomes a problem with direct oral route. In comparing Na pentobarbital consumed by oral (Meisch et al., 1981) and intragastric (Woolverton and Schuster, 1983) routes in self-administering rhesus monkeys, the amounts of the drug consumed were remarkably similar.

In rats, a gastric fistula method was developed by Nozaki et al. (1981), and has been used for evaluation of crosstolerance and physical dependence characteristics of barbiturates and benzodiazepines (Martin et al., 1982; McNicholas and Martin, 1982). The advantage of the gastric fistula preparation is the easy administration of large doses of drugs in a solid form. The preparation is relatively durable and permits efficient administration of the drug over months. However, repeated leakage of gastric fluid, when it occurs, may erode the tissues around the exit surface of the fistula. A similar preparation has been used for the chronic administration of barbiturates and benzodiazepines to dogs (Essig and Lam, 1968; McNicholas et al., 1983).

A gastric fistula in rats was made on either the left lateral or the ventral aspect of the abdomen using a sterile technique. A small incision was made on the left side of the abdomen beneath the rib margin or immediately below the xiphoid process. A passage was made through the abdominal muscle and the parietal peritoneum using blunt dissection. The great curvature of the stomach was sutured to the passage, and the stomach incised. The fistula was dilated with a glass rod daily after the surgery to prevent closure. One week postoperatively, all sutures were removed. At least 2 wk of recovery after the surgery were allowed before experiments were initiated. All drugs used in this study were put in #5 gelatin capsules and were administered through the gastric fistula. The control animals received a capsule containing lactose.

A gastric fistula method in the dog was originally developed by Essig and Lam (1968) and involved insertion of a modified Pavlovian cannula into the greater curvature of the stomach under aseptic conditions. All drugs were administered in #4 capsules beginning not less than 2 wk after surgery.

A method of implanting a chronic intragastric tube into the stomach as a mechanism of delivering the drug was used in cats (Rosenberg and Okamoto, 1974; Okamoto et al., 1975). The abdomen of the animal was opened by a midline incision, and a nonpyrogenic polyethylene cannula (modified from Butterfly infusion $19 \times 7/8$, 12-in tubing, Abbott, IL) with a flared lip at the end was implanted into the greater curvature of the stomach. A cloth jacket was placed around the abdomen and thorax, and tied around the trunk and neck to protect the tube from damage. The free end of the tube was attached to the animal's back. For the following week, daily injections of a few milliliters of tap water were made through the cannula to ensure its potency. After 1 wk of recovery, the animal started to receive chronic barbiturate treatment.

4.1.3. Systemic Route of Administration

Although a few studies report the effects of barbiturates after chronic sc injection in animals (Stanton, 1936; Schmidt and Kleinman, 1964), direct sc administration of water-soluble Na barbiturates is generally not recommended for their repeated administration owing to the high alkalinity of the salt solution. The minimum pH of the salt solution obtainable without precipitating commonly used barbiturates is with Na phenobarbital, the pH of which in solution is approx 9.1. Painful stresses and tissue necrosis at the site of injection might occur by repeated injection, and interfere with the outcome and the interpretation of the experimental results. Acid forms of barbiturates are not water-soluble.

As with the sc injection of barbiturates, strong alkalinity of the soluble salt form of barbiturates prevents the repeated im injections of barbiturates. A 10% aqueous solution of Na phenobarbital was injected by im route daily in monkeys for 6 wk in progressively increasing doses (Yanagita and Takahashi, 1970). Necrosis and scarring were observed at the site of injection in most of the animals, and the experiment had to be terminated.

Intraperitoneal and iv routes of repeated barbiturate administration have also been used for investigation of barbiturate tolerance and dependence studies. These studies are listed in Table 5. Intraperitoneal injection of barbiturates has been used frequently in rodents (mice and rat) for studies that do not involve direct investigation of severe barbiturate tolerance, physical dependence, and/or withdrawal. Accordingly, those studies often report their experimental results, i.e., biochemical and neurochemical alterations, and so forth, found under a chronic treatment schedule that was arbitrarily selected and not pharmacokinetically clarified. For example, many investigations (1) do not consider the presence of pharmacokinetic tolerance when they measure biochemical, neurochemical, and molecular changes in the brain in relation to tolerance and dependence development, and (2) do not report the degree of physical dependence produced by the treatment. As described, rodents in general eliminate barbiturates very efficiently, even long-acting barbiturates, such as phenobarbital and barbital. Arbitrarily designed treatment schedules, such as once-a-day systemic administration methods, do not permit production of consistent and reliable barbiturate blood concentrations necessary for production of reliable pharmacodynamic tolerance and physical dependence. Therefore, without measuring some of those parameters, the experimental results should not be overinterpreted, especially with regard to relevance to human studies.

The iv route has often been the preferred route of barbiturate administration in large-size animals, such as dogs and monkeys, because of relatively large-size veins, such as the sapenous vein and the cephalic vein, which are accessible for direct and repeated injections. However, repeated administration through this route tends to necrotize and cause hardening of the vein at the site of injection because of the high alkalinity of the Na barbiturate solution. On the other hand, iv administration of barbiturates into the jugular vein through an indwelling catheter is one of the major routes of administration in primates for self-administration studies of barbiturates.

Intraperitoneal administration of barbiturates has also been attempted through chronically implanted indwelling catheters utilizing Na pentobarbital (Yutrzenka et al., 1985). Rats were surgically prepared with chronically placed ip cannulae as described by Teiger (1974). A polyethylene-50 cannula was anchored into the peritoneal space with the free end advanced sc along the back, to exit at the neck. Rats were secured in specially designed harnesses and placed in individual cages. The indwelling cannula was connected to a flow-through swivel connected to a syringe. The drug was delivered through a Harvard infusion pump. Rats were initially infused with pentobarbital at a dose of 100 mg/kg/24 h and were maintained at this dose for

24 h. Subsequently, the pentobarbital dose was gradually increased daily by 100 mg/kg/d for a total 12 d. When the animals showed signs of profound CNS depression, the dose was increased only by 50 mg/kg/ d. By day 11, almost all the rats were receiving pentobarbital at an infused dose of 1000 mg/kg/24 h. Daily doses of pentobarbital were adjusted according to the degree of CNS depression exhibited by the individual rat. Rats were evaluated for signs of CNS depression at 9:00 AM each day using a 12-point rating scale modified from that reported by Rosenberg and Okamoto (1974). The rating system provided a reliable means of pentobarbital dosage adjustment sufficient to maintain effects on the CNS in the face of increased disposition of pentobarbital, while limiting the possibility of overdosage of pentobarbital and subsequent death of the rat. On day 12, pentobarbital was removed from the infusate, and rats were subjected to a drug-free period during which they received only 0.9% saline. It appeared that a minimum of 12 d of treatment was required to produce reproducible pentobarbital physical dependence in all rats.

4.1.4. Intracerebroventricular (icv) Administration

Because of the high alkalinity of the sodium form of barbiturates in solution, chronic icv administration of sodium barbiturates is not generally recommended. The first icv administration of barbiturates and the characteristics of tolerance developed were reported by Stolman and Loh (1975), Mycek and Brezenoff (1976), and Lyness et al. (1979).

A 9 mm length of 23-gage stainless-steel tubing (injection guide) was directed stereotaxically into the lateral ventricle in rat through a burr hole in the skull. The guide was fixed to the skull with dental cement. A 30-gage stilette was used as a stopper. A 10-μL vol of the drug solution was injected over a 30-s period into conscious animals through the guide via a 30-gage cannula. Control animals were injected with the saline vehicle adjusted to the same pH as the drug solution being tested. Since barbiturates are poorly soluble at neutral pH, solutions should be prepared daily in sterile 0.9% NaCl by adjusting with 0.1N NaOH to the minimal pH compatible with complete dissolution. This pH corresponded to, for example, 9.1 for phenobarbital and 9.6 for pentobarbital. The site of icv injection was confirmed by injecting methylene blue through the guide and by microscopical examination at the end of the experiment.

Acute icv injection of either Na barbital or Na phenobarbital to conscious rats produced a dose-related loss of consciousness as measured by the duration of loss of righting reflex. For example, Na phenobarbital doses of 2.4–3.9 μmol (i.e., 600–1000 μg) produced loss of righting reflex lasting from 3.5–19 min. The onset of actions was approx 1.8 min. Na barbital was found to be 1.3 times less potent than Na phenobarbital.

The chronic dosing regimen for Na phenobarbital was 800 μg (3.15 μmol) of the drug four times daily—twice in the morning, 30 min apart, and twice in the afternoon—for four d.

Chronic icv administration of Na phenobarbital resulted in tolerance to the loss of the righting reflex effect. Awakening phenobarbital brain levels of the tolerant rats were approx three times higher than those in nontolerant controls. Surprisingly, however, chronic Na barbital administration using an identical regimen did not produce tolerance nor were phenobarbital-tolerant rats crosstolerant to barbital. Underlying causes for this finding are not clear. Chronic icv Na pentobarbital resulted in an irreversible decrease in responsiveness to its own effects and to those of other barbiturates. This was attributed to the high alkalinity of the solution (pH 9.6), since icv injection of saline adjusted to the same pH also reduced responsiveness to icv barbiturates. However, rats tolerant to icv phenobarbital were tolerant to acute icv injection of Na pentobarbital. Similar crosstolerance was observed on systemic administration of the barbiturates. The efflux rates of icv phenobarbital or barbital from the CNS and their distribution in the tolerant rats were identical in tolerant and nontolerant rats. Although the icv dose of Na phenobarbital was approx 1/50th of the dose required for the peripheral routes, the amount of phenobarbital circulated by diffusion from the icv site was sufficient to induce the hepatic mixed-function oxidase system.

Although the technique affords the advantage of producing an anesthetic response to the barbiturates at a low dose compared to that required by peripheral routes, the high alkalinity of sodium solutions limits the usage of this route for chronic barbiturate administration. The only barbiturates that can feasibly be used without damaging the tissues are the ones that have a low pKa, and hence, the drug can be dissolved in saline solution at as low a pH as possible. It seems that the ventricular lining can retain reasonable permeability during chronic treatment while the pH of the injected solution is as high as 9.1 (Na phenobarbital), but not at 9.6 (Na pentobarbital). Otherwise, the anesthetic effect produced by the icv route is considerably shorter and, perhaps, more conveniently determined than that obtained employing the ip route. On the other hand, the icv injection procedure has been successful when it is used to infuse other pharmacological drugs that help elucidate the mechanism(s) of chronic barbiturate actions in animals made dependent on barbiturates administered by other peripheral routes.

4.1.5. Pellet Implantation

The technique of pellet implantation in the dorsal sc tissue of the mouse was originally developed by Huidobro and Maggiolo (1961) and Way et al. (1969) for studies of morphine dependence. This was later adapted for barbiturate studies (Ho et al., 1973,1975; Ho, 1976).

Huidobro and Maggiolo's original pellet (1961) released morphine rather slowly (25% of the dose in 8 d and 50% in 16 d) and took considerable time to develop morphine tolerance. Modification of the ingredients of the pellet, i.e., adding microcrystalline cellulose as a diluent, improved the drug release property of the pellet; the modified pellet now released 25–50% of the drug in the first 2 d (Way et al., 1969). The ingredients and the procedure for manufacturing this modified pellet were described in detail by Gibson and Tingstad (1970). The determining factor in selecting microcrystalline cellulose as the diluent was the desire to control pellet density or hardness. Because the drug is soluble in tissue fluid and cellulose is not, it was thought that, after some initial rapid release of the drug, the absorption rate would become constant as soon as the material at the surface of the pellet was dissolved. The microcrystalline cellulose was chosen because:

1. It is easily compressed to varying degrees of density while providing a physically stable pellet;
2. It proved a satisfactory vehicle for providing a reasonably constant absorption rate for the drug; and
3. Its physical characteristics are such that, after 5-d-implantation, the pellet remained as a semisolid palpable mass, facilitating easy removal from the animal.

The procedure for barbiturates involved the implantation of a specially formulated 75 mg/kg pellet of pentobarbital (containing 6.4 mg of pentobarbital acid form in a 130-mg weight pellet) in the sc tissues of a mouse. Each of the control animals had a placebo pellet implanted sc. Mice were divided into groups, and were implanted with pentobarbital in single or multiple numbers of tablets for 24, 48, and 72 h. At the end of each time period, the pellets were removed, and at 0, 6, 12, and 24 h after the removal of the pellet, the animals were examined.

The sc implantation of pellets of Na pentobarbital (6.4 mg) in the back of a conscious mouse results in measurable tolerance within 24–48 h. The development of tolerance to barbiturates was much more rapid than that produced by daily ip injection of 75 mg/kg Na pentobarbital in mice. The increased rate of tolerance development by pentobarbital pellet implantation was evident by a decrease in duration of loss of righting reflex when tested with either Na pentobarbital or Na barbital. The threshold for pentylenetetrazole-induced seizures was also significantly reduced compared to that of the Na pentobarbital daily injected and control groups. The extent of hepatic microsomal drug-metabolizing enzyme induction after pentobarbital pellet implantation also was found to be significantly higher than that produced by the injection technique. This reflected on the shorter half-life of pentobarbital elimination after the pellet implantation technique than after the daily injection. It was interesting to note that pellet implantation technique clearly demonstrated a greater reduction in mortality rate as compared to the daily injection technique.

4.1.6. Osmotic Minipump Method

The osmotic minipump (ALZET, ALZA Corp., Palo Alto, CA) has been used successfully for chronic treatment to demonstrate tolerance development to barbiturates (Siew and Goldstein, 1978). The osmotic minipump is a small capsule containing a collapsible reservoir surrounded by a sealed layer of salt solution, the osmotic driving agent. The whole is enclosed by a rate-controlling semipermeable membrane. The reservoir is prefilled to capacity with a concentrated barbiturate salt solution by means of a blunt-tip needle and syringe. When the device is implanted sc, it takes up body fluid from the tissues, forcing the drug solution out of the open end of the reservoir at a constant rate. This device has been widely and successfully utilized when a constant rate of delivery of a small amount of a relatively potent drug is required. The total delivery of the drug can be increased by implanting more than one pump at a time.

The pumps used in the study of Siew and Goldstein (1978) had flow rates of 1.3–1.4 µL/h and a nominal capacity of about 153 µL. About 90% of the volume was actually delivered in vivo. All these pumps were expected to function for 95–101 h. They were filled with sodium barbital, 180 mg/mL (which is close to the limit of solubility), so that each pump steadily delivered about 0.25 mg of drug/h until it was empty, about 4 d after implantation. Mice were given 300 mg/kg Na barbital ip. Just before the mice righted themselves from sleep, two, or up to five prefilled minipumps were implanted sc between the scapula of each mouse.

The experiments demonstrated that physical dependence could be readily developed in mice by 4 to 8 d of treatment by this method. The animals, however, did not show any abnormal behavioral patterns or spontaneous seizures after the pumps had emptied. However, withdrawal hyperexcitability was demonstrated by the incidence of pentylenetetrazole-induced seizures and the scores for convulsions elicited by handling. The osmotic minipumps are not ideal for sc administration of barbiturates. The amount of drug that can be administered is limited both by solubility and by the rather small capacity of the pumps. The blood levels achieved with this method were lower than those reported by Waters and Okamoto (1972) in a model using barbital administration in the drinking water. The withdrawal reactions were also mild under this condition.

The main advantage of this method is the stability of blood barbiturate levels during the intoxication period. Little or no metabolic tolerance develops to barbital, and the constant release rate of pumps assures that there is no day–night fluctuation. The continuous presense of the barbiturate in the brain probably accounts for the rapidity with which functional tolerance and physical dependence appear.

4.2. Self-Administration

The operant model that employs the self-administration technique (Weaks, 1962) was originally developed for studies of opiate dependence (Thompson and Pickens, 1969). This model assumes that the pharmacological effect of the drug is the primary determinant for both the cause and the maintenance of drug-taking behavior. Reinforcement has been demonstrated most often in studies employing the iv route. As a rule, the animal is first trained to press a bar in order to obtain food. It is then gradually trained to self-administer a drug solution through an indwelling catheter by pressing a bar that activates an injection pump. The effects of various environmental and physiological manipulations on the pattern of self-administration by the animal can be studied.

There are numerous review articles available on self-administration of psychoactive drugs, including barbiturates (Schuster and Thompson, 1969; Thompson and Pickens, 1969,1970; Woods and Schuster, 1970; Altschuler and Phillips, 1978; Ator and Griffiths, 1987). Studies on barbiturates are listed in Table 5. The general conclusion has been that different classes of drugs of abuse differ in their ability to act as primary reinforcers. For example, opiates, cocaine, and amphetamines are highly effective reinforcers, barbiturates and ethanol moderately effective, and mescaline and chlorpromazine relatively ineffective. When the drug reinforcement is continued, some animals will develop drug-taking patterns that resemble those of drug-dependent humans, leading to severe drug intoxication, physical dependence, and withdrawal.

4.2.1. Primates

Pentobarbital has been the most commonly used barbiturate and was reported to serve as a reinforcer for primates when it was made available by the oral (DeNoble et al., 1982; Lemaire and Meisch, 1984,1985; Meisch et al., 1981), intragastric (Altschuler et al., 1975; Woolverton and Schuster, 1983; Yanagita and Takahashi, 1973), and iv routes of administration (Bergman and Johanson, 1985; Deneau et al., 1969; Goldberg et al., 1971; Griffiths et al., 1981; Hoffmeister, 1977; Johanson, 1982,1987; Kubota et al., 1986; Winger et al., 1975; Yanagita and Takahashi, 1970,1973; Woods and Schuster, 1970). However, reinforcement has been demonstrated with virtually all the barbiturates that have intermediate/short elimination half-lives, except thiopental and hexobarbital. Not a great deal of work has been done with the very slowly eliminated barbiturates. Reinforcement was demonstrated with iv barbital (Winger et al., 1975) in primates; interestingly, however, there was a clear lack of reinforcement in studies with iv phenobarbital in rats (Collins et al., 1984).

For both the intragastric and iv self-administration, surgical implantation of an indwelling catheter is required. The surgeries are performed under appropriate anesthesia, and aseptic conditions comply to the accepted animal use guidelines.

For gastric self-administration (Altschuler et al., 1975; Woolverton and Schuster, 1983), a silicone catheter (0.08 cm id) was surgically implanted into the stomach. A 75-mm incision was made beginning 25 mm substernally. The stomach was exteriorized, and two concentric purse string silk (2–0) sutures approx 1–1.5 cm in diameter were placed in the fundus. The catheter was then inserted through a stab wound in the center of these sutures. The purse string sutures were closed around the catheter. Two nylon sutures that had been glued to the catheter were used to anchor it to the stomach. The stomach was then replaced in the peritoneal cavity and anchored to the ventral peritoneal lining with four silk sutures placed in a rectangular configuration. The catheter exited the peritoneal cavity through an opening in the lower left abdomen, was threaded sc to the back of the monkey, and exited via a small wound between the scapula. The abdominal incision was then closed.

For iv self-administration, the animals are surgically prepared with chronic indwelling internal jugular or femoral venous catheters. Silicone rubber catheters (0.08 cm id, 0.24 cm od) are implanted. The catheter is passed sc from the site of the incision to the middle of the animal's back where it is brought out through a stab wound.

Each animal is required to be fitted with a restraining jacket or stainless-steel harness for catheter protection. The experiments were done mostly in a specially designed primate cage (ex. Hoeltge no. HB-108; Labco ME-1305) for self-administration experiments in which the animal lived for the duration of the experiment.

The typical training of monkeys to lever press for injection of the drug follows. A piece of fruit was taped to the lever to increase the probability of a lever-press response (baiting) and initially the drug was delivered for every lever press, i.e., on a fixed-ratio one (FR-1) schedule. Additional fruit was used over the course of the 3-h session to assure exposure to behaviorally active drug doses. After a period of 1–2 wk of this procedure, the frequency of lever baiting was gradually reduced. If responding was maintained, the response requirement was gradually increased to 10 responses/injection (fixed ratio 10, FR 10). If, after several attempts, responding was not maintained, the animal was either removed from the study or training was attempted using another drug.

Because of the time-consuming nature of this method, a second, automated training method was also developed. Animals were initially trained to lever press for the delivery of a 1-g banana-flavored food pellet. Over the course of two or more sessions, response requirements were increased to FR 10. Subsequently, the frequency of food delivery was reduced to one pellet for every second or third FR 10. When responding was stable, self-administration training was begun. In some animals, sessions with food delivery alternated on a daily basis with sessions in which responding resulted in drug delivery. Gradually the number of consecutive sessions in which the drug was available was increased. In other animals, responding was maintained every day with drug delivery for every FR 10, and food delivery for every second or third FR 10. As responding stabilized, the frequency of food delivery was gradually reduced. These procedures were aimed at developing responding initially using food delivery and to gradually transferring to drug delivery as the maintaining event.

The schedule of reinforcement can be manipulated to determine the strength of the monkey's "drive" to self-administer the drug. Virtually all the self-administration studies with barbiturates have employed fixed-ratio schedules of reinforcement. Because most of the studies were concerned simply with whether a given drug would be self-administered at all, response requirements were chosen that would constrain drug-taking behavior as little as possible. In early studies utilizing the iv self-administration paradigm, monkeys frequently infused large amounts of the drug to become grossly intoxicated if they were allowed free access to the drug throughout the day. At the same time, they would spontaneously terminate their self-infusion of the drug. This pattern of self-administration resembles the "binge" in human alcoholic drinkers and has attracted some interest (Mello, 1978). However, this self-terminating pattern of drug administration made this experimental procedure inconvenient. In subsequent studies, Woods and colleagues (Woods et al., 1971; Kalory et al., 1978) found that, when these monkeys were given access to ethanol only through one session each day, the tendency to terminate self-administration could be eliminated. Accordingly, the majority of self-administration studies typically present animals' multiple opportunities to take the drug either throughout each day or within a limited experimental session of 2–3 h duration. Moreover, some studies, especially those using the iv route with monkeys, were designed to mimic the human condition and assured the sampling of barbiturate under study by first reinforcing lever pressing with a drug known to maintain regular self-administration (such as cocaine, codeine, or pentobarbital) and then substituting the barbiturate of interest (Bergman and Johanson, 1985; Griffiths et al., 1981). Other studies have focused on whether drug taking could be induced as a function of other environmental stimuli, such as a concurrent schedule of food reinforcement (Kodluboy and Thompson, 1971;

Tang et al., 1981) and delivery of electric shock (Davis et al., 1968; Davis and Miller, 1963).

Drug reinforcement has sometimes been difficult to show when response requirements were kept very low; however, when the response requirement was made high, the difference in intake between the drug and vehicle could be demonstrated (Winger et al., 1975). This was evident in studies of oral pentobarbital self-administration in rhesus monkeys. It was subsequently shown that, when the response requirement was increased, more drug than vehicle was consumed (DeNoble el al., 1982; Meisch et al., 1981). Additionally, when the linear relationship between the response requirement and the pentobarbital preference broke down at very high response requirements, it could be reinstated by increasing reinforcement (Lemaire and Meisch, 1984,1985).

An early iv pentobarbital study found, however, that the number of infusions per session established by pentobarbital in rhesus monkeys decreased when the response requirement was increased from FR 1 to FR 10, whereas cocaine did not affect this parameter of function. This was interpreted as being the result of the rate-suppressing effect of pentobarbital on lever pressing, which affected the higher response requirement more than the low FR (Goldberg et al., 1971). When this was corrected by imposing a 3-h timeout after each infusion, appropriate doses of pentobarbital, amobarbital, and secobarbital maintained the same number of injections per session as cocaine (Griffiths et al., 1981). Under these conditions, stable drug responding could be maintained for long periods of time, and the monkeys consistently infused enough drug to become grossly intoxicated.

The self-infusion of pentobarbital via the iv route of administration culminated in the development of tolerance and physical dependence (Deneau et al., 1969; Yanagita and Takahashi, 1970). Within a relatively brief period of time (around 10 d), tolerance appears to develop, and there is an increase in the response rate for the drug over time. Physical dependence can be readily demonstrated by abruptly terminated pentobarbital. Under these conditions, responding on the previously reinforced pentobarbital lever increases dramatically and is difficult to extinguish, giving some idea of the strength of pentobarbital as a reinforcer. Moreover, a full-blown withdrawal syndrome, consisting of tremors, hyperactivity, delirium, and convulsions, was demonstrated within a few hours after the drug withdrawal.

The advantages of the self-administration of barbiturates via indwelling iv cannulae are numerous. First, this technique provides a means of assessing the volitional aspects of barbiturate administration. Thus, a number of psychopharmacological studies, such as those determining the factors that give rise to and maintain the self-administration of drug or attempt to modify its administration, are possible with this method. In addition, the complexity of the monkey's behavior, relative to that of other species, permits substantially more sophisticated manipulations than would be possible with other species. It is likely that studies with primates can provide data with direct relevance to the psychological aspects of human behavior and drug dependence that would not be permitted with other species.

A disadvantage of the iv self-administration technique is that the drug is not given through the oral route, which is the likely route for barbiturate administration by humans. Also, not all laboratories are equipped to handle the physical requirements of maintaining monkeys, or have the relatively sophisticated surgical and technical expertise available to establish this method. In addition, it should be noted that not all monkeys can be trained to self-administer the drug; therefore, experimental procedures could be quite laborious, costly, and time-consuming. The technique is most suitable for investigation of underlying factors for induction and maintenance of self-administration of the drug.

Since most of the human population self-administers drugs by the oral route, it is beneficial to develop methods to study this aspect of drug dependence in animals. Meisch et al. (1981) have trained monkeys to drink solutions of pentobarbital by initially making the drug available with food, and then gradually separating feeding and drinking. In daily 3-h sessions, total intake of pentobarbital was directly related to dose using the oral route (approx 30–40 mg/kg/session). When food access sessions were separated from the drug access sessions, the consumption of pentobarbital in pharmacologically active amounts persisted. Hence, these investigators concluded that barbiturate became a reinforcer in its own right, apart from its previous relation to food consumption. During the early phase of the training, the animal presumably learns that the barbiturate has desirable pharmacological properties and sustains its consumption even when no food is made available. Meisch's work suggests that oral self-administration of pentobarbital in pharmacologically significant amounts is possible.

There are several disadvantages to the method. The monkeys or rodents are reduced to 80% of their prefeeding weight and are maintained in this partially reduced state for considerable periods of time. Since an animal model should permit an examination of barbiturate effects, any model that utilizes partial starvation or nutritional deprivation has a drawback.

Intragastric self-administration methods developed by Woolverton and Schuster (1983) deliver drugs to the same sites of absorption, but limit the disruptive effects of variables such as bitter taste, on behavior. In addition, humans rarely drink solutions of psychoactive drugs other than ethanol. It has been shown that there are notable similarities between oral and intragastric self-administration.

Perhaps the most important differences between drug delivery by the intragastric and iv routes are pharmacokinetic. These factors may have contributed to behavioral differences noted with the two routes of administration. For example, self-administration training was more protracted by the intragastric route than was found using the iv route. Rarely are more than one or two experimental sessions required to train for iv self-administration using a similar training procedure (baiting). However, a minimum period of 2 wk and usually longer is required for training intragastric self-administration. The principal reason for this effect is very likely the relatively slow onset of drug effects using the intragastric route, which makes the relationship between response and reinforcer more difficult to establish. In an attempt to make training more rapid, Woolverton and Schuster (1983) have used simultaneous food delivery to maintain responding initially and gradually reduced the frequency of food delivery. They have reported that reliable conditioning of self-administration could be achieved within approx 6 wk by this method.

The animals have been reported to maintain the responding above vehicle levels by pentobarbital at doses between 0.06–1.0 mg/kg/injection. Total drug intake ranged between 17.5 mg/kg at 0.12 mg/kg/injection and 41 mg/kg at 1.0 mg/kg/injection. Responding can be maintained under conditions of limited access (3- and 6-h sessions). Intragastric self-administration can be maintained under fixed-ratio schedules of reinforcement. Comparisons between results obtained by the iv and intragastric routes of administration are clearly of interest in assessing the usefulness of this method.

It is likely that pharmacokinetics are also an important determinant of the relative range of doses that maintain self-administration by intragastric and iv routes. For pentobarbital, responding was maintained over a range of 0.06–1.0 mg/kg/injection, by the intragastric route. Surprisingly, this dose range is similar to the range of doses that maintained self-administration in rhesus monkeys by the iv route (Johanson, 1982; Winger et al., 1975).

The advantages of the intragastric self-administration procedure are virtually the same as those described above for the iv route, as are the general disadvantages. However, the principal advantage of this procedure over the iv model is that the route of barbiturate administration is the normal one and, hence, this analog incorporates an important feature of the human hypnotic user. The problem with the intragastric technique, relative to the iv procedure, however, is an increased incidence of postsurgical complication.

Other possible consequences of intragastric self-administration are delayed drug onset and the possibility of overdose. By the iv route, the rapid onset of drug effects directly alters the rate of responding. Using a fixed-ratio schedule, these effects tend to decrease response rates and limit drug intake. However, by the intragastric route, the onset of these rate-decreasing drug effects is delayed, allowing the animal to accumulate potentially lethal doses of drug.

4.2.2. Rodents

Limited numbers of experiments have utilized rats for self-administration studies of barbiturates (Kaneto and Kosaka, 1983; Kosaka and Kaneto, 1982; Tang et al., 1981; Yanaura and Tagashira, 1975). Long-acting barbiturates, either barbital or phenobarbital, have been the drugs of choice. The route of administration has been primarily oral via drinking water. Most studies of oral self-administration in rodents have not, however, demonstrated reinforcement, despite a number of behavioral manipulations to induce drug intake (Tang et al., 1981).

Acknowledgment

The author gratefully acknowledges the assistance of Velicia Adams in preparing this manuscript.

References

Alexander E. J. (1965) Withdrawal effects of sodium amytal. *Dis. Nerv. Syst.* **12**, 77–82.

Altschuler H. L. and Phillips P. E. (1978) Intragastric self-administration of drugs by the primate, in *Drug Discrimination and State Dependent Learning* (Ho B. T., Richards D. W., and Chute D. L., eds.), Academic, New York, pp. 263–280.

Altschuler H., Weaver S., and Phillips P. (1975) Intragastric self-administration of psychoactive drugs by the rhesus monkey. *Life Sci.* **17**, 883–890.

Andrews P. R., Graham P. J., and Lodge D. (1979) Convulsant, anticonvulsant, and anaesthetic barbiturates. 5-ethyl-5-(3'-methyl-but-2'-eny)barbituric acid and related compounds. *Eur. J. Pharmacol.* **55**, 115–120.

Appel J. B. and Freedman D. X. (1968) Tolerance and cross-tolerance among psychotomimetic drugs. *Psychopharmacologia* **13**, 267–274.

Arvola A., Sammalisto L., and Wallgreen H. (1958) A test for level of alcohol intoxication in the rat. *Quart. J. Stud. Alc.* **19**, 563–572.

Aston R. (1965) Quantitative aspects of tolerance and posttolerance hypersensitivity to pentobarbital in the rat. *J. Pharmacol. Exp. Ther.* **150**, 253–258.

Aston R. (1966) Acute tolerance inducer for pentobarbital in male and female rats. *J. Pharmacol. Exp. Ther.* **152**, 350–353.

Ator N. A. and Griffiths (1987) Self-Administration of barbiturates and benzodiazepines: A review. *Pharmacol. Biochem. Behav.* **27**, 391–398.

Baker T. and Cannon D. (1979) Potentiation of ethanol withdrawal by prior dependence. *Psycopharmacologia (Berlin)* **60,** 105–110.

Barnes C. D. and Eltherington L. G. (1966) *Drug Dosage in Laboratory Animals. A Handbook.* University of Calififornia Press, CA.

Belknap J. K., Waddingham S., and Ondrusek G. (1973) Barbiturate dependence in mice induced by simple short-term oral procedure. *Physiol. Psychol.* **1,** 394–396.

Belknap J. K., Berg J. H., Ondrusek G., and Waddingham S. (1978) Barbiturate withdrawal and magnesium deficiency in mice. *Psychopharmacology* **59,** 299–303.

Belknap J. K., Danielson P. W., Lane M., and Crabbe J. C. (1988) Ethanol and barbiturate withdrawal convulsion are extensively codetermined in mice. *Alcohol* **5,** 167–171.

Bergman J. and Johanson C. E. (1985) The reinforcing properties of diazepam under several conditions in rhesus monkey. *Psychopharmacology* **86,** 108–113.

Blachly P. H. (1964) Procedure for withdrawal of barbiturates. *Am. J. Psychiat.* **120,** 894–895.

Boisse N. R. and Okamoto M. (1978a) Physical dependence to barbital compared to pentobarbital. I. "Chronically equivalent" dosing method. *J. Pharmacol. Exp. Ther.* **204,** 497–506.

Boisse N. R. and Okamoto M. (1978b) Physical dependence to barbital compared to pentobarbital. II. Tolerance characteristics. *J. Pharmacol. Exp. Ther.* **204,** 507–513.

Boisse N. R. and Okamoto M. (1978c) Physical dependence to barbital compared to pentobarbital. III. Withdrawal characteristics. *J. Pharmacol. Exp. Ther.* **204,** 514–525.

Boisse N. R. and Okamoto M. (1978d) Physical dependence to barbital compared to pentobarbital. IV. Influence of elimination kinetics. *J. Pharmacol. Exp. Ther.* **204,** 526–540.

Boissier J. R., Tardy J., and Diverres J. C. (1960) Une nouvelle methode simple pour explorer l'action "tranquilisante": le test de la cheminee. *Med. Exp.* **3,** 81–84.

Boyd E. M. (1960) Chlorpromazine tolerance and physical dependence. *J. Pharmacol. Exp. Ther.* **128,** 75–78.

Branch M. N. (1983) Behavioral tolerance to eliminating effects of pentobarbital: a within-subject determination. *Pharmacol. Biochem. Behav.* **18,** 25–30.

Broder S. (1936) Sleep induced by sodium amytal, an abridged method for use in mental illness. *Am. J. Psychiat.* **93,** 57–74.

Brodie B., Mark L., Lief P. A., Bernstein E. , and Papper E. M. (1951) Acute tolerance to thiopental. *J. Pharmacol. Exp. Ther.* **102,** 215–218.

Brodie B., Mark L., Papper E. M., Lief P. A., Bernstein E., and Rovenstine E. A. (1950) The fate of thiopental in man and a method for its estimation in biological material. *J. Pharmacol. Exp. Ther.* **98,** 85–96.

Buch H., Grund W., Buzello W., and Rummel W. (1969) Narkotische Wirksamkeit im Gewebsverteilung der optischen Antipoden des Pentobarbitals bei des Ratte. *Biochem. Pharmacol.* **18**, 1005–1009.

Bush M. T. (1963) Sedative and hypnotic. 1. Aborption, fate and excretion, in *Physiological Pharmacology*, vol. 1: *The Nervous System*, Part A. *Central Nervous System Drugs* (Root W. S. and Hoffmann F. G., eds.), Academic, New York, pp. 185–218.

Butler T. C., Mahaffee C., and Waddell W. J. (1954) Phenobarbital: Studies of elimination, accumulation, tolerance, and dosage schedules. *J. Pharmacol. Exp. Ther.* **111**, 425–435.

Carmichael E. B. and Posey L. C. (1933) Observations on effect of repeated administration of Nembutal in guinea pigs. *Proc. Soc. Exp. Biol. Med.* **36**, 1329–1330.

Carroll M. E., Stotz D. C., Kliner D. J., and Meisch R. A. (1984) Self-administration of orally-delivered methohexital in rhesus monkeys with phrencyclidine or pentobarbital histories: Effects of food deprivation and satiation. *Pharmacol. Biochem. Behav.* **20**, 145–151.

Cicero T. J. (1979) A critique of animal analogues of alcoholism, in *Biochemistry and Pharmacology of Ethanol*, vol 2 (Majchrowiz E. and Noble E. P. eds.) Plenum, New York, pp. 533–560.

Christensen H. D. and Lee I. S. (1973) Anesthetic potency and acute toxicity of optically active disubstituted barbituric acids. *Toxicol. Appl. Pharmacol.* **26**, 495–503.

Clemmesen L. and Hemmingsen R. (1984) Physical dependence on ethanol drug multiple intoxication and withdrawal episodes in the rat: Evidence of a potentiation. *Acta Pharmacol. Toxicol.* **55**, 345–350.

Collins R. J., Weeks J. R., Cooper M. M., Good P. I., and Russell R. R. (1984) Prediction of abuse liability of drugs using i.v. self-administration by rats. *Psychopharmacology (Berlin)* **82**, 6–13.

Conney A. H. (1967) Pharmacological implications of microsomal enzyme induction. *Pharmacol. Rev.* **19**, 317–366.

Conney A. H. and Burns J. J. (1962) Factors influencing drug metabolism. *Adv. Pharmacol.* **1**, 31–58.

Crabbe J. C., McSwigan J. D., and Belknap J. K. (1985a) The role of genetics in substance abuse, in *Determinants of Substance Abuse* (Galizio M. and Maistos A., eds.) Plenum, New York, pp. 13–64.

Crabbe J. C., Kosobud A., Young E. R., Tam B., and McSwigan J. D. (1985b) Bidirectional selection for susceptibility to ethanol withdrawal seizures in Mus muculus. *Behav. Genet.* **15**, 521–536.

Crawford J. S. (1966) Speculation: the significance of varying the mode of infection of a drug. With special reference to the brain and the placenta. *Brit. J. Anaesth.* **38**, 628–640.

Crossland J. and Leonard B. E. (1963) Barbiturate withdrawal convulsions in the rat. *Biochem. Pharmacol.* **12 (Suppl.)**, 103.

Crossland J. and Turnbull M. J. (1972) Gamma-aminobutyric acid and the barbiturate abstinence syndrome in rats. *Neuropharmacology* **11**, 733–738.

Curren M. A., Newman L. M., and Becker G. L. (1988) Barbiturate anesthesia and alcohol tolerance in a rat model. *Anesth. Anal.* **67**, 868–871.

Davis J. D. and Miller N. E. (1963) Fear and pain: Their effect on self-injection of amobarbital sodium by rats. *Science* **141**, 1286,1287.

Davis J. D., Lulenski G. C., and Miller N. E. (1968) Comparative studies of barbiturate self-administration. *Int. J. Addict.* **3**, 207–214.

Deneau G. A., Yanagita T., and Seevers M. H. (1969) Self-administration of psychoactive substances by the monkey—A measure of psychological dependence. *Psychopharmacolgia* **16**, 30–48.

DeNoble V. J., Svikis D. S., and Meisch R. A. (1982) Orally delivered pentobarbital and a reinforcer for rhesus monkeys with concurrent access to water: Effects of concentration, fixed-ratio size, and liquid positions. *Pharmacol. Biochem. Behav.* **16**, 113–117.

DeSalva S. J. (1956) Tolerance development to anticonvulsive drugs. *J. Pharmacol. Exp. Ther.* **116**, 15,16.

Dews P. B. (1955) Studies on behavior I. Differential sensitivity to pentobarbital of pecking performance in pigeons depending on the schedule of reward. *J. Pharmacol. Exp. Ther.* **113**, 393–401.

Dews P. B. (1978) Behavioral tolerance, in *Behavioral Tolerance: Research and Treatment Implications* (Kasnegor N. A., ed.), DHEW, Washington DC, pp. 18–26.

Diaz J. and Schain R. J. (1978) Phenobarbital: effects of long term administration on behavior and brain of artificially reared rats. *Science* **199**, 90,91.

Dorfman A. and Goldbaum L. R. (1947) Detoxification of barbiturates. *J. Pharmacol. Exp. Ther.* **90**, 330–337.

Dunham N. W. and Miya T. S. (1957) A note on a simple apparatus for detecting neurological deficit in rats and mice. *J. Am. Pharm. Assoc.* **46**, 208–209.

Ebert A. G., Yim G. K., and Miya T. S. (1964) Distribution and metabolism of barbital [^{14}C] in tolerant and intolerant rats. *Biochem. Pharmacol.* **13**, 1267–1274.

Eddy N. B., Halbach H., Isbell H., and Seevers M. H. (1965) Drug dependence: it significance and characteristics. *World Health Organ. Bull.* **32**, 721–733.

Ellis F. W. and Pick J. R. (1970) Experimentally induced ethanol dependence in Rhesus monkeys. *J. Pharmacol. Exp. Ther.* **175**, 88–93.

Epstein R. S. (1980) Withdrawal symptoms from chronic use of low-dose barbiturates. *J. Psychiat.* **137**, 107,108.

Essig C. F. (1962) Convulsive and shamrage behavior in decorticated dogs during barbiturate withdrawal. *Arch Neurol.* **7**, 471–475.

Essig C. F. (1963) Anticonvulsant effect of aminooxyacetic acid during barbiturate withdrawal in the dog. *Int. J. Neuropharmacol.* **2**, 199–204.

Essig C. F. (1964) Barbiturate withdrawal convulsions in decerebellate dogs. *Int. J. Neuropharmacol.* **3**, 453–456.

Essig C. F. (1966) Barbiturate withdrawal in white rats. *Int. J. Neuropharmacol.* **5**, 103–107.

Essig C. F. (1967) Clinical and experimental aspects of barbiturate withdrawal convulsions. *Epilepsia* **8**, 21–30.

Essig C. F. (1968) Possible relation of brain gamma-aminobutyric acid (GABA) to barbiturate abstinence convulsions. *Arch. Int. Pharmacodyn. Ther.* **176**, 97–103.

Essig C. F. (1970a) Barbiturate dependence, in *Drug Dependence* (Harris R. T., McIsaac W. M., and Schuztes C. R., eds.), Chapter 11, Texas Press, Austin.

Essig C. F. (1970b) Reduction of barbiturate dependence induced by repeated electroconvulsion. *Arch. Int. Pharmacodyn. Ther.* **188**, 387–391.

Essig C. F. (1972) Drug withdrawal convulsions in animals, in *Experimental Models of Epilepsy* (Purpura D. P., Penry J. K., Tower D. B., Woodbury D. M., and Walter R. D., eds.), Raven, New York, pp. 495–508.

Essig C. F. and Carter W. W. (1962) Failure of diphenylhydantoin in preventing barbiturate withdrawal convulsions in the dog. *Neurology* **12**, 481–484.

Essig C. F. and Flanary H. G. (1959) Convulsion in cats following withdrawal of barbital sodium. *Exp. Neurol.* **1**, 529–533.

Essig C. F and Flanary H. G. (1961) Convulsive aspects of barbital sodium in the cat. *Exp. Neurol.* **3**, 149–159.

Essig C. F. and Fraser H. F. (1958) Electroencephalographic changes in man during use and withdrawal of barbiturate in moderate dosage. *Electroencephalogr. Clin. Neurophysical* **10**, 649–656.

Essig C. F. and Fraser H. G. (1959) Convulsion in cats following withdrawal of barbital sodium. *Exp. Neurol.* **1**, 529–533.

Essig C. F. and Fraser H. F. (1966) Failure of chlorpromazine to prevent barbiturate-withdrawal convulsions. *Clin. Pharmacol. Ther.* **7**, 465–469.

Essig C. F. and Lam R. C. (1968) Convulsion and hallucinatory behavior following alcohol withdrawal in dogs. *Arch Neurol.* **18**, 626–632.

Ettinger G. H. (1938) The duration of anesthesia produced in the dog by repeated administration of Dial and Nembutal. *J. Pharmacol. Exp. Ther.* **63**, 82–87.

Flint B. A. and Ho I. K. (1980) Continuous administration of barbital by pellet implantation. *J. Pharmacol. Meth.* **4**, 127–139.

Fraser H. F. (1957) Tolerance to and physical dependence on opiates, barbiturates and alcohol. *Ann. Rev. Med.* **8**, 427–440.

Fraser H. F. and Isbell H. (1954) Abstinence syndrome in dogs after chronic barbiturate medication. *J. Pharmacol. Exp. Ther.* **112**, 261–267.

Fraser H. F., Wikler A., Essig C. F., and Isbell H. (1958) Degree of physical dependence induced by secobarbital or pentobarbital. *J. Am. Med. Assoc.* **166**, 126–129.

Fraser H. F., Wikler A., Isbell H., and Johnson N. K. (1957) Partial equivalence of chronic alcohol and barbiturate intoxications. *Quart. J. Stud. Alc.* **18**, 541–551.

Fraser H. F., Isbell H., Eisenman A. J., Wikler A., and Pescor F. T. (1954) Chronic barbiturate intoxication. Further studies. *Arch Intern. Med.* **94**, 34–41.

Freund G. (1971) Alcohol, barbiturate and bromide withdrawal syndrome in mice, in *Recent Advances in Alcoholism* (Mendelson J. and Bello N., eds.) GPO, Washington DC, pp. 453–471.

Frey H. H. and Kampman (1965) Tolerance to anticonvulsant drugs. *Acta Pharmacol. Toxicol.* **22,** 159–171.

Gates G. R. and Chen C. S. (1974) Effects of barbiturate withdrawal on audiogenic seizure susceptibility in BALB-c mice. *Nature* **249,** 449–451.

Gay M. H., Rayan G. P., Boisse N. R., and Guarino J. J. (1983) Phenobarbital tolerance and physical dependence, chronically equivalent dosing model. *Eur. J. Pharmacol.* **95,** 21–29.

Gibbins R. J., Kalant H., and LeBlanc A. E. (1968) A technique for accurate measurement of moderate degrees of alcohol intoxification in small animals. *J. Pharmacol. Exp. Ther.* **159,** 236–242.

Gibbins R. J., Kalant H., LeBlanc A. E., and Clark J. W. (1971) The effects of chronic administration of ethanol on startle-threshold in rats. *Psychopharmacologia* **19,** 95–104.

Gibson R. D. and Tingstad J. E. (1970) Formulation of morphine implantation pellet suitable for tolerance-physical dependence studies in mice. *J. Pharm. Sci.* **59,** 426,427.

Ginsburg B. E. (1958) Genetics as a tool in the study of behavior. *Prospect. Biol. Med.* **1,** 397–424.

Glauert H. P., Schwarz M., and Pitot H. C. (1986) The phenotypic stability of altered hepatic foci: effects of the short term withdrawal of phenobarbital and the long-term feeding of purified diets after the withdrawal of phenobarbital carcinogenesis. *Carcinogenesis* **7,** 117–121.

Gogerty J. H. (1976) Preclinical methodologies related to the development of hypnotic substances, in *Pharmacology of Sleep* (William R. L. and Karacan I., eds.), Wiley, New York, pp. 33–52.

Goldbaum L. R. and Smith P. K. (1954) The interaction of barbiturates with serum albumin and its possible relation to their disposition and pharmacological action. *J. Pharmacol. Exp. Ther.* **111,** 197–209.

Goldberg M. E., Marian A. A., and Efron D. H. (1967) A comparative study of certain pharmacologic responses following acute and chronic administration of chlordiazepoxide. *Life Sci.* **6,** 481–491.

Goldberg S. R., Hoffmeister F., Schlicting U. U., and Wuttke W. (1971) A comparison of pentobarbital and cocaine self-administration in rhesus monkeys: Effects of dose and fixed-ratio parameter. *J. Pharmacol. Exp. Ther.* **179,** 277–283.

Goldstein D. B. (1972) Relationships of alcohol dose to intensity of withdrawal signs in mice. *J. Pharmacol. Exp. Ther.* **180,** 203–215.

Gougos A., Khanna J. M., Le A. D., and Kalant H. (1986) Tolerance to ethanol and cross-tolerance to pentobarbital and barbital. *Pharmacol. Biochem. Behav.* **24,** 801–807.

Green M. W. and Koppanyi T. (1944) Studies on barbiturates. XXVII. Tolerance and cross tolerance to barbiturates. *Anesthesiology* **5,** 329–340.

Griffiths R. R., Lukar S. E., Bradford L. D., Brady J. V., and Snell J. D. (1981) Self-injection of barbiturates and benzodiazepines in baboons. *Psychopharmacology (Berlin)* **75**, 101–109.

Groh K. R., Ehret C. F., Peraino C., Meinert J. C., and Readey M. A. (1988) Circadian manifestations of barbiturate habituation, addiction and withdrawal in the rat. *Chronobiol. Int.* **5**, 153–166.

Gruber C. M. and Keyser G. F. (1946) A study on the development of tolerance to and physical dependence on barbiturates in experimental animals. *J. Pharmacol. Exp. Ther.* **86**, 186–196.

Guttman R., Lieblich I., and Naftali G. (1969) Variation in activity scores and sequences in two imbred mouse strains, their hybrids, and backcrosses. *Anim. Behav.* **17**, 374–385.

Haley T. J. and Gidley J. T. (1970) Pharmacological comparison of R(+) S(–) and racemic secobaribital in mice. *Eur. J. Pharmacol.* **9**, 358–361.

Harris R. A. and Snell D. (1980) Effects of acute and chronic administration of phenobarbital and *d*-amphetamine on schedule controlled behavior. *Pharmacol. Biochem. Behav.* **12**, 47–52.

Harrison J. W., Ambrus C. M., and Ambrus J. L. (1952) Tolerance of rats toward amphetamines. *J. Am. Pharm. Assoc.* **41**, 539–541.

Harvey S. C. (1985) Hypnotics and sedatives, in *The Pharmacological Basis of Therapeutics,* 7th ed., Chap. 17 (Goodman L. S., Gilman A., Rall T. W., and Murad F., eds.), McMillan Co., London, chapter 17, pp. 339–371.

Hillbom M. E. (1975) The prevention of ethanol withdrawal seizures in rats by dipropylacetate. *Neuropharmacology* **14**, 755–761.

Ho I. K. (1976) Systematic assessment of tolerance to pentobarbital by pellet implantation. *J. Pharmacol. Exp. Ther.* **197**, 479–487.

Ho I. K. and Harris R. A. (1981) Mechanism of action of barbiturates. *Ann. Rev. Pharmacol.* **21**, 83–111.

Ho I. K., Sutherland V. C., and Loh H. H. (1973) A model for the rapid development of tolerance to barbiturates. *Res. Commun. Chem. Pathol. Pharmacol.* **6**, 33–46.

Ho I. K., Yamamoto I., and Loh H. H. (1975) A model for the rapid development of dispositional and functional tolerance to barbiturates. *Eur. J. Pharmacol.* **30**, 164–171.

Hoffmeister F. (1977) Assessment of the reinforcing properties of stimulant and depressant drugs in the rhesus monkey as a tool for the prediction of psychic dependence-producing capability in man, in *Predicting Dependence Liability of Stimulant and Depressant Drugs* (Thompson T. and Unna R., eds.), Univerity Park Press, Baltimore, pp. 185–201.

Holck H. G., Riedesel C. C., and Robidoux F. A. (1950) Studies on tolerance and cross-tolerance to nostal (propallylonal isopropyl-beta-bromallyl barbituric acid). *J. Am. Pharm. Assoc.* **39**, 630–637.

Huang L-Y. M. and Barker J. L. (1980) Pentobarbital: Stereospecific actions of (+) and (–) isomers revealed on cultured mammalian neurons. *Science* **207**, 195–197.

Hubbard T. F. and Goldbaum L. R. (1949) The mechanisms of tolerance to thiopental in mice. *J. Pharmacol. Exp. Ther.* **97**, 488–491.

Huidobro F. and Maggiolo C. (1961) Some features of the absentinence syndrome to morphine in mice. *Acta Physiol. Latinoamer.* **11**, 201–209.

Idenstrom C. M. (1954) Flicker-fusion in chronic barbiturate usage. A quantitiative study in the pathophysiology of drug addiction. *Acta Psychiat. Neurol. Scand.* **91**, 1–93.

Irwin S. (1976) Drug screening and evaluation of new compounds in animals, in *Animal and Clinical Pharmacologic Techniques in Drug Evaluation* (Nodine J. H. and Siegler P. E., eds.), Medical Publishers Yearbook, Chicago, pp. 36–54.

Isbell H. (1950) Addiction to barbiturates and the barbiturate abstinence syndrome. *Ann. Intern. Med.* **33**, 108–121.

Isbell H. and Fraser H. F. (1950) Addiction to analgesia and barbiturates. *Pharmacol. Rev.* **2**, 355–397.

Isbell H., Fraser H. F., Wikler A., Belleville R. E., and Eisenman A. J. (1955) An experimental study of the etiology of "run fits" and delirium tremens. *Quart. J. Stud. Alc.* **16**, 1–33.

Isbell H., Altschul S., Kornetsky C. H., Eisenman A. J., Flanary H. G., and Fraser H. F. (1950) Chronic barbiturate intoxication: an experimental study. *Am. Med. Assoc. Arch. Neurol. Psychiat.* **64**, 1–28.

Jaffe J. H. (1985) Drug Addiction and drug abuse, in *The Pharmacological Basis of Therapeutics*, 7th ed. (Goodman L. S., Gilman A. G., Rall T. W., and Murad F., eds.), Macmillan, New York, pp. 532–581.

Jaffe J. H. and Sharpless S. K. (1965) The rapid development of physical dependence on barbiturates. *J. Pharmacol. Exp. Ther.* **150**, 140–145.

Jay G. E. (1955) Variation in response of various mouse strains to hexobarbital (evipal). *Proc. Soc. Exp. Biol. Med.* **90**, 378–380.

Johanson C. E. (1982) Behavior maintained under fixed-internal and record-order schedules of cocaine or pentobarbital in rhesus monkeys. *J. Pharmacol. Exp. Ther.* **221**, 384–393.

Johanson C. E. (1987) Benzodiazepine self-administration in rhesus monkeys: Estazolan, flurazepam and lorazepam. *Pharmacol. Biochem. Behav.* **26**, 521–526.

Jones B. E., Prada J. A., and Martin W. R. (1976) A method for bioassay of physical dependence on sedative drugs in dog. *Psychopharmacology* **47**, 7–15.

Jones B. L. and Roberts D. J. (1968) The quantitative measurements of motor incoordination in naive mice using an accelerating rota rod. *J. Pharm. Pharmacol.* **20**, 302–304.

Kaim S. C., Klitt C. J., and Rothfield B. (1969) Treatment of the acute alcohol withdrawal state: a comparison of four drugs. *Am. J. Psychiat.* **125**, 1640–1646.

Kalant H. (1977) Comparative aspects of tolerance to and dependence on alcohol, barbiturates and opiates. *Adv. Exp. Med. Biol.* **858**, 169–186.

Kalant H., LeBlanc A. E., and Gibbons R. T. (1971) Tolerance to and dependence on some non-opiate psychotropic drugs. *Pharmacol. Rev.* **23**, 135–190.

Kalinowsky L. B. (1942) Convulsions in nonepileptic patients on withdrawal of barbiturates, alcohol and other drugs. *Arch. Neurol. Psychiat.* **48**, 946–956.

Kalory A. J., Winger G., Ikomi F., and Woods J. H. (1978) The reinforcing property of ethanol in the rhesus monkey II. Some variables in ethanol-reinforced responding. *Psychopharmacology* **58**, 19–25.

Kaneto O. H. and Kosaka N. (1983) Determination of psychic dependence liability of drugs in small animal species (3). Selective drinking behavior for the solution of dependence liable drugs in mice. *Folia Pharmacol. Japn.* **81**, 267–274.

Kaneto H., Kawatuni S., and Kaneda H. (1986) Differentiation of alcohol and barbital physical dependence. *Yakubutso Seishin Kodo* **6**, 267–273.

Kato R. (1967) Analysis and differentiation of the mechanism in the development of drug tolerance. *Jap. J. Pharmacol.* **17**, 499–508.

Kato R., Chiesara L. E., and Vassanelli P. (1964) Further studies on the inhibition and stimulation of microsomal drug metabolizing enzymes of rat liver by various compounds. *Jap. J. Pharmacol.* **13**, 69–83.

Khanna J. M., Le A. D., Gougos A., and Kalant H. (1988) Effect of chronic pentobarbital treatment on the development of cross-tolerance to ethanol and barbital. *Pharmacol. Biochem. Behav.* **31**, 179–189.

Kodluboy D. W. and Thompson T. (1971) Adjunctive self-administration of barbiturate solution, in Proc. 79th Annual Convention APA, pp. 749,750.

Kosaka N. and Kaneto H. (1982) Determination of psychic dependence liability of the drugs in small animal species (4); Selective drinking behavior for barbital and ethanol solution in mice. *Japn. J. Pychopharmacol.* **2**, 61–68.

Kosman M. E. and Unna K. R. (1968) Effects of chronic administration of amphetamines and other stimulants on behavior. *Clin. Pharmcol. Ther.* **9**, 240–254.

Kramer J. C., Fischman V. S., and Littlefield D. C (1967) Amphetamine abuse. Pattern and effects of high doses taken intravenously. *J. Am. Med. Assoc.* **201**, 305–309.

Krop S. and Gold H. (1946) Comparative study of several barbiturates with observations on irreversible neurological disturbances. *J. Pharmcol. Exp. Ther.* **88**, 260–267.

Kubota A., Kuwahara A., Hakkei M., and Nakamura K. (1986) Drug dependence tests as a new anesthesia inducer, midazolam. *Folia Pharmacol. Jpn.* **88**, 125–158.

Kulig B. M. (1986) Attenuation of phenobarbital-induced deficits in coordinated locomotion during subacute exposure. *Pharmacol. Biochem. Behav.* **24**, 1805–1807.

Le A. D. and Khanna J. M. (1989) Dispositional mechanisms in drug tolerance and sensitization, in *Psychoactive Drugs* (Goudie A. J. and Emmett-Oglesby, ed.), Humana, Clifton, NJ, pp. 281–351.

Le A. D., Khanna J. M., Kalant H., and Grosi F. (1986) Tolerance to and crosstolerance among ethanol, pentobarbital and chlordiazepoxide. *Pharmacol. Biochem. Behav.* **24,** 93–98.

LeBlanc A. E., Kalant H., and Gibbins R. J. (1969) Acquisition and loss of tolerance to ethanol by the rat. *J. Pharmacol. Exp. Ther.* **168,** 244–250.

Lemaire G. A. and Meisch R. A. (1984) Pentobarbital self-administration in rhesus monkeys: Drug concentration and fixed-ratio size interaction. *J. Exp. Anal. Behav.* **42,** 37–49.

Lemaire G. A. and Meisch R. A. (1985) Interactions between drug amount and fixed-ratio size. *J. Exp. Anal. Behav.* **44,** 377–389.

Leonard B. E. (1965) Effect of barbitone sodium on the pituitary-adrenal system on some labile compounds in the brain of the rat. *Biochem. J.* **96,** 56P.

Leonard B. E. (1966a) The effect of chronic administration of barbitone sodium on labile compounds in the rat brain. *Biochem. Pharmacol.* **15,** 255–262.

Leonard B. E. (1966b) The effect of chronic administration of barbitone sodium on pituitary-adrenal function in the rat. *Biochem. Pharmacol.* **15,** 263–268.

Leonard B. E. (1967) The effect of chronic administration of barbitone sodium on the behavior of the rat. *Int. J. Neuropharmacol.* **6,** 63–70.

Leonard B. E. (1968) The effect of chronic administration of sodium barbitone on chemically and electrically induced convulsions in the rat. *Int. J. Neuropharmacol.* **7,** 463–468.

Lester D. and Freed E. X. (1972) Biological aspects of alcohol consumption. *Finnish Foundation of Alcohol Studies* **20,** 51–57.

Letz R. and Belknap J. K. (1975) Simple induction and assessment of barbiturate physical dependence in the rat. *Physiol. Psychol.* **3,** 249–252.

Lim R. K. (1967) Animal technique for evaluating hypnotics, in *Animal and Clinical Pharmacologic Techniques in Drug Evaluation* (Nodine J. H. and Siegler P. E., eds.), Medical Publishers Yearbook, Chicago, pp. 291–302.

Lim R. K., Pindell M. H., Glass H. G., and Rink K. (1956) The experimental evaluation of sedative agents in animals. *Ann. NY Acad. Sci.* **64,** 667–678.

Lutz J. (1929) Uber die dauernarkosebehandlung in der psychiatric. *Z. Gesamte Neurol. Psychiatr.* **123,** 91–122.

Lyness W. H., Brezenoff H. E., and Mycek M. J. (1979) Cross tolerance to centrally injected barbiturates. *Eur. J. Pharmacol.* **54,** 319–330.

Mackinnon G. L. and Parker W. A. (1982) Benzodiazepine withdrawal syndrome: a literature review and evaluation. *Am. J. Drug Alc. Abuse* **9,** 19–33.

Maier D. M. and Pohorecky L. A. (1989) The effects of repeated withdrawal episodes on subsequent withdrawal severity in ethanol-treated rats. *Drug and Alcohol Dependence* **23,** 103–110.

Majchrowicz E. (1975) Induction of physical dependence upon ethanol and the associated behavioral changes in rats. *Psychopharmacologia* **43**, 249–254.

Mark L. C. (1963) Metabolism of barbiturates in man. *Clin. Pharmacol. Ther.* **4**, 504–530.

Martin P. R., Kapier B. M., Whiteside E. A., and Seller E. M. (1979) Intravenous phenobarbital therapy in barbiturate and other hypersedative withdrawal reactions: a kinetic approach. *Clin. Pharmacol. Ther.* **26**, 256–264.

Martin W. R., McNicholas L. F., and Cherian S. (1982) Diazepam and pentobarbital dependence in the rat. *Life Sci.* **31**, 721–730.

Masuda M., Budde R. N., and Dille J. M. (1938) An investigation of acquired tolerance to certain short-acting barbiturates. *J. Am. Pharm. Assoc.* **27**, 830–836.

Maynert E. W. and Klingman G. I. (1960) Acute tolerance to intravenous anesthetics in dogs. *J. Pharmacol. Exp. Ther.* **128**, 192–200.

Maynert E. W. and VanDyke H. B. (1949) The metabolism of barbiturates. *Pharmacol. Rev.* **1**, 217–242.

McBride A. and Turnbull M. J. (1970) The brain acetylcholine system in barbitone-dependent and withdrawn rats. *Brit. J. Pharmacol.* **39**, 210,211.

McGee D. H. and Bourn W. M. (1978) Influence of treatment duration on audiogenic seizure susceptibility during barbiturate withdrawal in rats. *Experientia* **34**, 873,874.

McNicholas L. F. and Martin W. R. (1982) The effect of a benzodiazepine antagonist, R015-1788, in diazepam dependent rats. *Life Sci.* **31**, 731–737.

McNicholas L. F., Martin W. R., and Cherman S. (1983) Physical dependence on Diazepam and Lorazepam in the dog. *J. Pharmacol. Exp. Ther.* **226**, 783–789.

Mendelson J. H. (1964) Experimentally induced chronic intoxication and withdrawal in alcoholics. *Quart. J. Stud. Alc.* **25**, 1–52.

Meerloo A. M. (1930) Zur pathologie und psychopathologie der schlafmittelkuren. 2. *Gesamte Neurol. Psychiatr.* **127**, 168–187.

Meisch R. A., Kliner D. J., and Henningfield J. E. (1981) Pentobarbital drinking by Rhesus monkeys: Establishment and maintenance of pentobarbital-reinforced behavior. *J. Pharmacol. Exp. Ther.* **217**, 114–120.

Mello N. K. (1978) Psychopharmacology: A Generation of Progress (Lipton M., DiMascia A., and Killan K. F., eds.), Raven, New York, pp. 1619–1637.

Milner G. (1968) Modified confinement motor activity test for use in mice. *J. Pharm. Sci.* **57**, 1900–1902.

Moir W. M. (1937) The influence of age and sex on the repeated administration of sodium pentobarbital to albino rats. *J. Pharmacol. Exp. Ther.* **59**, 68–85.

Molinengo L. (1964) Azione di alcuni depressive del SNC sulls coorinazione motoria del ratto. *Arch Ital. Sci. Farmacol.* **14**, 288–291.

Morgan K. G. and Bryant S. H. (1977) Pentobarbital: Presynaptic effect in the squid giant synapse. *Experientia* **33**, 487,488.

Morgan W. W. (1976) Effect of 6-hydroxydopamine pretreatment on spontaneous convulsions induced by barbital withdrawal. *Experientia* **32**, 489–491.

Morgan W. W., Pfeil K. A., and Gonzales E. G. (1977) Catecholamine concentration in discrete brain areas following the withdrawal of barbital-dependent rats. *Life Sci.* **20**, 493–500.

Morgan W. W., Huffman R. D., Pfeil K. A., and Gonzales E. G. (1978) Effects of synthesis inhibition on the levels of brain catacholamines in barbital-dependent rats. *Psychopharmacology* **56**, 41–44.

Mott F. W., Woodhouse D. L., and Pickworth F. A. (1926) The pathological effects of hypnotic drugs upon the central nervous system of animals. *Br. J. Exp. Pathol.* **7**, 325–336.

Mycek M. J. and Brezenoff H. E. (1976) Tolerance to centrally administrated phenobarbital. *Biochem. Pharmacol.* **25**, 501–504.

Nakamura H. and Shimizu M. (1983) Persistence of drug experience in rats formally dependent or phenobarbital on meprobamate. *Neuropharmacol.* **22**, 923–926.

Newman H. W. and Abramson M. (1941) Relation of alcohol concentration to intoxication. *Proc. Soc. Exp. Biol. Med.* **48**, 509–513.

Noble E. P., Gillier R., Vigran R., and Mandel P. (1976) The modification of the ethanol withdrawal syndrome in rats by di-*n*-propylactate. *Psychopharmacology* **46**, 127–131.

Nordberg A. and Wahlstrom G. (1977) Effect of long-term forced oral barbital administration on endogenous acetylcholine in different regions of rat brain. *Eur. J. Pharmacol.* **43**, 237–242.

Nordberg A. and Wahlstrom G. (1979) Regional biosynthesis of acetylcholine in brain following forced chronic barbitone treatment to rat. *J. Neurochem.* **32**, 371–378.

Norton P. R. E. (1970) The effects of drugs on barbiturate withdrawal convulsions in the rat. *J. Pharm. Pharmacol.* **22**, 763–766.

Norton P. R. E. (1971) Some endocrinological aspects of barbiturate dependence. *Br. J. Pharmacol.* **41**, 317–330.

Nozaki M., Martin W. R., Driver M. B., Diringer M., and Wu K. M. (1981) Use of gastric fistula rats for the study of sedative, hypnotic and antianxiety drugs. *Drug and Alcohol Dependence* **7**, 221–231.

Okamoto M. (1985) Barbiturate tolerance and physical dependence: Contribution of pharmacological factors, in *Mechanisms of Tolerance and Dependence*. NIDA Research Monograph Service. vol. 54, pp. 333–347.

Okamoto M., Aaronson L., and Hinman D. (1983) Comparison of effects of diazepam on barbiturate and on ethanol withdrawal. *J. Pharmacol. Exp. Ther.* **225**, 589–594.

Okamoto M., Rao L. S., and Walewski J. L. (1986) Effect of dosing frequency on the development of physical dependence and tolerance to pentobarbital. *J. Pharmacol. Exp. Ther.* **238**, 1004–1008.

Okamoto M., Rosenberg H. C., and Boise N. R. (1975) Tolerance characteristics produced during the maximally tolerable chronic pentobarbital dosing in the cat. *J. Pharmacol. Exp. Ther.* **192**, 555–569.

Okamoto M., Rosenberg H. C., and Boisse N. R. (1976) Withdrawal characteristics following chronic pentobarbital dosing in cat. *Eur. J. Pharmacol.* **40**, 107–119.

Okamato M., Rosenberg H. C., and Boisse N. R. (1977) Evaluation of anticonvulsants in barbiturate withdrawal. *J. Pharmacol. Exp. Ther.* **202**, 479–489.

Okamoto M., Boisse N. R., Rosenberg H. C., and Rosen R. (1978) Characteristics of functional tolerance during barbiturate physical dependence production. *J. Pharmacol. Exp. Ther.* **207**, 906–915.

Okamoto M., Rao S. N., Reyes J., and Rifkind A. B. (1985) Recovery from dispositional and pharmacodynamic tolerance after chronic pentobarbital treatment. *J. Pharmacol. Exp. Ther.* **235**, 26–31.

Palermo-Neto J. and DeLima T. C. (1982) Effects of withdrawal from long-term barbital treatment on open-field behavior and seizure susceptibility of rats. *Neuropharmacology* **21**, 277–281.

Palmer H. and Braceland F. (1937) Six years experience with narcosis therapy in psychiatry. *Am. J. Psychiat.* **94**, 37–57.

Phillip B. M., Miya T. S., and Yim G. K. W. (1962) Studies on the mechanism of meprobamate tolerance in the rat. *J. Pharmacol. Exp. Ther.* **135**, 223–229.

Quinn G. P., Axelrod J., and Brodie B. B. (1958) Species, strain and sex differences in metabolism of hexobarnitone, amidopyrine, antipyrine and aniline. *Biochem. Pharmacol.* **1**, 152–150.

Rastogi S. K., Wenger G. R., and McMillan D. E. (1985) Effect of optical isomers of pentobarbital on behavior in rats maintained on either the d or 1 isomers of methadone. *Arch. Int. Pharmacodyn. Ther.* **276**, 247–262.

Raventos J. (1954) The distribution in the body and metabolic fate of barbiturates. *J. Pharm. Pharmacol.* **6**, 217–235.

Remmer H. (1962) Drug tolerance, in *Ciba Foundation Symposium on Enzyme and Drug Action* (Monger J. L. and de Reuch A. V. S., eds.), Churchhill, London, pp. 276–298.

Remmer H. (1964) Gewhnung au hexobarbital durch begchleunigten abbau. *Arch. Int. Pharmacodyn.* **152**, 346–359.

Remmer H. and Siegert M. (1960) Beschleunigung des Abbaus und Adaptation des ZNS wahrend der Gewohnung an Barbiturate. *Naunyn-Schmiedebergs Arch. Pharmakol. Exp. Pathol.* **240**, 22,23.

Richter J. A., Harris P. S., and Hanford P. V. (1982) Similar development of tolerance to barbital-induced inhibition of avoidance behavior and loss of righting reflex in rats. *Pharmacol. Biochem. Behav.* **16**, 467–471.

Rosenberg H. C. and Okamoto M. (1974) A method for producing maximal pentobarbital dependence in cats: dependency characteristics, in *Drug Addiction.* vol. 3 *Neurobiology and Influence on Behavior* (Singh J. M. and Lal H., eds.), Symposia Specialist, Miami, FL, pp. 89–103.

Roth-Schechter B. F. and Mandel P. (1976) A study on the development of barbiturate tolerance and dependence in hamster glial cells in culture. *Biochem. Pharmacol.* **25**, 563–571.

Rumke C. L., vanStrik R., De Jonge H., and Delver A. (1963) Experiments on the duration of hexobarbital narcosis in mice. *Arch. Int. Pharmacodyn.* **146**, 10–26.

Schmidt H., Jr. and Kleinman K. M. (1964) Effect of chronic administration and withdrawal of barbiturates upon drinking in the rat. *Arch. Int. Pharmacodyn.* **151**, 142–149.

Schuster C. R. and Thompson T. (1969) Self-administration of and behavioral dependence on drugs. *Ann. Rev. Pharmacol.* **9**, 483–502.

Seevers M. H. and Deneau G. A. (1963) Physiological aspects of tolerance and physical dependence, in *Physiological Pharmacology* (Root W. S. and Hoffman F. G., eds.), Academic, New York, pp. 565–640.

Seevers M. H. and Tatum A. L. (1931) Chronic experimental barbital poisoning. *J. Pharmacol. Exp. Ther.* **42**, 217–231.

Sereny G. and Kalant H. (1965) Comparative clinical evaluation of chlordiazepoxide and promazine in treatment of alcohol withdrawal syndrome. *Br. Med. J.* **1**, 92–97.

Sharpless S. K. (1970) Hypnotics and sedatives, in *The Pharmacological Basis of Therapeutics*, 4th ed. (Goodman L. S. and Gilman A., eds.), Macmillan, London, pp. 98–120.

Siemens A. J. and Chan A. W. K. (1976) Differential effects of phenobarbital and ethanol in mice. *Life Sci.* **19**, 581–590.

Siew C. and Goldstein D. B. (1978) Osmotic minipumps for administration of barbital to mice: Demonstration of functional tolerance and physical dependence. *J. Pharmacol. Exp. Ther.* **204**, 541–546.

Singh J. M. (1970) Clinical signs and development of tolerance to thiopental. *Arch. Int. Pharmacodyn.* **187**, 199–208.

Sjorgen J., Solvell L., and Karlsson I. (1965) Studies on the absorption rate of barbiturates in man. *Acta Med. Scand.* **178**, 553–559.

Smith C. M. (1977) The pharmacology of sedative/hypnotics, alcohol, and anesthetics: sites and mechanisms of action, in *Drug Addiction 1: Morphine, Sedative/Hypnotic and Alcohol Dependence* (Martin W. R., ed.), *Handbuch der Experimenteller Pharmakologie*, vol. 45, pf 1, Springer-Verlag, Berlin, pp. 413–587.

Smith D. E. (1969) Physical vs. psychological dependence and tolerance in high-dose methamphetamine abuse. *Clin. Toxicol.* **2**, 99–103.

Stanton E. J. (1936) Addiction and tolerance to barbiturates? The effects of daily administration and abrupt withdrawal of phenobarbital-sodium and pentobarbital-sodium in the albino rat. *J. Pharmacol. Exp. Ther.* **57**, 245–252.

Stevenson I. H. and Turnbull M. J. (1968) Hepatic drug-metabolizing enzyme activity and duration of hexobarbitone anesthesia in barbitone-dependent and withdrawn rats. *Biochem. Pharmacol.* **17**, 2297–2305.

Stevenson I. H. and Turnbull M. J. (1969) The effect of chronic barbitone administration and withdrawal on the sensitivity of the central nervous system to barbiturate. *Br. J. Pharmacol.* **37,** 502–503.

Stevenson I. H. and Turnbull M. J. (1970) The sensitivity of the brain to barbiturate during chronic administration and withdrawal of barbitone sodium in the rat. *Br. J. Pharmacol.* **39,** 325–333.

Stockhaus K. (1986) Physical dependence capacity of brotizalam in Rhesus monkey. *Arzneimittel-Forschung* **36,** 601–605.

Stolman S. and Loh H. H. (1975) Barbital induced cross-tolerance to barbiturates by the intracisternal route of administration. *Res. Commun. Chem. Pathol. Pharmacol.* **12,** 309–316.

Stump C. and Chiari I. (1965) Echte Gewhnung an Hexobarbital. *Naunyn-Schmiedebergs Arch. Pharmakol Exp. Pathol.* **251,** 275–287.

Suzuki T., Koike Y., Chida Y., and Misawa M. (1988) Cross-physical dependence of several drugs in methagualine-dependent rats. *Jpn. J. Pharmacol.* **46,** 403–410.

Svensson T. and Thieme G. (1969) An investigation of a new instrument to measure motor activity of small animals. *Psychopharmacologia* **14,** 157–163.

Swanson E. E., Weaver M. M., and Chen K. K. (1937) Repeated administration of amytal. *Am. J. Med. Sci.* **193,** 246–251.

Tabakoff B., Yanai J., and Ritzmann R. F. (1978) Brain noradrenergic systems as a prerequisite for developing tolerance to barbiturates. *Science* **200,** 449–451.

Tabashima T. and Ho I. K. (1981) Pharmacological responses of pentobarbital in different strains of mice. *J. Pharmacol. Exp. Ther.* **216,** 198–204.

Tagashira E., Hiramori T., Urano T., and Tanaura S. (1983) Formation of physical dependence on barbiturates and cerebral monoamines. *Jap. J. Pharmacol.* **33,** 415–422.

Tang M. et al. (1981) Barbiturate dependence and drug preference. *Pharmacol. Biochem. Behav.* **14,** 405–408.

Teiger D. G. (1974) Induction of physical dependence to morphine, codeine and meperidine in the rat by continuous infusion. *J. Pharmacol. Exp. Ther.* **190,** 408–415.

Thompson T. and Pickens R. (1969) Drug self-administration and conditioning, in *Scientific Basis of Drug Dependence* (Steinberg H., ed.), Churchhill, London, pp. 177–198.

Thompson T. and Pickens R. (1970) Behavioral variables influencing drug self-administration, in *Drug Dependence* (Harris R. T., McIsaac W. M., and Schuster C. R., eds.), Univ. of Texas Press, Austin, TX, pp. 143–157.

Turnbull M. J. and Watkins J. W. (1976) Acute tolerance to barbiturate in the rat. *Eur. J. Pharmacol.* **36,** 15–20.

Ulrichsen J., Clemmensen L., Flachs H., and Hemmingsen R. (1986) The effect of phenobarbital and carbamezepine on the ethanol withdrawal reading in the rat. *Psychopharmacology* **89,** 162–166.

Vesell E. S. (1968) Genetic and environmental factors affecting hexobarbital metabolism in mice. *Ann. NY Acad. Sci.* **151,** 900–912.

Victor M. (1966) Treatment of alcoholic intoxication and the withdrawal syndrome. *Psychosom. Med.* **28,** 636–650.

Victor M. and Adams R. D. (1953) The effect of alcohol on the nervous system, in *Metabolic and Toxic Diseases of the Nervous System. Assoc. Res. Nerv. Ment. Dir. Res. Publ.* **32,** 526–573.

Waddell W. J. and Baggett B. (1973) Anesthetics and lethal activity in mice of the stereoisomers of 5-ethyl-5-(1-methylbutyl) barbituric acid (pentobarbital). *Arch. Int. Pharmacodyn. Ther.* **205,** 40–44.

Waddell W. J. and Butler T. C. (1957) The distribution and excretion of phenobarbital. *J. Clin. Invest.* **36,** 1217–1226.

Wahlstrom G. (1966) Differences in anaesthetic properties between the optical antipodes of hexobarbital in the rat. *Life Sci.* **5,** 1781–1790.

Wahlstrom G. (1968) Hexobarbital (Enhexymalum NFN) sleeping times and EEG threshold doses as measurements of tolerance to barbiturates in the rat. *Acta Pharmacol. Toxicol.* **26,** 64–80.

Wahlstrom G. (1978) The effects of atropine on the tolerance and the convulsions seen after withdrawal from forced barbital drinking in the rat. *Psychopharmacology* **59,** 123–128.

Waters D. H. (1973) A quantitative evaluation of barbiturate physical dependence in rodents. Ph. D. Thesis, Cornell University, 1973 Graduate School of Medical Science.

Waters D. H. and Okamoto M. (1972) Increased control excitability in nondependent mice during chronic barbital dosing, in *Drug Addiction: Experimental Pharmacology* vol. 1 (Singh J. M., Miller L., and Lal H., eds.), Futura, New York, pp. 199–209.

Way E. L., Loh H. H., and Shen F. H. (1969) Simultaneous quantitative assessment of morphine tolerance and physical dependence. *J. Pharmacol. Exp. Ther.* **167,** 1–8.

Weaks J. R. (1962) Experimental morphine addiction: Method for automatic intravenous injections in unrestrained rats. *Science* **138,** 143,144.

Wenger G. R. (1986) Behavioral effects of the isomer of pentobarbital and secobarbital in mice and rats. *Pharmacol. Biochem. Behav.* **25,** 375–380.

Wenger G. R. (1988) Chronic administration of S-(–)-pentobarbital in pigeons and rats: Tolerance development. *Pharmacol. Biochem Behav.* **31,** 459–465.

Wenger G. R., Donald J. M., and Cunny H. C. (1986) Stereoselective behavioral effects of the isomers of pentobarbital and secobarbital in the pigeon. *J. Pharmacol. Exp. Ther.* **237,** 445–449.

Wessinger W. D. and Wenger G. R. (1987) The discriminative stimulus properties of barbiturate stereoisomers. *Psychopharmacology* **92,** 334–339.

Wikler A. and Essig C. F. (1970) Withdrawal seizure following chronic intoxification with barbiturates and other sedative drugs. *Epilepsy Mod. Problems Pharmaco-psychiat.* **4,** 170–184.

Wikler A., Fraser H. T., Isbell H., and Pescor F. T. (1955) Electroencephalogram during cycles of addiction to barbiturates in man. *E.E.G. J.* **7**, 1–13.

Williams R. T. and Parke D. V. (1964) The metabolic fate of drugs. *Ann. Rev. Pharmacol.* **4**, 85–114.

Winger G., Stitzer M. L., and Woods J. H. (1975) Barbiturate-reinforced responding in rhesus monkeys: Comparison of drugs with different duration of action. *J. Pharmacol. Exp. Ther.* **195**, 505–514.

Woods J. H. and Schuster C. R. (1970) Regulation of drug self-administration, in *Drug Dependence* (Harris R. T., McIsaac W. M., and Schuster C. R., eds.), University Texas Press, Austin, TX, pp. 158–169.

Woods J. H., Ikomi F., and Winger G. D. (1971) The reinforcing property of ethanol. In, *Biological Aspects of Alcoholism* (Roach M. K., McIsaac W. M., and Creaven P. J., eds.), University Texas Press, Austin, TX, pp. 371–388.

Woolverton W. L. and Schuster C. R. (1983) Intragastric self-administration in Rhesus monkeys under limited access condition: Methodological studies. *J. Pharmacol. Meth.* **10**, 93–106.

Wulff M. H. (1959) The barbiturate withdrawal syndrome. A clinical and electroencephalographic study. *Electroencephalogr. Clin. Neurophysiol.* **14**, 1–173.

Yamamoto I. and Ho I. K. (1978) Sensitivity to continuous administration of pentobarbital in different strains of mice. *Res. Commun. Chem. Pathol. Pharmacol.* **19**, 381–388.

Yanagita T. and Takahashi S. (1970) Development of tolerance to and physical dependence on barbiturates in Rhesus monkeys. *J. Pharmacol. Exp. Ther.* **172**, 163–169.

Yanagita T. and Takahashi S. (1973) Dependence liability of several sedative hypnotic agents evaluated in monkeys. *J. Pharmacol. Exp. Ther.* **185**, 307–316.

Yanai J., Guttman R., and Slern E. (1989) Genotype-treatment interaction in response of mice to early barbiturate administration. *Biol. Neonate* **56**, 109–116.

Yanaura S. and Tagashira E. (1975) Dependence on and preference for morphine (II). Comparison among morphine, phenobarbital and diazepam. *Folia Pharmacol. Jpn.* **71**, 285–294.

Young R., Glennon R. A., and Dewey W. L. (1984) Stereoselective stimulus effects of 3-methylflunitrazepam and pentobarbital. *Life Sci.* **34**, 1977–1983.

Yutrzenka G. J. and Kosse K. (1989) Dependence on phenobarbital but not pentobarbital using drug-adulterated food. *Pharmacol. Biochem. Behav.* **32**, 891–895.

Yutrzenka G. J., Patrick G. A., and Rosenberg W. (1985) Continuous intraperitoneal infusion of pentobarbital: A model of barbiturate dependence in the rat. *J. Pharmacol. Exp. Ther.* **232**, 111–118.

Benzodiazepine Tolerance and Dependence

Richard G. Lister

1. Introduction

In the late 1950s, researchers at Hoffmann-La Roche discovered that benzodiazepines were able to cause changes in animal behavior that resembled those caused by barbiturates and meprobamate. Shortly thereafter, chlordiazepoxide was introduced ino clinical practice as an anxiolytic, and other benzodiazepines soon followed. They were greatly favored over meprobamate and barbiturates not only because of their wide safety margin—death from benzodiazepine overdose alone is extremely rare—but also because benzodiazepines were not thought to produce dependence except following prolonged administration of very high doses. During the 1980s, this latter belief underwent considerable revision (*see* File, 1990), and there is now considerable interest in the development of drugs with less dependence-producing potential for the treatment of anxiety disorders.

Since their introduction, much has been learned of the behavioral pharmacology of the benzodiazepines. An exhaustive review of the abuse liability of these drugs was published recently (Woods et al., 1987), and File (1990) recently reviewed the animal literature on benzodiazepine dependence. This chap-

The views expressed herein are not necessarily those of the institute with which the author was affiliated.

From: *Neuromethods, Vol. 24: Animal Models of Drug Addiction*
Eds: A. Boulton, G. Baker, and P. H. Wu ©1992 The Humana Press Inc.

ter focuses on the changes that occur following the chronic administration of benzodiazepines to rodents. It begins with an overview of the different methods available for administering benzodiazepines chronically. It then considers the effects of chronic benzodiazepine administration, and finally considers some of the mechanisms that may account for the behavioral changes observed.

2. Methods of Chronic Benzodiazepine Administration

Animal studies examining the effects of chronic benzodiazepine treatment have generally used one of three different methods to administer the drug: injection, self-administration, or the surgical implantation of a device that releases the drug of interest at a regular rate. By far the most frequently used method of administering benzodiazepines chronically has been by injection (either po, ip, sc, or ig). The primary advantage of this method is that a known dose of drug is given at each time-point. Further, with this method it is possible to approximate closely the clinical situation in which fixed doses of drug are taken one or more times each day. A disadvantage is that this method can be labor-intensive if multiple doses have to be given every day over a long period of time. Further, since rodents generally metabolize benzodiazepines more rapidly than humans, even twice daily injections may be insufficient to achieve chronic brain concentrations of the drug.

An alternative procedure involves administering the drug in animals' food (e.g., Yanaura et al., 1975; Gallaher et al., 1986) or drinking water (Rosenberg and Chiu, 1981). The latter method clearly requires the use of a water-soluble benzodiazepine, such as flurazepam. Since animals' food and liquid intake varies considerably over a 24-h period, drug levels are also likely to vary during the course of each day. This variation will be most significant for benzodiazepines with short elimination half-lives. An example of this approach comes from the work of Rosenberg and Chiu (1981). They offered rats a flurazepam and saccharin solution as their only source of water. They prepared a solution containing 100 mg flurazepam/mL water by dissolving flurazepam dihydrochloride in distilled water and adjusting the pH to 5.8–6.0 with dilute sodium hydroxide. Rats were allowed to drink a 0.02% saccharin solution for 2 d. Based on the volume consumed over a 24-h period and the weight of each rat, a final drug concentration was adjusted so that each rat would receive approx 100 mg/kg daily. After a week, they increased the drug concentration to give the animals approx 150 mg/kg. This paradigm produces tolerance to a number of the behavioral effects of benzodiazepines, including their motor impairing and anticonvulsant actions (Rosenberg and Chiu, 1981; Rosenberg et al., 1985; Tietz and Rosenberg, 1988).

If the experimenter requires as constant a brain concentration of drug as possible during chronic exposure, then this is probably best achieved using an osmotic minipump (Theeuwes and Yum, 1976) or silastic pellet implant that slowly releases the drug at a relatively constant rate. For example, Miller et al. (1988) administered lorazepam to mice for 14 d by dissolving the drug in polyethylene glycol 400 and placing the solution in osmotic minipumps that were implanted sc under brief ether anesthesia. Different groups of mice received four different doses of lorazepam, and in each group, brain and plasma concentrations of lorazepam remained constant over the 14-d treatment period. Tolerance was observed to the motor-impairing effects of lorazepam over this time. Similarly Gallager et al. (1985) administered diazepam chronically to rats using silastic tubing. They initially implanted two pellets each containing 90 mg diazepam, and then 10 d later, implanted an additional 90-mg pellet. It was found that this protocol resulted in brain levels of diazepam that varied little over a 3-wk period after an initial surge during the 24 h that followed implantation. Tolerance to the anticonvulsant action of diazepam against bicuculline-induced seizures developed over the 3-wk period.

2.1. Benzodiazepine Pharmacokinetics

In all of the above methods of benzodiazepine administration, it is most important to consider the pharmacokinetics of the drug under investigation. In so doing, it may be noted that drug metabolism can vary considerably from one species to another. Garattini et al. (1973) reported on the concentrations of diazepam, desmethyldiazepam, and oxazepam in rats, mice, and guinea pigs following a single iv injection of 5 mg/kg diazepam. They found that, although the peak level of diazepam and its rate of disappearance were approximately the same in the three species, there were marked differences in the concentrations of the two metabolites. Only trace levels of metabolites were detectable in the rat, whereas guinea pigs had a sustained level of N-methyldiazepam lasting for approx 10 h, and in mice a peak of N-desmethyldiazepam followed the disappearance of diazepam. Further, in the mice, oxazepam concentrations increased after the N-desmethyldiazepam peak and were still detectable 10 h after the initial injection.

Data on the metabolism of lorazepam (Schillings et al. 1977; Lister et al., 1983b), chlordiazepoxide (Lister et al., 1983b), midazolam (Woo et al., 1981), and triazolam (Kitagawa et al., 1979) in the rat have also been reported. In all these cases, the benzodiazepine and metabolites have short elimination half-lives. It is, perhaps, for this reason that early studies in rats failed to produce reliable withdrawal syndromes when the drugs were administered by injection at doses that were considered to be clinically relevant. As will be seen later, many of the later studies in rats that have produced significant withdrawal following injection regimens have administered very high doses of benzodiazepines that maintain significant brain concentrations of the drug throughout the treatment period. For more detailed discussions of benzodiazepine pharmacokinetics, *see* Greenblatt et al. (1982) and Arendt et al. (1983).

3. Tolerance to the Effects of Benzodiazepines

Tolerance refers to the phenomenon in which a response to a drug is reduced as a result of prior exposure. Alternatively, a larger dose of drug is required to produce an equivalent effect in subjects previously exposed to the agent. Tolerance develops to most, if not all, of the effects of the benzodiazepines. Interestingly, however, the rate at which tolerance develops varies according to the behavioral effect (*see* File, 1985). Tolerance rapidly develops to benzodiazepines' sedative effects (e.g., Sansone, 1979; File, 1981), but somewhat more slowly to their anticonvulsant and anxiolytic effects (Goldberg et al., 1967; File, 1983; Haigh and Feely, 1988b; Stephens and Schneider, 1985; Vellucci and File, 1979; Davis and Gallager, 1988).

The variation in the rates at which tolerance develops to the different behavioral effects suggests that benzodiazepine tolerance has a pharmacodynamic rather than a pharmacokinetic basis. That is, the reduced behavioral response seen following chronic treatment is not simply the result of increased drug metabolism. This conclusion is supported by studies that have measured plasma and brain concentrations of benzodiazepines following acute and chronic treatment (e.g., Lister et al., 1983a; Rosenberg et al., 1985; Haigh et al., 1986; Miller et al., 1988).

3.1. Tolerance to the Effects of Other Benzodiazepine Receptor Ligands

There exists a wide spectrum of ligands with high affinity for central benzodiazepine receptors. Some of these are shown in Fig. 1. Drugs, such as the benzodiazepines diazepam and lorazepam, act as full agonists and produce the full range of behavioral effects associated with benzodiazepines, including sedation, ataxia, anxiolysis, anterograde amnesia, and anticonvulsant effects. Other drugs at the other end of the spectrum, such as the β-carboline DMCM, are termed inverse agonists and produce behavioral effects opposite to those of benzodiazepines (e.g., they cause anxiety and induce seizures). Partial agonists, such as Ro 16-6028, cause a number of the effects associated with benzodiazepines (e.g., anxiolyis and an anticonvulsant action), but are less likely to cause sedation. Partial inverse agonists, such as FG 7142, cause some of the effects associated with inverse agonists (e.g., they cause anxiety), but tend not to induce seizures (although they are proconvulsant). Finally, some drugs, such as flumazenil and the β-carboline ZK 93426, act primarily as antagonists, antagonizing the effects of both agonists and inverse agonists, but exerting few behavioral effects on their own. It is important to note that the classification of ligands is not based

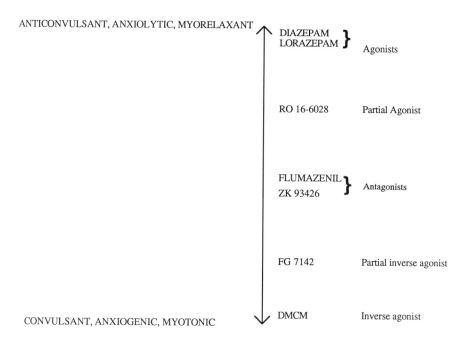

on their behavioral effects, but on the nature of each drug's interaction with benzodiazepine receptors. In general, however, the behavioral effects of each drug can be predicted from the classification of each ligand.

There has been considerable interest in the development of benzodiazepine receptor partial agonists for use in the treatment of anxiety disorders. It has been suggested that these drugs are less likely to lead to tolerance and physical dependence. For example, tolerance to the anticonvulsant action of Ro 16-6028 has not been observed in mice (Haigh and Feely, 1988a). More recently, Hernandez et al. (1989) compared the effects of 3 wk of treatment with diazepam, Ro 16-6028, or flumazenil. Although tolerance developed consistently to the anticonvulsant action of diazepam, the data suggested heterogeneity in the development of tolerance to the anticonvulsant action of Ro 16-6028, some animals showing tolerance and others not. There was similar heterogeneity in the development of subsensitivity to GABA (*see* Section 3.2.2.) in the dorsal raphe nucleus. It seems reasonable to speculate that these observations are causally linked.

Just as tolerance seems to develop less readily to the effects of a benzodiazepine partial agonist, it also seems difficult to induce physical dependence with these drugs (Moreau et al., 1990). Finally, it should be noted that a number of changes accompany chronic treatment with benzodiazepine receptor inverse agonists. These changes are beyond the scope of the present chapter, and the interested reader is referred elsewhere (Little et al., 1984,1987; Petersen and Jensen, 1987a,b; Corda et al., 1988; Stephens et al., 1988; Marley et al., 1991).

3.2. Mechanisms of Benzodiazepine Tolerance

Drug tolerance in general and benzodiazepine tolerance in particular have been discussed at many levels of analysis (e.g., *see* Goudie and Emmett-Oglesby, 1989). Explanations of benzodiazepine tolerance have come from investigators reporting alterations in benzodiazepine receptor binding, modification of GABA-benzodiazepine coupling, and changes in electrophysiological responses to benzodiazepines. Discussions of benzodiazepine tolerance have also taken place at the behavioral level by considering the role of classical and instrumental conditioning in tolerance development. These will be considered in turn.

3.2.1. Tolerance and Benzodiazepine Receptor Binding

There have been a number of investigations into the effects of chronic treatment with various benzodiazepines on benzodiazepine receptor binding using in vitro or ex vivo techniques. Some have observed no change in receptor number or affinity (e.g., Mohler et al., 1978; Braestrup and Nielsen, 1983; Gallager et al., 1984), whereas others have reported decreases in benzodiazepine receptor number (Rosenberg and Chiu, 1981; Crawley et al., 1982; Tietz et al., 1986). Recently, Miller et al. (1988) examined changes in benzodiazepine receptor number using in vivo methods. By measuring the specific uptake of the benzodiazepine receptor ligand [^{3}H]Ro 15-1788 (flumazenil), they observed downregulation of benzodiazepine receptors in cortex, hypothalamus, and hippocampus following 7 d of treatment with lorazepam. There were no significant changes in the cerebellum. The time-course of the change paralleled the development of tolerance to the ataxic effects of lorazepam assessed using a rotarod. Tietz et al. (1986) also observed regional variations in benzodiazepine receptor downregulation using autoradiography. For example, they found that 4 wk of flurazepam produced a 40% downregulation in the substantia nigra pars reticulata, but no change in the lateral hypothalamus (*see also* Rosenberg et al., 1988).

It is possible that benzodiazepine receptor downregulation may contribute to tolerance to some benzodiazepine effects, but not others. Rosenberg et al. (1985) noted that tolerance to the anticonvulsant effect of a benzodiazepine was present after receptor downregulation was no longer observable. It would appear, therefore, that benzodiazepine receptor downregulation is not critical for tolerance to benzodiazepine's anticonvulsant action.

3.2.2. Tolerance and Subsensitivity to GABA

Gallager and coworkers (1984) reported a decrease in postsynaptic sensitivity to GABA following 3 wk of diazepam administration assessed using iontophoretic application of GABA and serotonin on serotonergic cells in the midbrain dorsal raphe nucleus. More recently, a comparison has been made between the effects of chronic benzodiazepine administration on dorsal raphe neurons and neurons in the substantia nigra pars reticulata (Gallager and Wilson, 1988; Wilson and Gallager, 1989). The authors argued that dorsal raphe neurons became subsensitive to GABA with a time-course that paralleled the development of tolerance to the anticonvulsant action of benzodiazepines in contrast to the effects observed in reticulata neurons.

Marley and Gallager (1989) examined stimulation of $^{36}Cl^-$ influx by GABA in cortical and cerebellar membrane preparations in rats treated with diazepam for 3 wk. They observed a decreased responsiveness to stimulation of Cl^- influx by GABA in the cortical preparation as a result of chronic diazepam exposure, but no change was seen in the membrane vesicles prepared from cerebella. It seems likely that regional differences in the rates at which adaptation to the effects of benzodiazepines occurs may underlie the different rates at which tolerance develops to the various behavioral effects of benzodiazepines.

3.2.3. Tolerance and Response to Other Benzodiazepine Receptor Ligands

Little et al. (1987) and Petersen and Jensen (1987a) have examined the response of animals chronically treated with benzodiazepines (flurazepam or lorazepam) to a number of drugs that act at the benzodiazepine–GABA receptor complex. Both groups observed reduced response to benzodiazepine receptor agonists, and increased responses to benzodiazepine receptor partial inverse agonists and inverse agonists. Further, flumazenil and ZK 93426 behaved like partial inverse agonists in animals chronically treated with the benzodiazepines. The change in the effects of these drugs was not simply owing to their precipitating a withdrawal syndrome, since they could be observed when no benzodiazepine was present in the brain (Little, 1988). It was suggested that chronic treatment with benzodiazepines caused a "withdrawal shift" in the coupling at the chloride ionophore, moving all ligands towards the inverse agonist part of the spectrum shown in Fig. 1 (that is, the effects of inverse agonists are increased, the effects of agonists are decreased, and antagonists behave like partial inverse agonists). Increased sensitivity to a benzodiazepine receptor partial inverse agonist can be observed even after a single dose of a benzodiazepine (Lister and Nutt, 1986).

3.2.4. Psychological Accounts of Benzodiazepine Tolerance

A number of psychological accounts of drug tolerance have been proposed, based on classical conditioning, instrumental conditioning, and habituation (Siegel, 1976,1983; Baker and Tiffany, 1985; File, 1985; Goudie, 1989). It is difficult to provide a detailed analysis of benzodiazepine tolerance using any of these frameworks, because there have been few relevant studies, and the findings from these studies have been somewhat contradictory. The Pavlovian model has received the most attention. In this account, cues that predict the presence of a drug become associated with the development of a conditioned compensatory response that opposes the acute drug effect. As a result, if a drug is administered in the presence of drug-predictive cues (e.g., in an environment associated with drug experience) the drug response is reduced, and the subject appears to be tolerant to the effect of the drug. However, if the same dose of drug is given in the absence of relevant cues, tolerance will not be observed.

Greeley and Cappell (1985) reported some evidence of Pavlovian control of tolerance to the sedative and hypothermic effects of chlordiazepoxide. Chlordiazepoxide was paired with one environment and saline administration with another. Finally, tolerance to the effects of chlordiazepoxide was tested in the environment associated with drug, the environment associated with saline, and a novel environment. No associative control of tolerance was observed on a sleep-time measure, i.e., animals showed similar tolerance to chlordiazepoxide in all three environments. However, on an activity measure, tolerance was observed in the environments paired with drug or saline, but not in the novel environment. There was some slight evidence that tolerance was greatest in the environment paired with the drug. A similar pattern of results was obtained in experiments examining tolerance to the hypothermic effect of chlordiazepoxide. In these studies, a compensatory hyperthermia was observed when rats were given saline in the environment paired with chlordiazepoxide. In an extensive series of studies, however, Griffiths and Goudie (1986) failed to find evidence of context specificity of tolerance to the hypothermic effect of midazolam.

More recently, King et al. (1987) reported associative control of tolerance to the sedative effects of midazolam. In their experiments, tolerance to the sedative effect of midazolam was specific to the environment in which the drug had been given. Further, they found that tolerance was still present in the drug-associated context after 14 d without drug, but that tolerance could be extinguished by repeatedly exposing animals to the midazolam-associated environment in the absence of the drug. When challenged with saline, drug-tolerant animals tested immediately after injection showed a hyperactive response.

In summary, conditioning may play a role in some aspects of benzodiazepine tolerance. However, it seems unlikely to be able to account for tolerance to benzodiazepine effects that occur in circumstances where drug administration cues are minimized, such as following continuous release of benzodiazepines from osmotic minipumps or silastic tubing implants.

3.3. The Effects of Flumazenil on Benzodiazepine Tolerance

There have been a number of reports that a single dose of the benzodiazepine antagonist flumazenil can reverse the development of tolerance to benzodiazepine effects in rodents. Gonsalves and Gallager (1988) treated rats for 21 d with diazepam using silastic tubing implants as described earlier. Animals treated this way are tolerant to the anticonvulsant action of diazepam against bicuculline-induced seizures, and are subsensitive to the effects of GABA on dorsal raphe neurons. A single dose of flumazenil given either 24 h or 7 d before testing (i.e., 14 or 20 d after implantation) was able to restore both GABAergic sensitivity of dorsal raphe neurons and the anticonvulsant action of diazepam. Sannerud et al. (1989) found that an acute injection of midazolam caused a slight attenuation of the degree of tolerance to midazolam in baboons. Finally, Nutt and Costello (1988) reported that flumazenil also reversed the increased sensitivity to benzodiazepine receptor inverse agonists seen after chronic benzodiazepine treatment.

The exact mechanism that might account for this "resetting" of benzodiazepine receptor function by a single dose of flumazenil remains unclear. However, the possibility of reversing benzodiazepine tolerance, and as will be discussed later, withdrawal phenomena, is potentially of considerable clinical importance.

4. Benzodiazepine Withdrawal

Two basic methods have been used to examine benzodiazepine withdrawal. The first examines the behavioral changes that occur following the termination of benzodiazepine treatment, as brain and plasma concentrations of the drug decrease to zero. Such studies assess spontaneous benzodiazepine withdrawal.

The second uses a pharmacological antagonist, such as flumazenil, to displace the benzodiazepine from its receptors in the brain. This approach involves a precipitation of withdrawal. The two methods will be considered in turn. In each case, an overview of the data suggests that the higher the dose of drug used and the longer the duration of treatment, the more marked the withdrawal syndrome (Boisse et al., 1982; Woods et al., 1987). It should also be noted that different mechanisms may account for different aspects of the withdrawal syndrome. For example, increased anxiety associated with benzodiazepine withdrawal may have a different basis and respond differently to pharmacological treatment than tremor. Such dissociations have been observed in ethanol withdrawal (e.g., File et al., 1990), but to date, relatively little attention has been focused on different components of benzodiazepine withdrawal.

4.1. Spontaneous Benzodiazepine Withdrawal

Many behavioral changes have been observed in rodents undergoing spontaneous benzodiazepine withdrawal. These range from spontaneous seizures, increased sensitivity to audiogenic seizures, weight loss, tremor, and muscle rigidity to behavior consistent with increased anxiety in various animal models (Rastogi et al., 1976; Yoshimura and Yamamoto, 1979; Ryan and Boisse, 1983; Emmett-Oglesby et al., 1983; Baldwin and File, 1988; Boisse et al., 1986; Martin and McNicholas, 1982.

Marietta et al. (1990) examined cerebral glucose utilization in rats undergoing spontaneous withdrawal from diazepam. They observed decreased glucose utilization in some brain areas associated with motor system function (e.g., frontal sensorimotor cortex, globus pallidus), and in olfactory cortex and nucleus accumbens. Increased glucose utilization was seen in reticulata and compacta of substantia nigra, visual cortex, mamillary body, and dorsal hippocampus.

4.2. Benzodiazepine Withdrawal
Precipitated by Flumazenil

Following the discovery of benzodiazepine receptor antagonists, a number of laboratories successfully used these drugs to precipitate benzodiazepine withdrawal, in a way analogous to the precipitation of opiate withdrawal by naloxone. Withdrawal symptoms have been observed in rats, mice, cats, and primates (e.g., Cumin et al., 1982; Lukas and Griffiths, 1982; McNicholas and Martin, 1982; Rosenberg and Chiu, 1982; Patel et al., 1988). The reported symptoms included hypermotility, weight loss, increased muscle tone, hyperreactivity, tremor, hypersalivation, and seizures. A pentylenetetrazole-like stimulus can also be produced by administering a benzodiazepine antagonist to animals treated chronically with benzodiazepines (Harris et al., 1988; Idemudia and Lal, 1989).

It may be noted that the behaviors observed in the above studies of precipitated benzodiazepine withdrawal resemble the effects of benzodiazepine receptor inverse agonists. Indeed, it might be argued that these behaviors are not a result of precipitated withdrawal, but reflect the shift in the effects of benzodiazepine receptor ligands discussed in Section 3.2.3. on benzodiazepine tolerance (that is, flumazenil no longer behaves like an antagonist in animals tolerant to the effects of benzodiazepines, but behaves like an inverse agonist).

Ableitner et al. (1985) examined cerebral glucose utilization during benzodiazepine withdrawal precipitated by flumazenil. Their results differed somewhat from those of Marietta et al. (1990) reported above in that they found no decrease in glucose utilization in any of the areas studied. Increased glucose utilization was observed in a number of areas consistent with those reported by Marietta et al. (1990).

4.3. Attenuation of Benzodiazepine Withdrawal by Flumazenil

Although the above studies indicate that the administration of a benzodiazepine antagonist can precipitate a withdrawal syndrome, other investigations show that, in some cases, flumazenil may reduce benzodiazepine withdrawal (*see* File and Hitchcott, 1990). Lamb and Griffiths (1985) noted that flumazenil precipitated withdrawal signs in baboons chronically treated with benzodiazepines. However, when it was administered at 1- or 3-d intervals, withdrawal signs were attenuated. It may be noted that, in this study, in contrast to some of the studies discussed above in rodents, tolerance to the sedative effect of the benzodiazepines did not appear to be attenuated by intermittent flumazenil administration. Gallager et al. (1986) also found that periodic administration of flumazenil reduced the incidence of withdrawal symptoms in rhesus monkeys chronically treated with diazepam.

Baldwin and File (1989) administered chlordiazepoxide (10 mg/kg) ip to rats once daily for 27 d. Twenty-four hours after the last injection, animals showed an anxiogenic response in the social interaction test relative to animals chronically treated with the vehicle. This anxiogenic response was reversed by flumazenil given 20 min before the test. A similar pattern of results was observed in animals tested on a plus-maze (Baldwin and File, 1988). Moreover, a single dose of flumazenil given 6 d before the social interaction test also prevented the increase in anxiety associated with chlordiazepoxide withdrawal. The ability of flumazenil to reverse benzodiazepine withdrawal when given shortly before testing was suggested to be the result of an antagonism of an endogenous ligand for benzodiazepine receptors with inverse agonist effects. An increased turnover of such a ligand has been reported to accompany chronic benzodiazepine administration (Miyata et al., 1987). The reduction in benzodiazepine withdrawal caused by the administration of flumazenil at the earlier time-point may involve a mechanism similar to that underlying the reversal of benzodiazepine tolerance discussed in Section 3.3.

In summary, the effects of flumazenil in benzodiazepine withdrawal seem to depend on the time of its administration relative to behavioral testing, the duration of benzodiazepine treatment, whether or not a benzodiazepine is present at the time of administration, and, perhaps, the basal tone of the chloride ionophore (File and Hitchcott, 1990).

5. Conclusions

Many changes accompany the chronic administration of benzodiazepines, including the development of tolerance to the drugs' behavioral effects and an abstinence syndrome that follows the termination of drug treatment. Tolerance and dependence are both important clinically and can now be reliably produced in laboratory animals. This has allowed an examination of the mechanisms underlying benzodiazepine tolerance and dependence. Most studies have focused on alterations in the benzodiazepine/GABA chloride ionophore complex. There is some evidence that benzodiazepine receptor downregulation occurs following chronic benzodiazepine administration, although there are a number of negative reports. Subsensitivity to GABA seems to play a role in some aspects of benzodiazepine tolerance, and regional differences in the rate at which this occurs may account for the different rates at which tolerance develops to benzodiazepines' various behavioral effects. Chronic benzodiazepine treatment seems to modify the actions of all benzodiazepine receptor ligands, shifting their effects toward the inverse agonist end of the spectrum shown in Fig. 1. Conditioning mechanisms may play a role in benzodiazepine tolerance in some circumstances. However, no attempt has yet been made to account for such mechanisms at a neurobiological level.

Animal studies have also suggested some pharmacological interventions that may prove to be useful clinically for reversing tolerance and minimizing benzodiazepine withdrawal. In particular, a number of studies suggest that the intermittent use of flumazenil during a benzodiazepine therapy may have clinical value.

Finally, it may be noted that chronic treatment with benzodiazepines is just one way of reducing behavioral responses to these drugs. Other methods include the use of pharmacological antagonists (some of which were discussed above), manipulations of the testing environment, and varying the genetic background of the test subjects (Gallager et al., 1987). Comparisons between these different methods of antagonizing drug effects may cast further light on the mechanisms of benzodiazepine tolerance (*see* Lister, 1989).

References

Ableitner A., Wuster M., and Herz A. (1985) Specific changes in local cerebral glucose utilization in the rat brain induced by acute and chronic diazepam. *Brain Res.* **359,** 49–56.

Arendt R. M., Greenblatt D. J., DeJong R. H., Bonin J. D., Abernethy D. R., Ehrenberg B. L., Giles H. G., Sellers E. M., and Shader R. I. (1983) In vitro correlates of benzodiazepine cerebrospinal fluid upake, pharmacodynamic action and peripheral distribution. *J. Pharmacol. Exp. Ther.* **227,** 98–106.

Baker T. B., and Tiffany S. T. (1985) Morphine tolerance as habituation. *Psychol. Rev.* **92,** 78–108.

Baldwin H. A. and File S. E. (1988) Reversal of increased anxiety during benzodiazepine withdrawal: evidence for an anxiogenic endogenous ligand for the benzodiazepine receptor. *Brain Res. Bull.* **20,** 603–606.

Baldwin H. A. and File S. E. (1989) Flumazenil prevents the development of chlordiazepoxide withdrawal in rats tested in the social interaction test of anxiety. *Psychopharmacology* **97,** 424–426.

Boisse N. R., Ryan G. P., and Guarino J. J. (1982) Experimental induction of benzodiazepine physical dependence in rodents. *Natl. Insr. Drug Abuse Monogr. Ser.* **41,** 191–199.

Boisse N. R., Periana R. M., Guarino J. J., Kruger H. S., and Samoriski G. M. (1986) Pharmacological characterization of acute chlordiazepoxide dependence in the rat. *J. Pharmacol. Exp. Ther.* **239,** 775–783.

Braestrup C. and Nielsen M. (1983) Benzodiazepine receptors, in *Handbook of Psychopharmacology,* vol. 17 (Iversen L. L., Iversen S. D., and Snyder S. H., eds.), Plenum, New York, pp. 285–384.

Corda M. G., Giorgi 0., Longoni B., Fernandez A., and Biggio G. (1988) Increased sensitivity to inverse agonists and decreased GABA-stimulated chloride influx induced by chronic treatment with FG 7142. *Adv. Biochem. Psychopharmacol.* **45,** 293–306.

Crawley J. N., Marangos P. J., Stivers J., and Goodwin F. K. (1982) Chronic clonazepam administration induces benzodiazepine receptor subsensitivity. *Neuropharmacology* **21,** 85–89.

Cumin R., Bonetti E. P., Scherschlicht R., and Haefely W. E. (1982) Use of the specific benzodiazepine receptor antagonist, Ro 15-1788, in studies of physiological dependence on benzodiazepines. *Experientia* **38,** 833–834.

Davis M. and Gallager D. W. (1988) Continuous slow release of low levels of diazepam produces tolerance to its depressant and anxiolytic effects on the startle reflex. *Eur. J. Pharmacol.* **150,** 23–33.

Emmett-Oglesby M., Spencer J. R., Lewis M., Elmesallamy F., and Lal H. (1983) Anxiogenic aspects of diazepam withdrawal can be detected in animals. *Eur. J. Pharmacol.* **92,** 127–130.

File S. E. (1981) Rapid development of tolerance to the sedative effects of lorazepam and triazolam in rats. *Psychopharmacology* **73,** 240–245.

File S. E. (1983) Tolerance to the anti-pentylenetetrazole effects of diazepam in the mouse. *Psychopharmacology* **79,** 284–286.

File S. E. (1985) Tolerance to the behavioral actions of benzodiazepines. *Neurosci. Biobehav. Rev.* **9,** 113–121.

File S. E. (1990) The history of benzodiazepine dependence: a review of animal studies. *Neurosci. Biobehav. Rev.* **14**, 135–146.

File S. E. and Hitchcott P. K. (1990) A theory of benzodiazepine dependence that can explain whether flumazenil will enhance or reverse the phenomena. *Psychopharmacology* **101**, 525–532.

File S. E., Zharkovsky A., and Hitchcott P. R. (1990) Drug treatment of anxiety in alcohol withdrawal. *Clin. Neuropharmacol.* **13**, (suppl. 2). 510–511.

Gallager D. W. and Wilson M. A. (1988) Chronic benzodiazepine agonist exposure: comparison of electrophysiological changes by brain region and benzodiazepine receptor ligand. *Adv. Biochem. Psychopharmacol.* **45**, 325–336.

Gallager D. W., Heninger K., and Heninger G. (1986) Periodic benzodiazepine antagonist administration prevents benzodiazepine withdrawal symptoms in primates. *Eur. J. Pharmacol.* **132**, 31–38.

Gallager D. W., Lakoski J. M., Gonsalves S. F., and Rauch S. L. (1984) Chronic benzodiazepine treatment decreases postsynaptic GABA sensitivity. *Nature* **308**, 74–77.

Gallager D.W., Malcolm A. B., Anderson S. A., and Gonsalves S. F. (1985) Continuous release of diazepam: electrophysiological, biochemical and behavioral consequences. *Brain Res.* **342**, 26–36.

Gallaher E. J., Henauer S. A., Jacques C. J., and Hollister L. E. (1986) Benzodiazepine dependence in mice after ingestion of drug-containing food pellets. *J. Pharmacol. Exp. Ther.* **237**, 462–467 .

Gallaher E. J., Hollister L. E., Gionet S. E., and Crabbe J. C. (1987) Mouse lines selected for genetic differences in diazepam sensitivity. *Psychopharmacology* **93**, 25–30.

Garattini S., Mussini E., Marcucci F., and Guaitani A. (1973) Metabolic studies on benzodiazepines in various animal species. *The Benzodiazepines* (Garattini S., Mussini E., and Randall L. O., eds.), Raven, New York, pp. 75–97.

Goldberg M. E., Marian A. A., and Efron D. H. (1967) A comparative study of certain pharmacological responses following acute and chronic administration of chlordiazepoxide. *Life Sci.* **6**, 481–491.

Gonsalves S. F. and Gallager D. W. (1988) Persistent reversal of tolerance to anticonvulsant effects and GABAergic subsensitivity by a single exposure to benzodiazepine antagonist during chronic benzodiazepine administration. *J. Pharmacol. Exp. Ther.* **244**, 79–83.

Goudie A. J. (1989) Behavioral techniques for assessing drug tolerance and sensitization, in *Neuromethods*, vol. 13 (Boulton A. A., Baker G. B., and Greenshaw A. J., eds.), Humana Press, Clifton, NJ, pp. 565–622.

Goudie A. J. and Emmett-Oglesby M. W. (1989) *Psychoactive Drugs: Tolerance and Sensitization.* Humana Press, Clifton, NJ.

Greeley J. and Cappell H. (1985) Associative control of tolerance to the sedative and hypothermic effects of chlordiazepoxide. *Psychopharmacology* **86**, 487–493.

Greenblatt D. J., Shader R. I., Abernethy D. R., Ochs H. R., Divoll M., and Sellers E. M. (1982) Benzodiazepines and the challenge of pharmacokinetic taxonomy, in *Pharmacology of Benzodiazepines* (Usdin E., Skolnick P., Tallman J. F., Greenblatt D. J., and Paul S. M., eds.), MacMillan, London, pp. 257–269.

Griffiths J. W. and Goudie A. J. (1986) Analysis of the role of drug-predictive environmental stimuli in tolerance to the hypothermic effects of the benzodiazepine midazolam. *Psychopharmacology* **90,** 513–521.

Haigh J. R. M. and Feely M. (1988a) Ro 16-6028, a benzodiazepine receptor partial agonist, does not exhibit anticonvulsant tolerance in mice. *Eur. J. Pharmacol.* **147,** 283–285.

Haigh J. R. M. and Feely M. (1988b) Tolerance to the anticonvulsant effect of benzodiazepines. *Trends in Pharmacol. Sci.* **9,** 361–366.

Haigh J. R. M., Feely M., and Gent J. P. (1986) Tolerance to the anticonvulsant effect of clonazepam. *J. Pharm. Pharmac.* **38,** 931–934.

Harris C. M., Idemudia S. O., Benjamin D., Bhadra S., and Lal H. (1988) Withdrawal from ingested diazepam produces a pentylenetetrazole-like stimulus in rats. *Drug Dev. Res.* **12,** 71–76.

Hernandez T. D., Heninger C., Wilson M. A., and Gallager D. W. (1989) Relationship of agonist efficacy to changes in GABA sensitivity and anticonvulsant tolerance following chronic benzodiazepine ligand exposure. *Eur. J. Pharmacol.* **170,** 145–155.

Idemudia S. O. and Lal H. (1989) Pentylenetetrazole-like stimulus is produced in rats during withdrawal from ingested chlordiazepoxide. *Drug Dev. Res.* **16,** 23–29.

King D. A., Bouton M. E., and Musty R. E. (1987) Associative control of tolerance to the sedative effects of a short-acting benzodiazepine. *Behav. Neurosci.* **101,** 104–114.

Kitagawa H., Esumi Y., Kurosawa S., Sekine S., and Yokoshima T. (1979) Metabolism of 8-chloro-6-(o-chlorophenyl)-1-methyl-4H-s-triazolo[4,3-a][1,4]benzodiazepine, triazolam, a new central depressant. II. Identification and determination of metabolites in rats and dogs. *Xenobiotica* **9,** 429–439.

Lamb R. J. and Griffiths R. R. (1985) Effects of repeated Ro 15-1788 administration in benzodiazepine dependent baboons. *Eur. J. Pharmacol.* **110,** 257–261.

Lister R. G. (1989) Antagonizing the behavioral effects of drugs: a discussion with specific reference to benzodiazepines and alcohol. *J. Psychopharmacol.* **3,** 21–28.

Lister R. G. and Nutt D. J. (1986) Mice and rats are sensitized to the proconvulsant actions of a benzodiazepine receptor inverse agonist (FG 7142) following a single dose of lorazepam. *Brain Res.* **379,** 364–366.

Lister R. G., File S. E., and Greenblatt D. J. (1983a) Functional tolerance to lorazepam in the rat. *Psychopharmacolgy* **81,** 292–294.

Lister R. G., Abernethy D. R., Greenblatt D. J., and File S. E. (1983b) Methods for the determination of lorazepam and chlordiazepoxide and metabolites in brain tissue: a comparison with plasma concentrations in the rat. *J. Chromatog. Biomed. Applic.* **277**, 201–208.

Little H. J. (1988) Chronic benzodiazepine treatment increases the effects of inverse agonists. *Adv. Biochem. Psychopharmacol.* **45**, 307–323.

Little H. J., Nutt D. J., and Taylor S. C. (1984) Acute and chronic effects of the benzodiazepine receptor ligand FG 7142: proconvulsant proerties and kindling. *Br. J. Pharmacol.* **83**, 951–958.

Little H.J., Nutt D. J., and Taylor S. C. (1987) Bidirectional effects of chronic treatment with agonists and inverse agonists at the benzodiazepine receptor. *Brain Res. Bull.* **19**, 371–378.

Lukas S. E. and Griffiths R. R. (1982) Precipitated withdrawal by a benzodiazepine receptor antagonist (Ro 15-1788) after 7 days of diazepam. *Science* **217**, 1161–1163.

Marietta C. A., Eckardt M. J., Zbicz K. L., and Weight F. F. (1990) Cerebral glucose utilization during diazepam withdrawal in rats. *Brain Res.* **511**, 192–196.

Marley R. J. and Gallager D. W. (1989) Chronic diazepam treatment produces regionally specific changes in GABA-stimulated chloride influx. *Eur. J. Pharmacol.* **159**, 217–223.

Marley R. J., Heninger C., Hernandez T., and Gallager D. W. (1991) Chronic administration of β-carboline-3-carboxylic acid methylamide by continuous intraventricular infusion increases GABAergic function. *Neuropharmacology* **30**, 245–251.

Martin W. R. and McNicholas L. F. (1982) Diazepam and pentobarbital dependence in the rat. *Life Sci.* **31**, 721–730.

McNicholas L. F. and Martin W. R. (1982) The effect of a benzodiazepine antagonist, Ro 15-1788, in diazepam dependent rats. *Life Sci.* **31**, 731–737.

Miller L. G., Greenblatt D. J., Barnhill J. G., and Shader R. I. (1988) Chronic benzodiazepine administration. I. Tolerance is associated with benzodiazepine receptor downregulation and decreased gamma-aminobutyric acid, receptor function. *J. Pharmacol. Exp. Ther.* **246**, 170–176.

Miyata M., Mochetti I., Ferrarese C., Guidotti A., and Costa E. (1987) Protracted treatment with diazepam increases the turnover of putative endogenous ligands for the benzodiazepine/B-carboline recognition site. *Proc. Natl. Acad. Sci. USA* **84**, 1444–1448.

Mohler H., Okada T., and Enna S. J. (1978) Benzodiazepine and neurotransmitter receptor binding in rat brain after chronic administration of diazepam and phenobarbital. *Brain Res.* **156**, 391–395.

Moreau J-L., Jenck F., Pieri L., Schoch P., Martin J. R., and Haefely W. E. (1990) Physical dependence induced in DBA/2J mice by benzodiazepine receptor full agonists, but not by the partial agonist Ro 16–6028. *Eur. J. Pharmacol.* **190**, 269–273.

Nutt D. J. and Costello M. J. (1988) Rapid induction of lorazepam dependence and reversal with flumazenil. *Life Sci.* **43**, 1045–1053.

Patel J. B., Rinarelli C. A., and Malick J. B. (1988) A simple and rapid method of inducing physical dependence with benzodiazepines in mice. *Pharmacol. Biochem. Behav.* **29,** 753–754.

Petersen E. N. and Jensen L. H. (1987a) Chronic treatment with lorazepam and FG 7142 may change the effects of benzodiazepine receptor agonists, antagonists and inverse agonists by different mechanisms. *Eur. J. Pharmacol.* **133,** 309–317.

Petersen E. N. and Jensen L. H. (1987b) Lorazepam and FG 7142 induce tolerance to the DMCM antagonistic effect of benzodiazepine receptor ligands. *Brain Res. Bull.* **19,** 387–391.

Rastogi R. B., LaPierre Y. D., and Singhal R. L. (1976) Evidence for the role of brain norepinephrine and dopamine in "rebound" phenomena seen during withdrawal after repeated exposure to benzodiazepines. *J. Psychiat. Res.* **13,** 65–75.

Rosenberg H. C. and Chiu T. H. (1981) Tolerance during chronic benzodiazepine treatment associated with decreased receptor binding. *Eur. J. Pharmacol.* **70,** 453–460.

Rosenberg H. C. and Chiu T. H. (1982) An antagonist-induced benzodiazepine abstinence syndrome. *Eur. J. Pharmacol.* **81,** 153–157.

Rosenberg H. C., Tietz E. I., and Chiu T. H. (1985) Tolerance to the anticonvulsant action of benzodiazepines. Relationghip to decreased receptor density. *Neuropharmacology* **24,** 639–644.

Rosenberg H. C., Duggan J. M., Tietz E. I., and Chiu T. H. (1988) Nonuniformity of tolerance to the actions of benzodiazepine agonists. *Adv. Biochem. Psychopharmacol.* **45,** 355–366.

Ryan G. P. and Boisse N. R. (1983) Experimental induction of benzodiazepine tolerance and dependence. *J. Pharmacol. Exp. Ther.* **226,** 100–107.

Sannerud C. A., Cook J. M., and Griffiths R. R. (1989) Behavioral differentiation of benzodiazepine ligands after repeated administration in baboons. *Eur. J. Pharmacol.* **167,** 333–343.

Sansone M. (1979) Effects of repeated administration of chlordiazepoxide on spontaneous locomotor activity in mice. Psychopharmacology **66,** 109–110.

Schillings R. T., Sisenwine S. F., and Ruelius H. W. (1977) Disposition and metabolism of lorazepam in the male rat. *Drug Metabol. Dispos.* **5,** 425–435.

Siegel S. (1976) Morphine analgesic tolerance: its situation specificity supports a Pavlovian conditioning model. *Science* **193,** 323–325.

Siegel S. (1983) Classical conditioning, drug tolerance, and drug dependence, in *Research Advances in Alcohol and Drug Problems,* vol. 7 (Smart R. G., Glaser F. B., Israel Y., Kalant H., Popham R. E., and Schmidt W., eds.), Plenum, New York, pp. 207–246.

Stephens D. N. and Schneider H. H. (1985) Tolerance to the benzodiazepine diazepam in an animal model of anxiolytic activity. *Psychopharmacology* **87,** 322–327.

Stephens D. N., Schneider H. H., Weidmann R., and Zimmermann L. (1988) Decreased sensitivity to benzodiazepine receptor agonists and increased sensitivity to inverse agonists following chronic treatments: evidence for separate mechanisms. *Adv. Biochem. Psychopharmacol.* **45,** 337–354.

Theeuwes F. and Yum S. I. (1976) Principles of the design and operation of general osmotic pumps for the delivery of semisolid or liquid formulations. *Ann. Biomed. Eng.* **4,** 343–353.

Tietz E. I. and Rosenberg H. C. (1988) Behavioral measurement of benzodiazepine tolerance and GABAergic subsensitivity in the substantia nigra pars reticulata. *Brain Res.* **438,** 41–51.

Tietz E. I., Rosenberg H. C., and Chiu T. H. (1986) Autoradiographic localization of benzodiazepine receptor downregulation. *J. Pharmacol. Exp. Ther.* **236,** 284–292.

Vellucci S. V. and File S. E. (1979) Chlordiazepoxide loses its anxiolytic action with long-term treatment. *Psychopharmacology* **62,** 61–65.

Wilson M. A and Gallager D. W. (1989) Responses of substantia nigra pars reticulata neurons to benzodiazepine ligands following acute and prolonged diazepam exposure: I. Modulation of GABA sensitivity. *J. Pharmacol. Exp. Ther.* **248,** 879–885.

Woo G. K., Williams T. H., Kolis S. J., Warinsky D., Sasso G. J., and Schwartz M. A. (1981) Biotransformation of [^{14}c]midazolam in the rat *in vitro* and *in vivo. Xenobiotica* **11,** 373–384.

Woods J. H., Katz J. L., and Winger G. (1987) Abuse liability of benzodiazepines. *Pharmacol. Rev.* **39,** 251–413.

Yanaura S., Tagashira E., and Suzuki T. (1975) Physical dependence on morphine, pentobarbital, and diazepam in rats by drug-admixed food ingestion. *Jpn. J. Pharmacol.* **25,** 453–463.

Yoshimura K. and Yamamoto K. I. (1979) Neuropharmacological studies on drug dependence. I. Effects due to the difference in strain, sex, and drug administration time on physical dependence development and characterization of withdrawal signs in CNS-affecting drug dependent rats. *Folia Pharmacol. Jpn.* **75,** 805–828.

Self-Administration of Psychomotor Stimulants Using Progressive Ratio Schedules of Reinforcement

David C. S. Roberts and Nicole R. Richardson

1. Introduction

With the pioneering work of Weeks (1961,1962) and others (Thompson and Schuster, 1964; Collins and Weeks, 1965; Deneau et al., 1969; Yanagita and Takahashi, 1973), which showed that laboratory animals would voluntarily ingest psychoactive drugs, an enormous field of research has developed that uses "self-administration" techniques to study drug reinforcement. This chapter deals with some of the methods employed in self-administration studies involving rats, with special emphasis placed on the schedules of reinforcement that govern drug delivery.

Drug self-administration techniques have been developed for a variety of routes of administration, including iv (Weeks, 1962), intragastric (Gotestam, 1973), sc (Mucha, 1980), inhalation (Wood et al., 1977; Hatsukami et al., 1990), and intracranial (Goeders and Smith, 1987). Only the iv route will be dealt with here.

2. Schedules of Reinforcement

2.1. Fixed Ratio Schedules

A wide variety of schedules of reinforcement have been used in self-administration studies using primates as subjects. By contrast, studies using rats have typically been limited to simple schedules of reinforcement, most commonly a fixed ratio (FR).

From: *Neuromethods, Vol. 24: Animal Models of Drug Addiction*
Eds: A. Boulton, G. Baker, and P. H. Wu ©1992 The Humana Press Inc.

An FR schedule requires that an animal complete a preset number of lever responses in order to receive a drug infusion. Although FR schedules have the advantage of being easily implemented and readily learned by rodents, there is one major difficulty associated with their use. Rate of self-administration is the only dependent variable that can be derived from simple schedules, and it is now becoming clear that rate is an inadequate measure for studying changes in drug reinforcement.

Controversy exists with respect to what controls the rate of drug intake. Rate of responding is extremely sensitive to changes in the unit injection dose (Pickens and Thompson, 1968). Past some minimal level, small doses of cocaine or amphetamine will produce high infusion rates. Postreinforcement pauses, which occur immediately after each infusion, are also directly related to dose. As the unit dose is increased, pauses becomes longer with the result that fewer infusions are taken through the session. Some have suggested that postreinforcement pauses are the result of a general disruption of behavior that renders the animal incapable of responding for a short time following drug delivery (Wilson et al., 1971). Hence, rate of drug intake may, in fact, be a function of this incapacity, rather than a change in the reward value of the drug being self-administered. It has been reported, however, that rats are capable of pressing a lever at very high rates for intracranial stimulation immediately after self-injecting either amphetamine (Wise et al., 1977) or heroin (Gerber et al., 1985). This would seem to indicate that rate of self-injection is a matter of choice, rather than incapacitation.

2.2. Difficulty in Interpreting Changes in Rate

What does a change in rate of drug intake mean? The self-administration of psychomotor stimulants has been shown to be affected in a predictable manner by dopamine (DA) receptor antagonist drugs. Low doses of the DA antagonists haloperidol (Gill et al., 1975; De La Garza and Johanson, 1982), perphenazine (Johanson et al., 1976), pimozide (Yokel and Wise, 1975,1976; Risner and Jones, 1976; DeWit and Wise, 1977), butaclamol (Yokel and Wise, 1976), chlorpromazine (Wilson and Schuster, 1972; Risner and Jones, 1976), α-flupenthixol, sulpiride, metoclopramide, thioridazine (Roberts and Vickers, 1984), spiperone, and SCH 23390 (Koob et al., 1987) all produce an increased intake of psychomotor stimulants (*see also* Woolverton, 1986). The most compelling explanation for this phenomenon has been offered by Yokel and Wise (1975,1976), who have drawn an analogy between the increase in drug intake following DA receptor blockade and the increase in injection rate seen in animals after a reduction in the unit dose. Recalling that rate of cocaine self-administration is dependent on injection dose (Pickens and Thompson, 1968) and that a reduction in injection dose produces a compensatory increase in injection rate, Yokel and Wise suggest that DA antagonist-induced increases in self-administration also represent a compensatory response. They argue that partial antagonism of DA receptors produces a partial block of the stimulant's reinforcing effect and that the increase in drug intake represents an attempt by the animal to overcome this blockade. By this logic, if DA receptor blockade is the same as a reduction in dose, then it follows that DA receptor stimulation is critical for cocaine reward.

The analogy goes further. Lowering the unit injection dose of cocaine will increase responding only to a point, past which the self-administration behavior will extinguish. This is normally characterized by a burst of responding followed by response cessation. A similar phenomenon occurs with DA antagonists, such that pretreatment of animals self-administering cocaine with DA receptor blockers causes a dose-dependent increase in stimulant intake until, at some point, the behavior degenerates into a pattern reminiscent of behavioral "extinction" (Yokel and Wise, 1975,1976; DeWit and Wise, 1977). Thus, there are striking parallels between DA blockade and reductions in unit drug dose. These data bolster the hypothesis that DA receptor stimulation is the essential mechanism of action for cocaine reward.

A problem of interpretation develops, however, when lesion data are considered. Lesions to dopaminergic systems will often reduce cocaine or amphetamine intake (Roberts et al., 1977; Lyness et al., 1979). If one accepts the idea that increases in rate reflect a decrease in the reinforcing strength of the drug, then lesion-induced reductions in drug intake should be interpreted as reflecting an *increase* in the drug's reinforcing effects. On the contrary, lesion-induced decreases in drug intake have typically been interpreted as reflecting a *decrease* in drug reinforcement.

It is apparent that the lesion and pharmacological literature have interpreted changes in rate of self-administration differently. A decrease in the reinforcing value of a drug is inferred from *increases* in drug intake following DA receptor blockade, whereas the same conclusion is reached following DA lesion-induced *decreases* in drug intake. Although both observations are robust and interesting, it is clearly an unacceptable situation when opposite results can be used to support the same conclusion. We have offered *post hoc* explanations that address these inconsistencies (*see* Roberts and Zito, 1987), but it remains clear that a change in rate of drug intake has limited value as an indicator of change in level of drug reinforcement. Most pharmacologists have long recognized the difficulties with interpreting changes in rate of self-administration, yet more complex schedules have seldom been employed in self-administration studies using rats.

2.3. Progressive-Ratio Schedules

Of the few studies using more complex methodologies, a number have employed a progressive-ratio (PR) schedule, although the actual implementation has been somewhat varied. Hodos (1961) originally suggested that a PR schedule could be employed to assess the relative strength of food reward. This method has been applied more recently to the study of drug reinforcement in rhesus monkeys (Hoffmeister, 1979), baboons (Griffith et al., 1975,1978,1979), and dogs (Risner and Silcox, 1981; Risner and Goldberg, 1983; Risner and Cone, 1986). In these studies, a fixed ratio is imposed during the daily session, and if an animal performs to some specified criterion, the ratio is increased on the following day. Bedford et al. (1978), in accordance with Hodos (1961), use a PR schedule in which the ratios increase following each infusion. In their tests with rhesus monkeys, if a breaking point is not reached within a session on one day, then the ratio series continues on the following day where the animal left off.

The PR schedule that we employ with rats is similar to that used by Bedford et al. (1978) in that the response requirements to earn a drug injection escalate with each reinforcement, with the added requirement that a breaking point is established each day. Specifically, the first injection of each test session requires only one response on the lever; however, the number of responses required to earn subsequent injections follows through an exponential series (*see* Section 2.5.) until the self-administration behavior extinguishes.

Figure 1 shows the pattern of responding of one animal on a PR schedule reinforced on different days by three different doses of cocaine. During the early part of each test session, infusions of cocaine are regularly spaced. Each infusion is followed by a remarkably consistent postreinforcement pause, after which the animal quickly responds until the requirements for the next infusion are met. At some point in the session, the response requirements exceed the reinforcing efficacy of the drug dose, and the lever pressing behavior extinguishes. The last ratio completed before extinction is defined as the "breaking point," which serves as the dependent measure. In general, higher doses of cocaine produce higher breaking points, although some animals might show an inverted U-shaped dose/response curve (*see also* Bedford et al., 1978).

2.4. Rate of Self-Administration vs Breaking Point

It has generally been assumed that, if the reinforcing value of a drug were to change, then rate of drug intake would be expected to change as well. Results from three recent experiments have established that this is simply not the case. The examination of self-administration behavior in female rats over their 4-d estrous cycle reveals a dissociation between rate and breaking point. We had reported earlier that rate of cocaine self-administration remained unaltered over the estrous cycle. In addition, it was neither affected by ovariectomy nor by estrogen replacement (Roberts et al., 1987). Reexamination of this effect with the PR schedule shows that breaking points are in fact sensitive to the hormonal manipulations earlier deemed inconsequential to rate of drug intake (Roberts et al., 1989a).

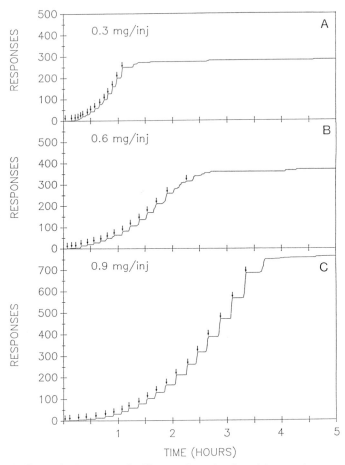

Fig. 1. Cumulative records illustrating the breaking points reached by the same animal self-administering cocaine at three different unit doses (0.3, 0.6, and 0.9 mg/kg/injection). The points represent the number of responses made by the animal throughout a 5-h session. The arrows indicate when injections were delivered. Note that, as the unit dose increases, the breaking point increases (from Roberts et al., 1989b).

A second example of the failure of rate to predict changes in reinforcement value can be found in our investigations of apomorphine self-administration. We previously showed that rats would not self-administer cocaine following bilateral 6-hydroxydopamine (6-OHDA) lesions of the nucleus accumbens, but would self-administer apomorphine at prelesion rates

(Roberts et al., 1977). This was surprising, since DA receptor supersensitivity should have developed in the accumbens, and therefore, some change in rate of apomorphine self-administration would be expected. Reexamination of this phenomenon with the PR schedule shows that the reinforcing effects of the drug are indeed augmented by such lesions. Although maintaining consistent rates of apomorphine self-administration, the lesioned rats responded to higher breaking points each successive day following 6-OHDA treatment as supersensitivity presumably developed (Roberts, 1989).

The effect of 5,7-dihydroxytryptamine (5,7-DHT) lesions on cocaine self-administration provides a third example of the discrepancy between rate and breaking point (Loh and Roberts, 1990). Lesions of the serotonin innervation of the forebrain in rats caused a dramatic increase in breaking points, despite the fact that no changes in the rate of cocaine self-administration were observed.

In summary, the PR schedule provides a sensitive measure for evaluating the effects of lesions and pharmacological pretreatments on drug reinforcement. The PR schedule has shown that, even though rate of drug intake may remain unchanged, there may indeed be changes in the motivation to self-administer a drug. Hence, no conclusion should be drawn from a failure to affect rate of drug intake.

2.5. The Progressive-Ratio Scale

The PR schedule, by definition, demands that the response requirements escalate during the experiment. The shape of the escalating scale, however, varies from lab to lab. Although some researchers use an arithmetic function (*see* Bedford et al., 1978; Koob et al., 1987), we prefer to use an exponential function. Examples of two exponential scales used in our lab are shown in Fig. 2. Specifically, the scale marked PR 1 was calculated using the spreadsheet equation:

$$\text{ROUND } [5^* \text{ EXP}(0.2^* \text{ infusion number}) - 5] \qquad (1)$$

which yields the values: (1, 2, 4, 6, 9, 12, 15, 20, 25, 32, 40, 50, 62, 77, 95, 118, 145, 178, 219, 268, 328, 402, 492, 603, 737, 901). The

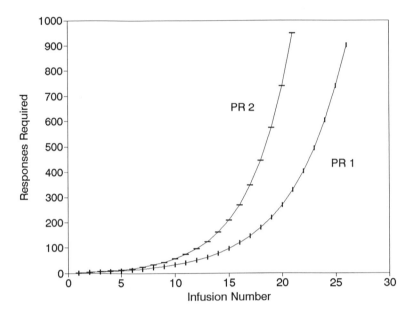

Fig. 2. Two different PR scales (PR 1 and PR 2). The points represent the number of responses an animal must make in order to receive a drug injection. Note that these two scales are derived from exponential functions (from Roberts, 1989; Roberts et al., 1989ab).

scale with the steeper slope, marked PR 2, is derived from the equation:

$$\text{ROUND} [5* \text{EXP}(0.25* \text{ infusion number}) - 5] \qquad (2)$$

We have found, in preliminary studies, that animals maintained on equivalent doses of cocaine will respond to approximately the same final ratios regardless of whether the PR 1 or PR 2 series of ratios is used (unpublished data).

Normally, animals respond to final ratios that range from approx 40 to 95 (or 11–15 infusions); however, neurotoxic lesions and hormonal fluctuations can cause animals to respond to much higher breaking points (i.e., 20–25 infusions) (Roberts, 1989; Roberts et al., 1989a,b).

All of the exponential scales used in our lab, including the two addressed here, were developed with a primary goal in mind—to allow self-administration behavior to extinguish in each animal each day. It was found, through pilot work, that these two particular series of response requirements (PR 1 and PR 2) satisfied this goal, since animals would self-administer only during the first few hours of the session, yet would fail to self-administer into the fifth and final hour. Consequently, all animals would display clear breaking points each session.

In our laboratory, breaking point is defined as the final ratio of responses successfully completed by a subject prior to a 1-h period of nonreward (Roberts, 1989; Roberts et al., 1989a,b). Bedford et al. (1978) have suggested the more dynamic criterion of three times an animal's longest baseline interresponse time (IRT).

Statistical analysis of final ratios is problematic. Since final ratios are derived from an escalating exponential function, they violate one of the assumptions of Analysis of Variance (ANOVA), namely, that there be homogeneity of variance within cells. Clearly, since final ratios are escalating in nature, values at the high end of the scale will be associated with much larger variances than will values at the low end. In order to avoid this problem, a log transformation of the final ratio data can be performed. In essence, such a transformation would simply yield values equivalent to the ordinal values of the final ratio (or number of infusions). It is appropriate therefore to skip all of the intermediate steps, and simply collect and analyze the number of infusions from the start.

2.6. What Do Breaking Points Measure?

Breaking point is a dependent measure used to quantify an animal's motivation to self-administer a drug. It is assumed that an animal's breaking point reflects the degree to which that animal will "work" to obtain an infusion of drug. Presumably, the higher the abuse liability of a drug, the higher the breaking points that the drug maintains (e.g., Griffiths et al., 1975).

It is important to note, however, that the manner in which the PR schedule is implemented can drastically alter the results. According to the PR schedule suggested here, the breaking point is established at the end of every session after the animal has self-administered many drug injections. Consequently, animals would have *high* systemic drug levels when their breaking points were established. There is good reason to believe that this would affect the results. Bennett and Roberts (unpublished data) examined whether rats would respond to high ratios for the first cocaine injection of the day. They discovered that rats would not. Although they responded to very high breaking points after being "primed" with an injection of cocaine, they failed to emit large numbers of responses in order to attain the first cocaine infusion of the day. Hence, it would appear that the motivation to *initiate* cocaine self-administration may not be the same as the motivation to *continue* a self-administration bout.

Some authors have described PR schedules in which a fixed ratio is maintained during an entire session (or during a number of trials on one day) (Griffith et al., 1975,1978,1979; Hoffmeister, 1979; Risner and Silcox, 1981; Risner and Goldberg, 1983; Risner and Cone, 1986). Specifically, if a subject performs to some preset criterion, then the ratio is increased the next day. In contrast to the PR schedule implementations that determine breaking points daily, these schedules establish breaking points when animals have *low* systemic drug levels.

Consequently, these two PR schedule implementations may be addressing fundamentally different aspects of drug-related motivation—the motivation to initiate drug-seeking behavior and the motivation to continue a drug binge. Clearly, these are two distinct phenomena, and various drug treatments may have differential effects on the two motivations.

The implementation used in our lab has proven effective for measuring an animal's motivation to continue a cocaine "binge." In fact, breaking points derived from our schedule are clearly sensitive to pretreatments with haloperidol (Roberts et al., 1989b), fluoxetine (Richardson and Roberts, 1991), and (+)-AJ 76 (Richardson et al., submitted for publication); as well as to 6-OHDA lesions of the nucleus accumbens (Koob et al., 1987), 5,7-DHT lesions of both the medial forebrain bundle and the amygdala (Loh and Roberts, 1990), fluctuations in hormonal status (Roberts et al., 1989a), and changes in unit dose (Yanagita and Takahashi, 1973; Roberts et al., 1989b).

The magnitude and time-course of psychomotor stimulant "craving" in a drug-free state have not yet been adequately addressed experimentally. Perhaps they could be studied through a PR implementation in which the ratios at the beginning of the session are adjusted.

2.7. Other Schedules of Reinforcement

Complex schedules of reinforcement, other than PR, have yet to be employed in self-administration studies involving rodents. Several have, however, been used extensively with primates. Specifically, concurrent (Iglauer and Woods, 1974; Catania, 1976; Iglauer et al., 1976; Llewellyn et al., 1976) and second-order schedules (Kelleher and Goldberg, 1977; Johanson, 1982), choice procedures (Johanson and Schuster, 1975; Johanson and Aigner, 1981), and discrete trials (Stretch, 1977) have contributed elegantly to our understanding of the contingencies that control self-administration behavior in primates. Applying such schedules to rodents should prove to be a valuable endeavor for future research.

Researchers who study rodents do not always turn to schedules of reinforcement when they discover the need for more complex methods to investigate drug reinforcement. In fact, Suzuki et al. (1990) have abandoned schedules for another, more original technique. Specifically, they have adopted a weight-pulling method for rats. In order to reach drug-admixed food, rats are required to "work" by pulling a weight attached to the end of their harness (Suzuki et al., 1990).

3. Apparatus

3.1. Operant Chambers

Operant boxes are available from a number of commercial suppliers, although it has been our experience that such test chambers are difficult to clean. Since we choose to house our animals in the test chambers for the duration of the experiment, it is important to us that our cages be easily sanitized. Prior to assembly, holes are precut in the separate Plexiglass™ pieces. These pieces are then glued together with dichloromethane to form the operant boxes. The boxes are equipped with retractable or removable levers, stimulus lights (Spectro, Scarborough, ONT), and a water bottle holder. A food pellet dispenser can also be attached. A counterbalance arm is attached to support structure, also fashioned from Plexiglass™.

3.2. Fluid Swivels and Syringe Pumps

Two types of fluid swivels are generally used—industrially manufactured swivels and in-house manufactured swivels. A number of commercial suppliers (e.g., Stoelting Co., Chicago, IL) now sell very durable swivels, whose only major drawback is the price. The alternative is to manufacture one's own swivel according to the method of Amit et al. (1976). This swivel is made from common lab supplies (syringes, and so on), is very inexpensive, and is easy to manufacture. It is most efficient to assemble a dozen or so at one time, and although one can expect to discard one or two because of leakage or rigidity, the remaining should perform well for many months.

The simplest method to deliver a fixed quantity of drugs is to activate a syringe driver for a precise amount of time. Razel syringe drivers (Razel Scientific Instruments Inc., Stamford, CT) are relatively inexpensive and, in our lab, have proven to be reliable for many years. The rate of injection speed can be chosen by selecting one of a number of different motors. A 5-rpm motor will deliver 0.1 mL in 4 s through a 10-mL syringe. The alternative is to use an injection pump. Weeks (1977,1981) has described an elegant design for a pneumatic pump that allows for very accurate injections.

3.3. Intravenous Cannulae

Cannula designs vary enormously across self-administration labs, although the essential features remain the same. Basically, a small Silastic tube runs from the jugular vein sc to a point of exit, usually in the back. From there, it extends to a fluid swivel suspended over the cage. A number of designs have been described (*see* Weeks, 1962). The following is a brief description of a simple and inexpensive system used at Carleton University.

Based on the assumption that the fewer the number of connections in the system the better, we opted for a cannula that was continuous from the vein to the fluid swivel. A schematic diagram illustrating the construction of this cannula is shown in Fig. 3. A small length of Silastic tubing (Dow Corning Corp., Midland, MI, Cat No. 602-105) is fitted inside another, wider piece of Silastic tubing (Dow Corning Corp., Midland, MI, Cat. No. 602-155) and is held there with Silastic glue. A second bump of glue is then applied about 1 cm from the junction. (These two bumps will serve to ensure that the cannula remains in place.) The larger piece of tubing is first pushed through a small hole in a 1.5×3.0 cm piece of felt and then through a similar hole in a 2.0 $\times 3.0$ cm piece of Marlex mesh (Bard Corp., Billerica, MA) until 8 cm extend from the mesh to the small-diameter tubing. These materials are used at the point of exit. The Marlex mesh adheres very rapidly to the skin, and the felt serves to absorb any wound excretions. The Silastic tubing is then led through a protective spring (Plastics One Inc., Roanoke, VA), and a piece of PE 260 is used to stiffen the leash assembly. It is important that the spring be firmly attached to the Marlex. This is accomplished by winding the end of the spring two or three turns through the mesh. The joint is then covered with RTV Silicone sealant so that the Silastic tubing can make a smooth 90° turn and not crimp. The distal end of the Silastic tubing is held in place on the fluid swivel by a 2–3 cm length of "heat-shrink" tubing. This tubing is shrunk down over the Silastic tubing with a soldering iron. (For further details, *see* Roberts and Goeders, 1989.)

This cannula assembly is useful for experiments in which chronic (unlimited) access to a drug is desired, or for any other situation where the animal need not be disconnected from the infusion apparatus. This system is advantageous in that it is extremely simple to construct, and implantation can be accomplished in <10 min.

If the cannula is to be routinely disconnected, then a joint must be made in the system. One option is to disconnect the tubing above the swivel, and transfer the animal along with the leash assembly to and from the home cage. Another

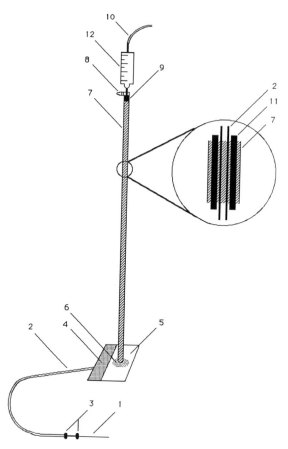

Fig. 3. A schematic diagram of the iv cannula assembly used at Carleton University. The components are as follows: 1, small-diameter Silastic tubing (id 0.012 in), 3 cm in length; 2, large-diameter Silastic tubing (id 0.025 in), 8 cm from vein to point of exit; 3, bumps of Silastic glue; 4, Marlex mesh (Bard Corp. cat. no. 011267), 2.0 × 3.0 cm; 5, piece of felt, 1.5 × 3.0 cm; 6, RTV silicone sealant; 7, protective spring (Plastics One Inc., cannula spring); 8, torque arm; 9, small piece of "heat-shrink" tubing; 10, polyethylene tubing (PE 90); 11, Polyethylene tubing (PE 260); 12, fluid swivel.

alternative is to disconnect the tubing at the point of exit from the skin. An example of this technique is the back-mounted connector used by W. Corrigal (Addiction Research Foundation, Toronto). (For a detailed description and illustration of this technique, *see* Roberts and Goeders, 1989). Again, Marlex mesh is used to anchor the cannula to the back, but in this case, a receptacle end is made with a piece of "heat-shrink" tubing passed through a plastic bolt. A 20-gage needle is then inserted into the heat-shrink tubing to form the connection. (If this connection becomes loose after several days, a touch with a soldering iron will reshrink the tubing.) The protective spring is equipped with a nut that fits on the plastic bolt to form a solid joint.

In some laboratories, the connection is made on the skull with commercially available fittings (e.g., Plastics One Inc., Roanoke, VA). After the cranial surface has been cleaned and dried, and several anchor screws placed in the skull, the plugs are cemented in place with dental cement. This has the advantage of having a firm base on which to mount the connector, namely, the skull. The extra time required for mounting the animals in a stereotaxic apparatus and cementing the plugs is, however, a disadvantage to this technique. If the experimental design calls for intracerebral injections or lesions, then a guide cannula must be implanted at the time of iv cannulation.

3.4. Computer Interface

A number of computer-controlled lab systems are on the market that might easily be adapted for self-administration experiments. Here we describe a way to interface an IBM-compatible computer to the test apparatus using standard electronic components. Central to the design are optically isolated Potter and Brunfield input or output (I/O) modules. Modules are available as either input sensors or output drivers. Figure 4 illustrates how banks of eight modules can be configured to switch dc stimulus lights, ac pumps, or detect switch closures (lever responses).

Each bank of eight modules is cabled to an interface port on the computer. Generally, interface cards, such as the Metrabyte PIO-12, are based on the Intel 8255 chip, which is able to control 24 I/O lines. Sometimes, several of these chips are mounted on the same board, such as the Contec PIO-96 board, which has four 8255 chips and can therefore control up to 96 I/O lines.

The Intel 8255 chip has four addresses: three I/O addresses and a control address. The first three addresses correspond to

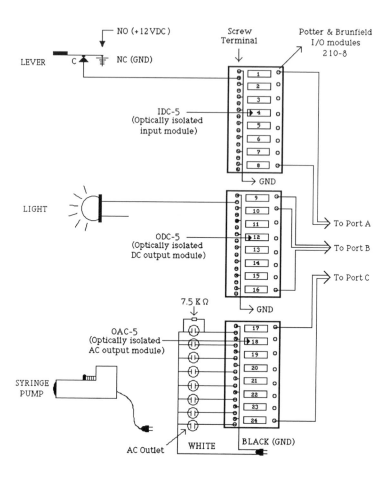

Fig. 4. A schematic diagram of the electronic components used to interface an IBM-compatible computer to a self-administration apparatus. Note: 7.5K ohm 3-W resistors are required across each ac outlet to provide minimum load current.

Ports A, B, and C, respectively. If a port is configured as input, *reading* the contents of the port will show which of the eight switches is open or closed. If the port is configured as output, then *writing* a number to that address will turn on or off the corresponding switches. The fourth address on the chip is the control address; whether Port A, B, or C is configured as input or output will depend on the data written to the control address (*see* Section 4.1.).

Figure 4 illustrates how eight input lines and 16 output lines can be sent to a single 8255 chip. The input switches are wired to Port A, the stimulus lights to Port B, and the ac switches (pumps) to Port C. This example would require that Port A be configured as "input," and Ports B and C as "output." How these ports are manipulated through software is discussed in Section 4.1.

4. Programming

4.1. Programming I/O Ports in Pascal

As stated earlier, switching an output on or off is done by writing data to the address corresponding to the appropriate port. Each port address is one byte (8 bits), and each bit controls one I/O module. Therefore, one port (A, B, or C) can control one bank of 8 Potter and Brumfield modules (*see* Fig. 4). Since output modules are *off* if the signal line is at +5 VDC and *on* if it is 0 VDC (ground), the task is to assign the correct binary code that will turn the desired modules on or off. Table 1 shows the binary code that turns on each switch. The hexadecimal and integer codes corresponding to the binary code are also given. The column under "Bit" gives the value of the bit being switched. It is clear from the table that the integer value required to turn on any particular bit is 255 minus the bit value.

Two or more switches can be turned on simultaneously by subtracting their combined bit values from 255. A value of 0 would turn all switches on; a value of 255 would turn all switches off.

A method exists for toggling individual switches by taking advantage of "bit-wise" procedures. Such procedures allow a single bit to be manipulated, leaving the other bits unchanged. Suppose that one wanted to turn on switch number 4 at address $300 (the factory set value for the Metrabyte PIO-12). This can be accomplished with the Pascal instruction:

$$\text{Port}[\$300] := \text{Port}[\$300] \text{ and } 247 \qquad (3)$$

The value 247 serves as a "mask" for switch 4, allowing only this bit to be changed. The instruction sets the 4th switch to 0 whether it is initially at 0 or 1. The following illustrates that only the 4th switch is changed by this instruction.

Table 1[a]
Binary Codes for Switching Individual Bits

Switch	Binary code	Hex	Integer	Bit
1	1111 1110	$FE	254	1
2	1111 1101	$FD	253	2
3	1111 1011	$FB	251	4
4	1111 0111	$F7	247	8
5	1110 1111	$EF	239	16
6	1101 1111	$DF	223	32
7	1011 1111	$BF	191	64
8	0111 1111	$7F	127	128

[a]This table shows the binary code, integer values, and hexadecimal equivalents that would cause switches 1–8 to be turned on if these data were written to the I/O port address.

```
        1 0 1 0 |1|0 1 0      <- Old value on port
And     1 1 1 1 |0|1 1 1      <- Mask
Equals  1 0 1 0 |0|0 1 0      <- New value on port      (4)
```

The masks for switching *1–8 on* are (254, 253, 251, 247, 239, 223, 191, 127). (Note that these are the integer values in Table 1.)

Turning individual bits *off* uses the OR command and a different set of masks. The masks for switching 1–8 *off* are the bit values in Table 1 (1, 2, 4, 8, 16, 32, 64, 128). The following Pascal instruction:

$$Port[\$300] := Port[\$300] \text{ or } 16 \qquad (5)$$

would set the 5th switch to 1 and leave all other bits unchanged.

4.2. Decoding Inputs

Switch closures (lever responses) cause the bit values on the input port to change from 1 to 0. Decoding the byte to indicate which switches are closed also takes advantage of bit-wise instructions. A small program that would decode an input port is shown in Fig. 5.

4.3. A Progressive-Ratio Schedule in Pascal

The complete listing of a Pascal program that will run eight operant chambers on a PR schedule is given in Appendix A. The

```
PROGRAM Intest;
USES     Dos, Crt;
CONST    Exponent : array [1..8] of integer = (1,2,4,8,16,32,64,128);
VAR      PortValue, Lever : integer;
         NewResponse : array [1..8] of boolean;

BEGIN
   ClrScr; {Clear Screen}
   Port[$303] := $9B;   {Set Control value so that Port A,B & C are Input}
   for lever := 1 to 8 do NewResponse[lever] := True;
   Write ('Press any key to Quit');
   Writeln;
   PortValue := 255;
   Repeat
    Begin
        if Port[$300] <> PortValue then
        Begin
            PortValue := Port[$300];
            for lever := 8 downto 1 do
            if PortValue AND exponent [lever] = 0 then
            Begin
                if NewResponse[lever] then
                Begin
                    NewResponse[lever] := False;
                    writeln(lever);
                End;
            End else NewResponse[lever] := True
        End;
    End;
   Until Keypressed;
End.
```

Fig. 5. A Pascal program for decoding an input to indicate which switches are closed. Note: This program utilizes bit-wise instructions.

program was written using Turbo Pascal 5.5 with each box represented as an "Object." The program repeatedly scans for a lever response (or a keyboard entry), and the timing is handled by interrupts (18.2/s). The scheme for this program is depicted in Fig. 6.

4.4. Graphic Representation of the Data

On-line graphic representation of the data is advised so that subject performance can be monitored, and equipment failure

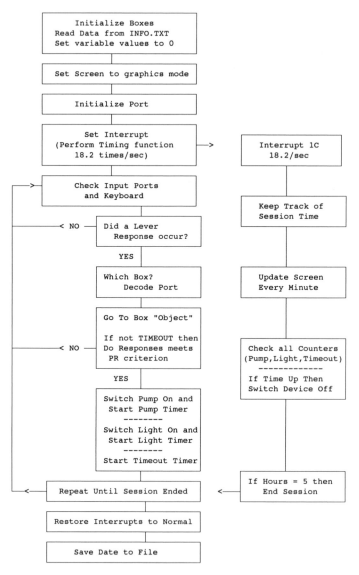

Fig. 6. The schema for a Pascal program capable of running eight operant chambers on a PR schedule of reinforcement.

and/or software bugs can be detected. The program listed in Appendix A creates an "event record" on the screen similar to that shown in Fig. 7. The data are stored to disk, and this event record or a cumulative record (e.g., Fig. 1) can then be reconstructed following the session.

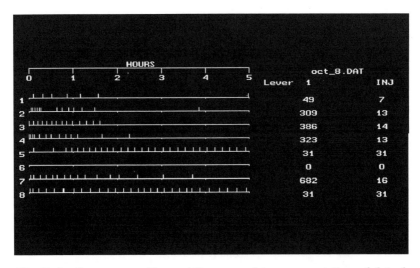

Fig. 7. An "event record" providing a graphic representation of data from eight rats self-administering cocaine during a 5-h session. Each dash represents the delivery of an injection of drug. The column of numbers on the extreme right shows the total number of injections received by each animal. The column immediately to the left of this INJ column depicts the total number of responses made by each animal throughout the session. Note: Animals 1, 2, 3, 4, and 7 are on a PR schedule, and animals 5 and 8 are on an FR 1 schedule. Box 6 was empty.

5. Surgery

Implantation: With the rat suitably anesthetized, two incisions are made in the skin—one located in the back slightly caudal to the scapula, and the other over the jugular vein, extending from the cheek to the clavicle. The Silastic tubing is then led sc over the shoulder to the ventral incision, where the small-diameter tubing is inserted into the vein. (Some laboratories prefer to lead the tubing under the arm, but this seems to add an unneccesary curve to the cannula.) Two sutures, tied between the "bumps" of Silastic glue, hold the cannula in place. The ventral wound is then closed with sutures. Next, the Marlex mesh is placed under the skin through the dorsal wound, and is held in place with two sutures that penetrate the skin, mesh, and muscle. The incision is then closed with two or three sutures, and Vetbond™ tissue adhesive is applied around the cannula at the point of exit to reduce wound seeping. This assembly will be tolerated quite well for many weeks providing the swivel turns freely and the leash is counterbalanced.

Appendix 1
Pascal Program

```
PROGRAM ONELEVER;
USES Dos, Crt, Graph;
CONST
    BitMask : array [1..16] of integer =
        (1,2,4,8,16,32,64,128,1,2,4,8,16,32,64,128);

    PR_Sequence : array [1..30] of integer =
    (  1,   2,   4,   6,   9,  12,  15,  20,  25,  32,
      40,  50,  62,  77,  95,118,145,178,219,268,
     328,402,492,603,737,901,999,999,999,999);

    LastBox = 8;
    FiveSeconds = 91;           TenSeconds = 182;
    TwentySeconds = 364;        OneMinute  = 1092;

    Lever_Port = $303;     {InPort1}
    Pump_Port = $301;
    Light_Port = $302:

    Black       : integer = 0;
    Blue        : integer = 1;
    LightGrey   : integer = 7;
    LightBlue   : integer = 9;
    LightGreen  : integer = 10;
    Yellow      : integer = 14;
    White       : integer = 15;


TYPE
    State = (Time_In, Reinforce_TO, Session_Over);
    BOX = OBJECT
      Light_On,   Pump_On   : Boolean;
      Mask, Negative_Mask : integer;
      Pump_Ticks, Time_Out_Ticks : integer;
      Box_Minute : integer;
      Screen_Row : integer;
    { Program  Variables }
      PR_Schedule : Boolean;
      PR_Pointer, Step_No, Criterion : Integer;
      Criterion_Str : String[4];
    { Dependent Variables }
      RAT_ID, Schedule, Drug, Dose : string; {define string length}
      Total_Responses    , Trial_Responses    , Injections : integer;
      Total_Responses_Str, Trial_Responses_Str, Injections_Str : string[4];
      Wrong_Responses : integer; Wrong_Responses_Str : string[4];
      Min_Responses  : array [0..299] of byte;
      Min_Injections : array [0..299] of byte;
      Phase : State;

    Procedure Init (Box_Number  : integer;
                    Init_Rat_ID, Init_Schedule : string );
    Procedure Time_Line;
    Procedure Timer_Tick;
    Procedure Go_To_Reinforce_TO;
```

```
      Procedure Go_To_Time_In;
      Procedure Detect_Lever;
      Procedure Switch_PumpOn;
      Procedure Switch_PumpOff;
      Procedure Switch_Light_On;
      Procedure Switch_Light_Off;
      Procedure Display_Wrong_Response;
   end; {Object BOX}


VAR
    New_Response : array [1..Lastbox] of boolean;
    Interrupts, ThisSecond, Second, Minute, TotalMinute, Hour : integer;
    Int1CSave, Int1CSet : pointer;
    Box_Number : integer;
    Diskfile : text;
    Session_Ended : boolean;
    Ch : char;
    Boxes : array [1..lastbox] of Box;
    InPort1_Data, InPort2_Data, InPort3_Data, InPort4_Data : integer;
    GraphDriver, GraphMode, HighMode, Errorcode : integer;
    Start_Hour, Start_Min, Start_Sec, Start_Hun : Word;
    Finish_Hour, Finish_Min, Finish_Sec, Finish_Hun : Word;


Procedure BOX.Init (Box_Number  : integer;
                    Init_RAT_ID, Init_Schedule : string );
    CONST  Exponent : array [1..8] of integer = (1,2,4,8,16,32,64,128);
    VAR Index : integer;
    Begin
     {Defined in BOXUNIT}
       Mask :=  Exponent[Box_Number];
       Negative_Mask := 255 - Mask;
       Light_On := False;
       Pump_On    := False;
       Pump_Ticks := 0; Time_Out_Ticks := 0 ;
       Box_Minute := 0 ;
       Screen_Row := 20 * Box_Number + 20;
        Schedule := Init_Schedule;
        If Schedule = 'PROG  ' then
           begin
             PR_Schedule := TRUE; PR_Pointer := 1 ;
             Criterion := PR_Sequence[PR_Pointer] ;
             STR(Criterion, Criterion_Str);
           end;
         If Schedule = 'FIXED ' then
           begin
             PR_Schedule := FALSE ;
             Criterion   := 1 ;
             STR(1,Criterion_Str);
           end;
         RAT_ID := Init_RAT_ID;
        { Dose := InitDose;}
        { Drug := InitDrug;}
         Total_Responses := 0;
         Str(Total_Responses, Total_Responses_Str);
         Trial_Responses := 0;
```

(continued)

Appendix 1 *(continued)*

```
        Str(Trial_Responses , Trial_Responses_Str);
        Injections := 0;
        Str(Injections, Injections_Str);
        Wrong_Responses := 0;
        Str(Wrong_Responses,Wrong_Responses_Str);
        Box_Minute := 0;
        for index := 0 to 299 do
            begin
                Min_Responses [index]   := 0;
                Min_Injections[index]   := 0;
            end;
        Phase := Time_In ;
End;


Procedure Box.Go_To_Reinforce_TO;
Begin
        Phase := Reinforce_TO;
        Switch_Light_ON;
      { Retract_Lever; }
        Pump_Ticks := FiveSeconds; Switch_PumpOn;
        Time_Out_Ticks := Twentyseconds;
        SetColor(Black);
        OutTextXY(150, Screen_Row, Injections_Str);
        SetColor(Yellow);
        Inc(Injections);
        Str(Injections, Injections_Str);
        OutTextXY(150, Screen_Row, Injections_Str);
        Inc(Min_Injections[Box_Minute]);
        line (Box_Minute + 210, Screen_Row, Box_Minute + 210, Screen_Row + 5);
End;


Procedure Box.Go_To_Time_In;
    Begin
        Phase := Time_In;
        { Insert Lever }
        Switch_Light_Off;
        If PR_Schedule then
        Begin Inc(PR_Pointer);
            SetColor(Black);
            OutTextXY(110,Screen_Row,Criterion_Str);
            SetColor(Yellow);
            Criterion := PR_Sequence[PR_Pointer];
            Str(Criterion,Criterion_Str);
            OutTextXY(110,Screen_Row,Criterion_Str);
        End;
    End;


Procedure BOX.Timer_Tick;
Begin
        if Pump_Ticks > 0 then
            begin
                Dec(Pump_Ticks);
                if Pump_Ticks = 1 then Switch_PumpOff;
            End;
        if Time_Out_Ticks > 0 then
```

```
                    begin
                      Dec(Time_Out_Ticks);
                      if Time_Out_Ticks = 1 then Go_To_Time_In;
                    End;
        End;

        Procedure Box.Detect_Lever;
        Begin
          If Phase = Time_In then
          begin
            SetColor(Black);
            OutTextXY(40,Screen_Row,Total_Responses_Str);
            OutTextXY(80,Screen_Row,Trial_Responses_Str);
            SetColor(Yellow);
            Inc(Total_Responses);
            Str(Total_Responses, Total_Responses_Str);
            OutTextXY(40, Screen_Row, Total_Responses_Str);
            Inc(Trial_Responses);
            Inc(Min_Responses[Box_Minute]);
            If Trial_Responses = Criterion then
            begin
                Trial_Responses := 0;
                Go_To_Reinforce_TO;
            end;
            Str(Trial_Responses, Trial_Responses_Str);
            OutTextXY(80, Screen_Row , Trial_Responses_Str);
          end;
        End;

        Procedure Box.Time_Line;
        Begin
            SetColor(Yellow);
            PutPixel(Box_Minute + 210, Screen_Row + 5, 15);
            Inc(Box_Minute);
        End;

        Procedure BOX.Switch_PumpOn;
        Begin
          Pump_On := True;
          Port[Pump_Port] := Port[Pump_Port] AND Negative_Mask;
          SetColor(Black);
          OutTextXY (620, Screen_Row, 'off');
          SetColor(Yellow);
          OutTextXY (620, Screen_Row, 'on');
        End;

        Procedure BOX.Switch_PumpOff;
        Begin
          Pump_Ticks := 0;
          Pump_On := False;
          Port[Pump_Port] := Port[Pump_Port] OR Mask;
          SetColor(Black);
          OutTextXY (620, Screen_Row, 'on');
          SetColor(Yellow);
          OutTextXY (620, Screen_Row, 'off');
```

(continued)

Appendix 1 *(continued)*

```
End;

Procedure BOX.Switch_Light_On;
Begin
  Light_On := True;
  Port[Light_Port] := Port[Light_Port] AND Negative_Mask;
  SetColor(Black);
  OutTextXY (540, Screen_Row, '.');
  SetColor(Yellow);
  OutTextXY (540, Screen_Row, '1');
End;

Procedure BOX.Switch_Light_Off;
Begin
  Light_On := False;
  Port[Light_Port] := Port[Light_Port] OR Mask;
  SetColor(Black);
  OutTextXY (540, Screen_Row, '1');
  SetColor(Yellow);
  OutTextXY (540, Screen_Row, '.');
end;

Procedure Box.Display_Wrong_Response;
Begin
  SetColor(Black);
  OutTextXY (590, Screen_Row, Wrong_Responses_Str);
  Inc(Wrong_Responses);
  Str(Wrong_Responses,Wrong_Responses_Str);
  SetColor(White);
  OutTextXY (590, Screen_Row, Wrong_Responses_Str);
End;

PROCEDURE InitPorts;
 VAR BoxNumber : integer;
    {The number that is placed in the CONTROL port will       }
    {determine whether the other three ports are INPUT        }
    {or OUTPUT.                                               }
    {The sequence of addresses is PortA, PortB, PortC, Control}
    {    PortA   PortB   PortC      Control                   }
    {    --------------------       Value                     }
    {    Out     Out     Out        128                       }
    {    In      Out     Out        144                       }
    {    Out     In      Out        130                       }
    {    Out     Out     In         137                       }
    {    In      In      Out        146                       }
    {    Out     In      In         139                       }
    {    In      Out     In         153                       }
    {    In      In      In         155                       }

Begin
  Port [$303] := 137;   {Configure $300, $301 as output & $303 as input}
  for BoxNumber := 1 to LastBox do
  begin
    New_Response[Box_Number] := FALSE;
  end;
```

```
End;

PROCEDURE GraphModeInit;
   VAR LowMode, HighMode : integer;
   Begin
      GraphDriver := Detect;
      InitGraph (GraphDriver, GraphMode, 'C:\TP\');
      ErrorCode := GraphResult;
      if ErrorCode <> grOK then
        begin
           Clrscr;
           Writeln ('Unable to initialize graphics screen '); Writeln;
           Case Errorcode of -1 : Writeln ('BGI graphics not installed');
                             -2 : Writeln ('Graphics hardware not detected');
                             -3 : Writeln ('Device driver file not found');
                             -4 : Writeln ('Invalid device driver file');
           end; {Case}
           Writeln ('Press return to HALT Program');
           Readln; Halt(1);
         end;
      SetGraphMode(1);
      SetTextJustify(CenterText,RightText);
   End;

Procedure ScreenInit;
VAR  S : string; B : integer;
Begin
  GraphModeInit;
  OutTextXY (300, 1, 'Q to quit');
  OutTextXY (550,20, 'LIGHT');
  OutTextXY (590,20, 'L2');
  OutTextXY (620,20, 'Pump');
  OutTextXY (210, 20, '0');
  OutTextXY (270, 20, '1');
  OutTextXY (330, 20, '2');
  OutTextXY (390, 20, '3');
  OutTextXY (450, 20, '4');
  OutTextXY (510, 20, '5');
  MoveTo(210, 29);
  LineTo(510, 29);
  OutTextXY (30, 25, 'Tot');
  OutTextXY (70, 25, 'Resp.');
  OutTextXY (110, 25, 'FR');
  OutTextXY (150, 25, 'Inj.');
  SetColor(Yellow);
  for B:= 1 to lastbox  do
  begin
    Str(B, S);
    OutTextXY(8,Boxes[B].Screen_Row, S);
    OutTextXY(40,  Boxes[B].Screen_Row, Boxes[B].Total_Responses_Str);
    OutTextXY(80,  Boxes[B].Screen_Row, Boxes[B].Trial_Responses_Str);
    OutTextXY(110, Boxes[B].Screen_Row, Boxes[B].Criterion_Str);
    OuttextXY(150, Boxes[B].Screen_Row, boxes[B].Injections_str);
    Boxes[B].Switch_Light_Off;
  end;
```

(continued)

Appendix 1 *(continued)*

```
    End;

{------------------------------------------------}
{          Clock's Method Implementation         }
{------------------------------------------------}

  Procedure Clock_Init;
  Begin
    Interrupts  := 0; ThisSecond  := 0; Second      := 0;
    Minute      := 0; TotalMinute := 0; Hour        := 0;
  End;

  Procedure Clock_Tick;
  Begin
    Inc(Interrupts);
    ThisSecond := Trunc (Interrupts/18.2); {18.2 interrupts/sec}
    if ThisSecond <> Second then
    begin
      Second := ThisSecond;
      if Second = 60 then
      begin
        Interrupts := 0; Second := 0; Inc(Minute); Inc(TotalMinute);
        for Box_Number := 1 to LastBox do Boxes[Box_Number].Time_Line;
        if Minute = 60 then
        begin
          Minute := 0; Inc(Hour);
        end;
      end;
    end;
  end;

{----------------------------------------------------------}
{----------          Port    Procedures       ------------}
{----------------------------------------------------------}

Procedure CheckInPort;
Begin
  if Port[Lever_Port] <> InPort1_Data then      {if something new then}
  begin
    InPort1_Data := Port[Lever_Port];
    for Box_Number:= 8 downto 1 do
    if InPort1_Data AND Bitmask[Box_Number] = 0 then
    begin
        if New_Response[Box_Number] then
        begin
          New_Response[Box_Number] := False;
          Boxes[Box_Number].Detect_Lever;
        end;
    end
    else New_Response[Box_Number] := True;
  end;
End;

PROCEDURE GetTick (Flags,CS,IP,AX,BX,CX,DX,SI,DI,DS,ES,BP: Word);
        Interrupt; {every 1/18th sec}
```

```
            CONST DispSeg = $B000; {Monochrome value}
            BEGIN
                Clock_Tick;
                CheckInPort;
                for Box_Number:= 1 to LastBox do boxes[Box_Number].Timer_Tick;
            END; { GetTick }

PROCEDURE SetInterrupt;
            BEGIN
                GetIntVec($1C, Int1CSave);
                SetIntVec($1C, @GetTick);
            END;

PROCEDURE ResetInterrupt;
            BEGIN
                SetIntVec($1C, Int1CSave);
            END;

Procedure SaveData;
Const
    months : array [1..12] of string[3] =
    ('JAN','FEB','MAR','APR','MAY','JUN','JUL','AUG','SEP','OCT','NOV','DEC');
VAR Min, Final_Ratio, Last_Injection : integer;
    Y,M,D,DOfW : word;
    filename : string;
    month, day : string[3];
Begin
    Write ('Save Data?');
    Repeat CH := Readkey until CH in ['Y','y','N','n'];
    If CH in ['N','n'] then exit;  Writeln;
    GetDate(Y,M,D,DOfW); month := Months[M]; STR(D,day);
    filename := concat('D:',Month,'_',day,'.DAT');
    Write ('Save data to: ',filename,' ? :');
    Repeat CH := Readkey until CH in ['Y','y','N','n'];
    if CH in ['N','n'] then
    begin
        Writeln;
        Write ('Enter filename:');
        Readln (filename);
    end;
    Writeln;
    Writeln ('Saving to ',filename);
    assign (diskfile,filename);
    rewrite (diskfile);
    GetDate(Y,M,D,DOfW); month := Months[M]; STR(D,day);
    writeln (diskfile,Month,' ',day,' ',Y);
    Write(diskfile,'Start Time :',
        Start_Hour:2,':',Start_Min:2,':',Start_Sec:2,':',Start_Hun:2);
    Writeln(diskfile,'   Finish at:',
        Finish_Hour:2,':',Finish_Min:2,':',Finish_Sec:2,':',Finish_Hun:2);
    writeln
        (diskfile,'Box   ID   Inactive  Active   Inj.  Final Ratio');
    for Box_Number:= 1 to LastBox do
    begin
        if Boxes[Box_Number].injections > 0 then
```

(continued)

Appendix 1 (*continued*)

```
       begin
          Last_Injection := boxes[Box_Number].injections;
          Final_Ratio := boxes[Box_Number].Criterion;
       end
       else Final_Ratio := 0;
       write (diskfile,Box_Number:3);
       Write (diskfile, ' ',boxes[Box_Number].RAT_ID,' ',
                          boxes[Box_Number].Wrong_Responses:6,' ',
                          boxes[Box_Number].Total_Responses:6,' ',
                          boxes[Box_Number].Injections:4,'    ',
                          Final_Ratio:4);
       writeln(diskfile);
    end;
    writeln (diskfile);
    for Min := 0 to 299 do
    begin
       write (diskfile, Min:3);
       for Box_Number:= 1 to LastBox do
       Begin
              write (diskfile, ' ');
          write (diskfile, boxes[Box_Number].Min_Responses[Min]:5);
          write (diskfile, boxes[Box_Number].Min_Injections[Min]:3);
       End;
       writeln(diskfile);
    end;
    close(diskfile);
    Writeln;
    Writeln('Data saved to file: ',filename);
end;

Procedure Init_Boxes_with_INFO;          {From INFO.TXT}
   Var
       INFO_ERROR : Boolean;
       line : string;
       Box_Number: integer;
       Rat_ID, Schedule : string[6];

   Procedure ErrorCheck;
       Begin
       If (Schedule <> 'PROG  ') And (Schedule <> 'FIXED ')
              Then INFO_ERROR := TRUE;
       End;

   Begin
       INFO_ERROR := FALSE;
       assign(diskfile,'INFO.TXT');
       reset(diskfile);
       Writeln ('The following info has been passed from INFO.TXT');
       Readln(diskfile,line); Writeln(line);
       Readln(diskfile,line); Writeln(line);
       Readln(diskfile,line); Writeln('  ',line);
       for Box_Number:= 1 to lastbox do
       begin
          readln(diskfile,Rat_ID, Schedule) ;
          ErrorCheck;
```

```
          Writeln(box_number:2,' ',Rat_ID, Schedule);
          Boxes[Box_Number].Init(Box_Number,Rat_ID,Schedule);
        end;
        readln;
        close(diskfile);
        If INFO_ERROR then
          begin
            Window(1,14,60,16);
            Textcolor(Black);
            TextBackground(White);
            Writeln ('******ERROR IN READING FILE****');
            Textcolor(White);
            Textbackground(Black);
            readln;
          end;
    end;

Procedure TurnBoxesOff;
Begin
for Box_Number:= 1 to LastBox do
    begin
        boxes[Box_Number].Switch_PumpOff;
        boxes[Box_Number].Switch_Light_Off;
    end;
end;

BEGIN { Program Body }
    ClrScr; {Clear Screen}
    Init_Boxes_With_INFO;
    ScreenInit;
    Clock_Init;
    InitPorts;
    Ch := '0';
    Session_Ended := False;
    For Box_Number := 1 to LastBox do boxes[Box_Number].Retract_Lever;
    SetInterrupt;
    GetTime(Start_Hour,Start_Min,Start_Sec,Start_Hun);
    Repeat
        if KeyPressed then
        begin
          Ch := Upcase (ReadKey);
          case Ch of '1'..'8' : Boxes[ORD(Ch)-48].Detect_Lever; {Test 1..8}
                       'Q' : Session_Ended := True;
              end; {case}
        end;
        if Hour = 1 then Session_Ended := True;
    Until Session_Ended;
    GetTime(Finish_Hour,Finish_Min,Finish_Sec,Finish_Hun);
    ResetInterrupt;
    TurnBoxesOff;
    CloseGraph;
    SaveData;
End.
```

Appendix 2
List of Supplies

Manufacturer	Item
Becton-Dickinson (BD) Rutherford, NJ 07070	Disposable syringe (3 mL) catalog # 5585; quantity = 100
	Disposable syringe (1 mL) catolog # 9602; quantity = 100
	Needles (22 gage 1-1/2 in) catalog # 5156; quantity = 100
Clay Adams Division of BD Parsippany, NJ 07054	PE 260 tubing catalog # 7456; quantity = 200 ft
Plastics One Inc. Roanoke, VA 24022	Stainless-steel springs 0.156 od × 0.120 id quantity = 24 × 4 ft
	Cannula tubing PE50 HVYWL catalog # C313CT; quantity = 200 ft
Dow Corning Corp. Medical Products Midland, MI 48640	Silastic medical-grade tubing 0.025 id × 0.047 od catalog # 602-155; quantity = 100 ft 0.012 id × 0.025 od catalog # 602-105; quantity = 50 ft
	Silastic medical adhesive catalog # 891; quantity = 2 tubes
BARD Cardiosurgery Division C. R. Bard Inc. Billerica, MA 01821	Marlex mesh catalog # 011267; quantity = 2 boxes
Razel Scientific Instruments Inc. Stamford, CT 06907	Syringe pump (5 rpm motor) model A; quantity = 1
Spectro Scarborough, ONT M1P2L6	LED indicator assembly (stimulus light) #1820; quantity = 10
Potter and Brumfield Relays Princeton, IN 47671	Mounting board for I/O modules # 21024; quantity = 1
	I/O optically isolated modules ODC-5 and IDC-5; quantity = 1
Contec Microelectronics USA Inc. San Jose, CA 95131	PIO-96W interface board quantity = 1

Acknowledgments

Grant support has been provided by the Medical Research Council of Canada.

References

Amit Z., Brown Z. W., and Sklar L. S. (1976) Intraventricular self-administration of morphine in naive laboratory rats. *Psychopharmacology* **48,** 291–294.

Bedford J. A., Bailey L. P., and Wilson M. C. (1978) Cocaine reinforced progressive ratio performance in the rhesus monkey. *Pharmacol. Biochem. Behav.* **9,** 631–638.

Catania A. C. (1976) Drug effects and concurrent performances. *Pharmacol. Rev.* **27,** 385–394.

Collins R. J. and Weeks J. R. (1965) Relative potency of codeine, methadone and dihydromorphine to morphine in self-maintained rats. *Naunyn Schmiedeberg's Arch. Pharmacol.* **249,** 509–514.

De La Garza R. and Johanson C. E. (1982) Effects of haloperidol and physostigmine on self-administration of local anaesthetics. *Pharmacol. Biochem. Behav.* **17,** 1295–1299.

Deneau G., Yanagita T., and Seevers M. H. (1969) Self-administration of psychoactive substances by the monkey. *Psychopharmacology* **16,** 30–48.

DeWit H. and Wise R. A. (1977) Blockade of cocaine reinforcement in rats with the dopamine receptor blocker pimozide, but not with noradrenergic blockers phentolamine and phenoxybenzamine. *Can. J. Psychol.* **31,** 195–203.

Gerber G. J., Bozarth M. A., Spindler J. E., and Wise R. A. (1985) Concurrent heroin self-administration and intracranial self-stimulation in rats. *Pharmacol. Biochem. Behav.* **23,** 837–842.

Gill C. A., Holz W. C., and Zinkle C. (1975) Reported to the NAS-NRC Committee in Problems of Drug Dependence.

Goeders N. E. and Smith J. E. (1987) Intracranial self-administration methodologies. *Neurosci. Biobehav. Rev.* **11,** 319–330.

Gotestam K. G. (1973) Intragastric self-administration of medazepam in rats. *Psychopharmacology* **28,** 87–94.

Griffiths R. R., Bradford L. D., and Brady J. V. (1979) Progressive ratio and fixed ratio schedules of cocaine-maintained responding in baboons. *Psychopharmacology* **65,** 125–136.

Griffiths R. R., Brady J. V., and Snell J. D. (1978) Progressive-ratio performance maintained by drug infusions: comparison of cocaine, diethylpropion, chlorphentermine, and fenfluramine. *Psychopharmacology* **56,** 5–13.

Griffiths R. R., Findley J. D., Brady J. V., Gutcher K., and Robinson W. W. (1975) Comparison of progressive-ratio performance maintained by cocaine, methylphenidate and secobarbitol. *Psychopharmacology* **43,** 81–83.

Hatsukami D., Keenan R., Carroll M., Colon E., Geiske D., Wilson B., and Huber M. (1990) A method for delivery of precise doses of smoked cocaine-base to humans. *Pharmacol. Biochem. Behav.* **36**, 1–7.

Hodos W. (1961) Progressive ratio as a measure of reward strength. *Science* **134**, 943–944.

Hoffmeister F. (1979) Progressive-ratio performance in the rhesus monkey maintained by opiate infusions. *Psychopharmacology* **62**, 181–186.

Iglauer C. and Woods J. H. (1974) Concurrent performances: Reinforcement by different doses of intravenous cocaine in rhesus monkeys. *J. Exp. Anal. Behav.* **22**, 179–196.

Iglauer C., Llewellyn M. E., and Woods J. H. (1976) Concurrent schedules of cocaine injection in rhesus monkeys: Dose variations under independent and non-independent variable-interval procedures. *Pharmacol. Rev.* **27**, 367–383.

Johanson C. E. (1982) Behavior maintained under fixed-interval and second-order schedules of cocaine or pentobarbital in rhesus monkeys. *J. Pharmacol. Exp. Ther.* **221**, 384–393.

Johanson C. E. and Aigner T. (1981) Comparion of the reinforcing properties of cocaine and procaine in rhesus monkeys. *Pharmacol. Biochem. Behav.* **15**, 49–53.

Johanson C. E. and Schuster C. R. (1975) A choice procedure for drug reinforcers: Cocaine and methylphenidate in the rhesus monkey. *J. Pharmacol. Exp. Ther.* **193**, 676–688.

Johanson C. E., Kandel D. A., and Bonese K. F. (1976) The effects of perphenazine on self-administration behavior. *Pharmacol. Biochem. Behav.* **4**, 427–433.

Kelleher R. T. and Goldberg S. R. (1977) Fixed-interval responding under second-order schedules of food presentation or cocaine injection. *J. Exp. Anal. Behav.* **28**, 221–231.

Koob G. F., Le H. T., and Creese I. (1987) The D1 dopamine receptor anatagonist SCH 23390 increases cocaine self-administration in the rat. *Neurosci. Lett.* **79**, 315–320.

Koob G. F., Vaccarino F. J., Amalric M., and Bloom F. E. (1987) Positive reinforcement properties of drugs: Search for neural substrates, in *Brain Reward Systems and Abuse* (Engel J. and Oreland L., eds.), Raven, New York. pp. 35–50.

Llewellyn M. E., Iglauer C., and Woods J. H. (1976) Relative reinforcer magnitude under a non-independent concurrent schedule of cocaine reinforcement in rhesus monkeys. *J. Exp. Anal. Behav.* **25**, 81–91.

Loh E. A. and Roberts D. C. S. (1990) Break-points on a progressive ratio schedule reinforced by intravenous cocaine increase following depletion of forebrain serotonin. *Psychopharmacology* **101**, 262–266.

Lyness W. H., Friedle N. M., and Moore K. E. (1979) Destruction of dopaminergic nerve terminal in nucleus accumbens: Effects of d-amphetamine self-administratlon. *Pharmacol. Biochem. Behav.* **11**, 553–556.

Mucha R. F. (1980) Indwelling catheter for infusions into subcutaneous tissue of freely-moving rats. *Physiol. Behav.* **24,** 425–428.

Pickens R. and Thompson T. (1968) Cocaine reinforced behavior in rats: Effect of reinforcement magnitude and fixed ratio size. *J. Pharmacol. Exp. Ther.* **161,** 122–129.

Richardson N. R. and Roberts D. C. S. (1991) Fluoxetine pretreatment reduce breaking points on a progressive ratio schedule reinforced by intravenous cocaine self-administration in the rat. *Life Sci.* **49,** 833–840.

Richardson N. R., Piercey M. F., Svensson K., Collins R. J., Myers J. E., and Roberts D. C. S. (1991) An assessment of the reinforcing and anti-cocaine effects of the preferential dopamine autorector antagonist, (+)-AJ 76 (submitted for publication).

Risner M. E. and Cone E. J. (1986) Intravenous self-administration of fencamfamine and cocaine by beagle dogs under fixed-ratio and progressive-ratio schedules of reinforcement. *Drug Alcohol Dep.* **17,** 93–102.

Risner M. E. and Goldberg S. R. (1983) A comparison of nicotine and cocaine self-administration in the dog: Fixed ratio and progressive-ratio schedules of intravenous drug infusion. *J. Pharmacol. Exp. Ther.* **224,** 319–326.

Risner M. E. and Jones B. E. (1976) Role of noradrenergic and dopaminergic processes in amphetamine self-administration. *Pharmacol. Biochem. Behav.* **5,** 477–482.

Risner M. E. and Silcox D. L. (1981) Psychostimulant self-administration by beagle dogs in a progressive-ratio paradigm. *Pychopharmacology* **75,** 25–30.

Roberts D. C. S. (1989) Breaking points on a progressive ratio schedule reinforced by intravenous apomorphine increase daily following 6-hydroxydopamine lesions of the nucleus accumbens. *Pharmacol. Biochem. Behav.* **32,** 43–47.

Roberts D. C. S. and Goeders N. (1989) Drug self-administration: Experimental methods and determinants, in *Neuromethods* (Boulton A. A., Baker G. B., and Greenshaw A. J., eds.), Humana Press, Clifton, NJ, pp. 349–398.

Roberts D. C. S. and Vickers G. J. (1984) Atypical neuroleptics increase self-administration of cocaine: An evaluation of a behavioral screen for antipsychotic drug activity. *Psychopharmacology* **82,** 135–139.

Roberts D. C. S. and Zito K. A. (1987) Interpretation of lesion effects on stimulant self-administration, in *Methods of Assessing the Reinforcing Properties of Abused Drugs* (Bozarth M. A., ed.), Springer-Verlag, New York, pp. 87–103.

Roberts D. C. S., Bennett S. A. L., and Vickers G. J. (1989a) The estrous cycle affects cocaine self-administration on a progressive ratio schedule in rats. *Psychopharmacology* **98,** 408–411.

Roberts D. C. S., Loh E. A., and Vickers G. J. (1989b) Self-administration of cocaine on a progressive ratio schedule in rats: Dose-response relationship and effect of haloperidol pretreatment. *Psychopharmacol.* **97,** 535–538.

Roberts D. C. S., Corcoran M. E., and Fibiger H. C. (1977) On the role of ascending catecholamine systems in intravenous self-administration of cocaine. *Pharmacol. Biochem. Behav.* **6**, 615–620.

Roberts D. C. S., Dalton J. C. H., and Vickers G. J. (1987) Increased self-administration of cocaine following haloperidol: Effect of ovariectomy, estrogen replacement, and estrous cycle. *Pharmacol. Biochem. Behav.* **26**, 37–43.

Stretch R. (1977) Discrete-trial control of morphine self-injection in squirrel monkeys: Effects of naloxone, morphine, and chlorpromazine. *Can. J. Physiol. Pharmacol.* **55**, 615–627.

Suzuki T., Masukawa Y., Yoshii T., Kawai T., and Yanaura S. (1990) Preference for cocaine by the weight pulling method in rats. *Pharmacol. Biochem. Behav.* **36**, 661–669.

Thompson T. and Schuster C. R. (1964) Morphine self-administration, food-reinforced, and avoidance behaviors in rhesus monkeys. *Psychopharmacology* **5**, 87–94.

Weeks J. R. (1961) Self-maintained "addiction" a method for chronic programmed intravenous injection in unrestrained rats. *Fed. Proc.* **20**, 397.

Weeks J. R. (1962) Experimental morphine addiction: Method for automatic intravenous injections in unrestrained rats. *Science* **138**, 143,144.

Weeks J. R. (1977) The pneumatic syringe: A simple apparatus for self-administration of drugs by rats. *Pharmacol. Biochem. Behav.* **7**, 559–562.

Weeks J. R. (1981) An improved pneumatic syringe for self-administration of drugs by rats. *Pharmacol. Biochem. Behav.* **14**, 573–574.

Wilson M. C. and Schuster C. R. (1972) The effects of chlorpromazine on psychomotor stimulant self-administration in the rhesus monkey. *Psychopharmacol.* **26**, 115–126.

Wilson M. C., Hitomi M., and Schuster C. R. (1971) Psychomotor stimulant self-administration as a function of dosage per injection in the rhesus monkey. *Psychopharmacology* **22**, 271–281.

Wise R. A., Yokel R. A., Hanson P. A., and Gerber G. J. (1977) Concurrent intracranial self-stimulation and amphetamine self-administation in rats. *Pharmacol. Biochem. Behav.* **7**, 459–461.

Wood R. W., Grubman J., and Weiss B. (1977) Nitrous oxide self-administration by squirrel monkey. *J. Pharm. Exp. Ther.* **202**, 491–499.

Woolverton W. L. (1986) Effects of a D1 and D2 dopamine antagonist on the self-administration of cocaine and piribedil by rhesus monkeys. *Pharmacol. Biochem. Behav.* **24**, 531–535.

Yanagita T. and Takahashi S. (1973) Dependence liability of several sedative-hypnotic agents evaluated in monkeys. *J. Pharm. Exp. Ther.* **185**, 307–316.

Yokel R. A. and Wise R. A. (1975) Increased lever pressing for amphetamine after pimozide in rats: Implication for a dopamine theory of reward. *Science* **187**, 547–549.

Yokel R. A. and Wise R. A. (1976) Attenuation of intravenous amphetamine reinforcement by central dopamine blockade in rats. *Psychopharmacol.* **48,** 311–318.

Opiate
Withdrawal-Produced Dysphoria

A Taste Preference Conditioning Model

Ronald F. Mucha

1. Introduction

The present chapter deals with a model of opiate abstinence in which the primary event is motivation, as indexed by the ability of withdrawal to produce avoidance behavior. The propensity of rodents to learn rapidly to avoid novel flavors paired with illness (Garcia et al., 1974; Rozin and Kalat, 1971) is the basis of this model, which should be viewed as analogous to that of simple place and taste preference conditioning models of the positive motivating effects of opiate administration (Rossi and Reid, 1976; Mucha and Herz, 1985). In the last few years, these models have resulted in a number of new and fruitful lines of anatomical (Zito et al., 1988; Mucha et al., 1985; Spyraki et al., 1988), behavioral (Lett and Grant, 1989; Martin et al., 1988; Vezina and Stewart, 1987), biochemical (Finlay et al., 1988), comparative (Mucha and Walker, 1987), genetic (Dymshitz and Lieblich, 1987), ontogenic (Kehoe and Blass, 1986; Schenk et al., 1985), and pharmacological (Acquas et al., 1989; Herz and Shippenberg, 1989; Mucha and Herz, 1985; Shippenberg and Herz, 1988) research.

The primary emphasis will be on the description of taste aversion produced by precipitated opiate withdrawal and the data needed to apply this efficiently as a simple test of withdrawal motivation. However, to place the model accurately in

From: *Neuromethods, Vol. 24: Animal Models of Drug Addiction*
Eds: A. Boulton, G. Baker, and P. H. Wu ©1992 The Humana Press Inc.

the context of the literature also presented, a short historical account of withdrawal motivation and a brief critical evaluation of existing research models will also be provided.

2. Background and Historical Perspective

That opiate withdrawal may be described as dysphoric, or otherwise unpopular, can be discerned from descriptions of the experiences of addicts. Often noted are discomfort, restlessness, agitation, anxiety, vomiting, diarrhea, neurogenic pain, abnormal sensations, and no interest in work. A variety of other effects connoting unpleasantness can also be found: tormenting sneezing and yawning, hallucinations, suicidal tendencies, sweating, shivering, complaints concerning room temperatures, tremors, and weakness (Sections 3 and 8 of Erlenmeyer, 1887). Together these intuitively suggest that withdrawal is to be avoided. From the fact that morphine can reverse the somatic symptoms and the sensory qualities of withdrawal in addicts, it can be further appreciated that withdrawal has long been used to explain why addicts seek out and self-administer more of the drug despite social and legal pressures not to do so.

There are, however, surprisingly few studies that have investigated the mechanisms of motivation produced by withdrawal (Goudie, 1990; Katz and Valentino, 1986; Mucha et al., 1986). The most recent thorough summary of a new body of data was almost 15 years ago (Goldberg, 1976).* The dearth of interest in this area may be explained partially by the complexity and expense of the existing models; this will be argued in Section 3. A further reason may come from the general assumption that all studies on opiate dependence and withdrawal are believed to have relevance for motivation, regardless of whether or not motivation is measured. This thesis is viewed as particularly relevant here, since it may help:

*This refers specifically to motivation produced by the withdrawal of the drug in the chronically treated organism. It must not be confused with the wealth of other literature to be found under the title drug dependence (eg., Wise, 1987a). This work is primarily concerned with the incentive or positive reinforcing properties of opiates (see Section 3.5.), and the term is a general one proposed by the WHO (Eddy et al., 1964) to refer to a state of psychic and physical dependence, or both, that can arise in the affected individual.

1. Explain why so little work has been done on withdrawal motivation even with the inefficient models available;
2. Clarify why little work and no discussions necessary to stimulate the development of new models of withdrawal motivation are found in the animal literature; and
3. Point out some of the directions that should be followed to stimulate the necessary work.

Thus, the views on motivation and opiate withdrawal have changed little since the start of systematic scientific investigations. In the early clinical studies, motivation produced by withdrawal was indicated through verbal reports, and the clinician/scientist of the 1920s and 1930s was confronted with a discrepancy between the intensity of withdrawal as described subjectively and its objective manifestations (Kolb and Himmelsbach, 1938). There was not the appreciation for subjective reports as data that exists today (Ericsson and Simon, 1980), and reliable ways to extract accurate information from the verbal reports have only been recently developed (Undeutsch, 1967; Yuille, 1989). Accordingly, the discrepancy was explained as the result of secondary gain to the addict of falsifying reports of withdrawal severity. What emerged was a view that only "objective measurements" of opiate withdrawal should be used (Kolb and Himmelsbach, 1938).

Advanced with this was an explanation for the relation between observable signs of withdrawal and an aversive state. Himmelsbach (1943) proposed that a withdrawal sign and morphine's "euphoria" in the addict are related mechanistically. In the spirit of Cannon's notion of homeostasis popular at the time (Cannon, 1929), the motivation for renewed morphine use was construed to be the result of the measurable signs of withdrawal in the way that thirst or hunger were thought to be the result of a parched mouth and stomach pangs (Cannon, 1929). This explanation has remained unchallenged. Unlike the situation for food and water intake (cf Montgomery, 1931; Wangensteen and Carlson, 1931), no extensive testing of the "Local Signs Theory" of withdrawal motivation is to be found. In 1949, Hebb drew attention to the need to understand processes of food and water intake in order to understand opiate

motivation; however, such developments of the natural reinforcement literature have had little impact on motivation associated
with opiate withdrawal. It may be that at the time of
Himmelsbach, and until very recently, morphine was accepted
to have significance only as a foreign toxin or poison (Grinker,
1938; Meggendorfer, 1928). The homeostatic readjustment to
morphine was then an acquired one, closely associated with prior
exposure to this toxin, and not to be compared with that of food
and water.

Therefore, the present literature on withdrawal relies almost
exclusively on somatic indices of withdrawal. This is the case
even in clinical areas concerned with drug intake and relapse
(*see* Turkington and Drummond, 1989). Only rarely does it men-
tion the sensory descriptions that were so rich in earlier reports
(Kolb, 1925). Accordingly, the study of withdrawal motivation
has become synonymous with the study of abstinence signs, as
a number of theoretical accounts suggest (e.g., Siegel et al., 1975;
Wikler, 1980; Bozarth and Wise, 1984).

3. A Critical Analysis of Models
   of Withdrawal Motivation

This section has a principal purpose of overviewing the
many potential models. A very functional approach is taken by
addressing the models with regard to the following two questions: First, what is needed to conclude that the measure used
reflects an underlying motivational state? Second, how practical
is the approach as a research tool for investigating the neurobiology of withdrawal motivation?

A difficulty was encountered with the first question. Controversy still rages, as it has for decades, on what constitutes a
motivational state (*see* Blackburn and Pfaus, 1988; Wise, 1987b),
and a definition is not clear even for thirst, hunger, or interest in
sex. This was considered to be much more problematic in the
case of drug withdrawal, which unlike food, as a example of a
motivating stimulus, does not have the equivalent to the organs
of taste and smell that can be used as the defining point. The
discovery of opioid receptors has made it only slightly easier to

find the "sensory organs" of opiate withdrawal motivation. Regardless of their nature and location, the abrupt cessation of chronic opiate exposure appears to result in an offset response (Fry et al., 1980; Schulz, 1988). Some controversy may exist on this point for the mouse vas deferens (Schulz and Herz, 1984), but it likely stems from the choice of measure used (North and Vitek, 1980). Accordingly, withdrawal after chronic opiate treatment may be a characteristic of most or all opiate receptors, which also happen to be widely distributed in the body.

Therefore, to avoid making any assumptions as to what constitutes motivation, the various models were categorized into two groups. Direct tests involve the demonstration of behaviors, such as conditioned avoidance of withdrawal. Indirect methods must rely on a direct measure to be validated.

3.1. Unconditioned Consequences of Withdrawal

The most commonly used method remains the observation and measurement of unlearned responses in a subject after withdrawal from chronic treatment with opiates. The termination of drug effect can be produced spontaneously, or it can be precipitated with a competitive antagonist, such as naloxone (Bläsig et al., 1973; Mucha et al., 1979; Wei, 1973). The signs of withdrawal range from various behavioral and autonomic signs to an almost exhaustive array of changes in electrical, biochemical, or physiological activity; the reader is directed to other reviews (Redmond and Krystal, 1984; Katz and Valentino, 1986) or any reasonable literature retrieval system under the keywords opiate withdrawal sign.

It will be noted that these models offer the advantage of a simple experimental design that can be applied to a wide range of investigations. The late Harry Collier and his colleagues provided a classic example. They enumerated a series of criteria derived from studies of physical dependence in intact animals (Collier et al., 1981). They then proceeded to use these as "benchmarks" for the presence of similar relevant processes in a purely physiological preparation. They found that incubation of segments of the isolated ileum in an organ bath with morphine resulted in naloxone-precipitated contracture of the gut. The magnitude of the effect showed a clear time-course of development related to the duration of the incubation and to the amount of morphine present. In addition, the sign was resulted from activity of the treatment drug on stimulation of the opioid receptors.

Indeed, there is a clear trend to the use of simple measures and individual signs in the study of mechanisms. Each laboratory tends to have a favorite single measure of withdrawal, such as jumping, weight loss, or shaking, which is measured alone. The reason is that the entire syndrome is difficult to quantify. Some individual signs of withdrawal show complex changes with variations in the severity of the morphine treatment (Bläsig et al., 1973; Mucha et al., 1979). However, two questions follow from applying the simplicity of this experimental design in the present context. The first concerns the relevance of a contracting gut preparation, or some other single sign, to withdrawal motivation. To the extent that stomach pains and diarrhea are well-known signs of withdrawal, an isolated gut preparation may be more than expedient. However, the second question is whether it has been validated as a criterion of withdrawal motivation or whether it has only been assumed to be so based on the theoretical notions of Himmelsbach, as outlined above.

Only in the case of the jumping response in rodents has there been any extended consideration of its motivational significance. This was first noted by the use of the term escape response to label the jumping (Wei, 1973). Siegel et al. (1975) showed that jumping was only seen when animals had the opportunity to escape and concluded that this was indicative that jumping was an index of an internal motivational state. However, this interpretation was subjected only recently to critical testing (Mucha, 1987). In using jumping (or any other single measure) to study withdrawal motivation, one would expect that some quantitative parameter of jumping should correlate with the magnitude of the motivational effect. However, jumping was not correlated with the place aversion of withdrawal across different conditions for precipitating withdrawal.

Various other commonly measured signs of withdrawal, such as weight loss, writhing, wet shakes, and diarrhea, were also poor predictors of withdrawal aversion (Mucha, 1987). Ten years earlier, Manning and Jackson (1977) drew a similar conclusion on the basis of a taste conditioning test, but their findings seem to have gone unnoticed in the literature. They described that naloxone produced a taste aversion 21 d after morphine pellet implantation without showing any measurable somatic signs of withdrawal.

3.2. Changes in Behavior
Produced by Other Reinforcers

A related set of indirect measures involves withdrawal-produced changes in operant responding for various reinforcers. Thus, animals that exhibit a stable baseline of responding for food, water, brain stimulation, saccharin, or any other reinforcer are put through withdrawal and allowed to engage further in the operant behavior. The literature is vast, and the behavior studies have ranged from simple lever pressing and normal consummatory behavior to complex schedules in a variety of species (e.g., Bergman and Schuster, 1985; France and Woods, 1985; Mello et al., 1981; Parker and Radow, 1974; Schaefer and Michael, 1983; Young and Thompson, 1979). Typically, the baseline is disrupted when the drug is withdrawn, and this recovers as the effects of withdrawal subside or soon after the treatment drug is given again.

These are thought to represent better measures than the somatic signs, despite the considerable time needed for operant training. The baselines are usually very stable and permit small effects to be seen (Pilcher and Stolerman, 1976a). Many data can be collected in the same subjects. Also, the use of different reinforcers and conditions for producing the baseline behavior can be used to look for the substrate of withdrawal. For example, with electrical stimulation in different regions as the reinforcer, there were different withdrawal-produced changes (Schaefer and Michael, 1983).

Finally, since the performance seen in withdrawal is usually described as disrupted (e.g., Carroll and Lac, 1987; Koob et al., 1989), the model is often interpreted as relevant for motivation of withdrawal. This is understandable given how addicts describe how they feel when in withdrawal. It is also known that they tend *not* to engage in favored activities during frank withdrawal. These can include eating, sleeping, sexual activity, and smoking (Sections 3 and 8 of Erlenmeyer, 1887).

However, whether a change in a baseline of operant responding can be interpreted as a sign of particular motivational change without other information should be seen as doubtful. How closely such changes parallel any negative motivational effects of withdrawal has not yet been analyzed. The increases and decreases in bar-pressing rates for electrical brain stimulation after morphine injections, for example, were only explained by Rossi and Reid (1976) after they used the conditioned place preference paradigm to look at the motivational states associated with the rate changes. The interpretation that changes in bar-pressing rates reflect specific motivation effects of morphine (Lorens and Mitchell, 1973; Olds and Travis, 1960; Olds, 1976) proved in fact to be wrong! This problem with rate measures may have been overcome with electrical brain stimulation as a reinforcer by using the thresholds for self-stimulation (Kornetsky and Esposito, 1979). However, this has not been regularly applied in assessing the effects of opiate withdrawal on a baseline of self-stimulation, and it is not readily possible with other reinforcers. This aspect of the model is particularly problematic for exploratory work, as exemplified in an attempt to extend the finding of a decreased baseline consumption of a saccharin-flavored solution seen in opiate withdrawal (Parker and Radow, 1984) to amphetamine withdrawal. The amphetamine animals showed, in contrast, an increase in saccharin preference (Mucha et al., 1990). There may be many interesting motivational interpretations of amphetamine's effect, but any conclusion requires the application of direct measures of motivation.

3.3. Conditioned Suppression

Occasionally noted in the literature is another model of withdrawal motivation that uses a baseline of responding. Instead of assessing the effect of putting animals through withdrawal, the effect of a stimulus previously paired with withdrawal is measured. Goldberg and Schuster (1967), for example, exposed opiate-dependent monkeys to a pairing of a light/tone stimulus with nalorphine infusions and showed that subsequently the stimulus suppressed operant behavior. For a comprehensive review, *see* Goldberg (1976).

This model has the advantage over studies of acute effects of withdrawal in that whatever is suppressing behavior on testing acts also as a stimulus for learning, and it can be considered a more direct measure of withdrawal aversion. The fact that testing is not during withdrawal itself, but sometime later, also precludes any direct motor activity effects of the withdrawal syndrome from confounding the test measure. This is particularly true if a reduction in responding is to be interpreted as a reflection of a withdrawal-produced aversive state.

However, the model is not entirely free of this and other criticisms. As Schuster (1976) queried in the context of the conditioned suppression produced by nalorphine, it not certain whether the operant behavior is affected through incompatible autonomic responses or because of a direct influence of the aversive properties of the withdrawal syndrome. From the literature on conditioned suppression with shock (Estes and Skinner, 1941), it is thought that, because of the aversive properties that had been acquired by the stimulus, the animals elicit defense behavior that is incompatible with a continuation of the operant responding. However, in following the thinking from the opiate–dependence literature, one can also argue that, through pairing with the opiate-abstinence syndrome, neutral stimuli can come to elicit some of the physiological changes associated with the unconditioned withdrawal (Wikler, 1948) and that these may then interfere with the baseline performance.

An additional criticism of this model is that conditioned suppression is not restricted to aversive stimuli. It has also been seen with positive motivating stimuli, such as food (Azrin and Hake, 1969; Stolerman and D'Mello, 1981). It might be further added that this work was performed mostly in monkeys; no data could be found on rodents. Also, most of the work was carried out almost 20 yr ago; therefore, problems can be expected in finding an empathetic reviewer for evaluation of the work, even if all the necessary controls were carried out.

3.4. Discriminative-Stimulus Tests

This model also falls somewhere between the indirect and the direct models of withdrawal motivation. The paradigm involves training animals to make a correct choice between two or more differentially reinforced outcomes when the information for choice is provided during a prior injection of drug or vehicle. The choice may involve pressing a left or right bar in an operant box to avoid electric shock, turning left or right in a T-maze for food, or predicting that a saccharin solution is or is not paired with an emetic (Colpaert, 1978; Overton, 1982; Jaeger and Mucha, 1990). In the study of withdrawal, opiate-dependent animals learn to discriminate the presence or absence of naloxone or some other antagonist (Gellert and Holtzman, 1979). This has been confirmed in various ways to reflect the withdrawal precipitated by the antagonist. For example, it is possible to show that a reduction in the morphine treatment dosing can be interpreted by the subject as similar to an injection of naltrexone (Gellert and Holtzman, 1979; *see* a review by Katz and Valentino, 1986).

Although these models require considerable time expenditures for initial training, they have the important advantage of studying a learned behavior associated with a sensory aspect of withdrawal. Therefore, it is possible to argue that what the addict recognizes as uncomfortable in withdrawal is actually being studied in this model.

Drug discriminative (DD) properties of a withdrawal in animals, however, should not always be taken as a clear indicator for its motivating effects. The relation between discriminative effects and motivation is not yet fully understood. A good example that discriminability and motivation can be unrelated is the fact that naloxone in opiate-naive animals has no DD properties, yet produces clear aversions in both operant and classical conditioning tests (Hoffmeister, 1986; Mucha and Walker, 1987). Also, DD properties of a drug are dependent on the procedures used to generate them (cf Jaeger and Mucha, 1990; Overton, 1979).

A slightly different DD model has been applied by Emmett-Oglesby and colleagues. Pentylenetetrazol (PTZ) injections are used to train naive animals, and state of withdrawal from various drugs, including opiates, is recognized as similar to the PTZ cue (Emmett-Oglesby et al., 1983,1984). Motivational interpretations of data with this technique also have to be validated with other models. Evidence that the PTZ may be producing the anxiety or discomfort seen in drug withdrawal comes from the fact that PTZ responds positively in screening tests for anxiety (Pellow and File, 1984; Lal and Emmett-Oglesby, 1983). However, in testing for the aversive effects of PTZ, we found that the effective doses for the DD were higher than those needed to produce a conditioned taste aversion (Mucha and Fassos, unpublished data).

3.5. Negative Reinforcement

Negative reinforcement is the main technique that springs to mind in the context of measuring the motivational effect of withdrawal. This term is usually applied to work with operant conditioning techniques and refers to the increased probability of a behavior produced by the termination of some event (Mackintosh, 1974); In the case of the opiate-dependent organism, the escape or avoidance of withdrawal is thought to produce opiate self-administration. Surprisingly little work, however, has been done to understand negative reinforcement, and the actual role that this process plays in self-administration of opiates is not clear (Katz and Valentino, 1986).

There are several variations of this model. In the standard paradigm, animals in withdrawal self-administer more opiate that those that are not (e.g., Mello et al., 1981). Animals can also be trained to emit a response in order to terminate infusions of opiate antagonist, and this effect is greatly potentiated in opiate-dependent animals as compared to opiate-naive controls (Downs and Woods, 1976). A third variation involves the demonstration that a neutral environmental stimulus paired with opiate withdrawal increases self-administration of morphine (Goldberg, 1976). In general, much of the work is reported in self-administering monkeys; however, various studies using rats and oral drug intake methods have been reported (Cappell and LeBlanc, 1981).

Little further attention will be given to these techniques here, since with regard to the present interests, they are complex and costly, which is in fact why attention was directed to simpler models of withdrawal motivation in the first place. This may also account for why so little work has been done with this model. They are also well described in the literature, and the reader is referred to Katz and Valentino (1986).

There are a number of additional problems with negative reinforcement of withdrawal as a model of withdrawal motivation. First, the standard test for negative reinforcement is confounded with other reinforcing effects of an opiate agonist, termed incentive or positive reinforcing effects. It has long been suspected that withdrawal may not account for all the reinforcing effects of opiates (*see* a review by Cappell and LeBlanc, 1981). Thus, it was finally demonstrated with the place preference paradigm that a positive motivating effect of opiates can be seen after a single injection, without any opportunity for withdrawal to occur (Mucha et al., 1982). Since the negative reinforcement model of withdrawal motivation uses a baseline of opiate self-administration, many controls must be incorporated to extract the contribution of responding for morphine produced by withdrawal over that of the positive reinforcing effect.

An additional problem with this model concerns the use of the iv self-administration method. Oral self-administration is hampered by the flavor of some of the opiate agonists (Stolerman and D'Mello, 1981) and by the pharmacological effects of opiates on ingestion processes (Reid, 1985). Because the iv catheters are very delicate, animals are often housed relatively immobile for extended periods or continually attached to infusions pumps (e.g., van Ree et al., 1978). Therefore, the work is further restricted to well-equipped and specialized laboratories. In addition, the stress of these manipulations or of the isolation on the subject has not been fully assessed, but has surfaced as a possible confound of animal studies of drug intake (Alexander et al., 1985). A related point is that in recent years the question of animal ethics has surfaced (McGregor, 1986). Therefore, some investigators believe that many questions can be worked out totally, or at least partially, in simpler and less traumatic experimental preparations.

3.6. Conditioned Aversions

Various reports can be found where exposure to different flavors while animals are in opiate withdrawal can lead to an avoidance of the flavor (Mucha et al., 1990; Parker et al., 1973; Parker and Radow, 1974; Zellner et al., 1984). This has also been seen with caffeine withdrawal (Vitello and Woods, 1977). The well-known fact that spontaneous withdrawal of opiates tends to be a prolonged process of up to several days, and the fact that withdrawal has consequences on normal ingestive processes, such as reduction of drinking and eating, make this phenomenon not well suited as an efficient model of withdrawal motivation for neurobiological studies.

Ternes (1975) was the first to report that naloxone injections (8 mg im) produced a strong conditioned aversion to a sucrose solution in a small group of rats dependent on morphine and not in nondependent controls. In the light of the present knowledge about opiate antagonists, this offered a vast improvement in the control of the withdrawal syndrome. Antagonists can be injected or infused to produce the local withdrawal of different opiate receptor subtypes (Paterson, 1988; Schulz, 1988) or anatomical populations of receptors (Brown and Goldberg, 1985), at the discretion of the experimenter and with a relatively short and predictable time-course.

Only a small number of additional studies have been carried out on conditioned aversions produce by precipitated opiate withdrawal, but the potential of these results was revealed in several ways. Pilcher and Stolerman (1976b) showed Ternes' effect with a lower dose of naloxone (10 mg/kg ip) and saccharin as the flavor. The degree of aversion produced was related to the doses of morphine used to produce the dependence, and the technique was found to respond to morphine treatment as low as 1 mg/kg twice per day. Frumkin (1976) also demonstrated an aversion to saccharin with the same dose of naloxone in rats implanted with two morphine pellets. Manning and Jackson (1977) showed this effect with 5 mg/kg at periods of 3–21 d after implantation of a single pellet. More recently, it was reported that three conditioning trials with 0.004 mg/kg naloxone (iv) produced a conditioned place aversion in rats treated chronically with a morphine pellet. In contrast, the lowest dose to produce any conditioning in animals implanted with a placebo pellet was 0.1 mg/kg iv naloxone. The basic effect was also replicated in placebo and morphine pellet-implanted mice (Mucha and Walker, 1987). In addition, variations of this paradigm have been used to study the relation between somatic signs and motivational indices as indicators of withdrawal and the central or peripheral locations of receptors responsible for the aversive effects of precipitated withdrawal (Hand et al., 1988; Mucha, 1987,1989).

A number of advantages over the operant paradigms have been enumerated on several occasions as reasons why variations of place and taste preference conditioning are useful for full-scale mechanistic studies of the motivation produced by opiates and other drugs of abuse (Bozarth, 1987; Cappell and LeBlanc, 1975; Carr et al., 1988; Mucha and Herz, 1985; Mucha et al., 1982; Spyraki, 1987; van der Kooy, 1987; van der Kooy et al., 1982). Included are precise experimental control over the conditions of training and testing, low costs for setting up and running, and high sensitivity. The fact that very little training is required also makes the procedure useful for anatomical mapping studies. Added to this is the general flexibility of the classical conditioning paradigm, which can be applied to yield comparable data in diverse types of subjects (*see* a study in infant rats by Kehoe and Blass, 1986). However, for balance, these can be contrasted with a review in this series (Swerdlow et al., 1988) that takes a less optimistic view of such models.

It was first formally suggested by Ternes (1975), Pilcher and Stolerman (1976b), and Manning and Jackson (1977) that an antagonist-induced conditioned aversion could be useful as a routine test for withdrawal, but little work followed. The long periods required for training and preparation of the animals in several of the procedures may have tended to make them appear complex and of no advantage over other established models of withdrawal. Also, the method seemed to be presented as an indicator of cumulative general malaise produced by withdrawal—a type of summary effect of the already well-known effects of withdrawal. Therefore, it may not have appeared particularly interesting theoretically. In addition, the doses originally used were very high (5–10 mg/kg) and in a range in which, as was pointed out shortly afterward, effects may not have been the result of activity on opioid receptors (Sawynok et al., 1979). Finally, it was also noted about this time that naloxone and other antagonists produce detectable conditioned aversions in nontreated animals (Stolerman et al., 1978), an effect that, at the time, could not be incorporated into the traditional view of dependence as an adjustment of homeostasis to the foreign toxin, morphine.

Ten years later, the preference conditioning methods were proposed again as a model of withdrawal motivation (Mucha et al., 1986). Evidence was now presented that the motivational response of precipitated withdrawal measured by conditioned place and taste aversions could be seen with doses of opioid antagonists that were in the ranges used to elicit withdrawal in humans (Zilm and Sellers, 1978; Peachy and Lei, 1989). It was further noted that the magnitudes of common somatic signs of abstinence appeared to be independent of one another and that the conditioned aversion procedures could be made very efficient (Mucha et al., 1986). In the meantime, research had also indicated that the naloxone-produced aversion seen in the naive animal may have involved blockade of endogenous opioid activity on opioid receptors (Mucha et al., 1982,1985; Mucha and Walker, 1987; Hoffmeister, 1986), and could be comparable to the larger and more potent aversion found in morphine-dependent animals.

4. Description of a Taste Aversion Model

The purpose of this section will be to describe further a two-flavor, unbiased, taste preference conditioning procedure for the study of opiate withdrawal motivation in morphine pellet-implanted rats. Basically, it involves the pairing of a distinctive taste with an unconditioned stimulus (UCS), in this case antagonist-precipitated withdrawal. This is followed by a test where the subject is given a choice between this flavor and a comparable control flavor. It is well known that animals will avoid the paired taste when it was previously paired with an UCS known to produce avoidance learning in other paradigms. Recent work suggestes that aversions seen with this and related paradigms predict dysphoria in humans (Pfeiffer et al., 1986).

Taste preference conditioning was chosen for its expediency and potential for broad application. Only the most basic rodent vivarium is required, and the time for economically running large numbers of taste conditioning subjects is also much less than that for place conditioning, the second choice in this regard. It should be noted, nevertheless, that most of the parameters studied here can be applied to a place procedure (*see* Mucha and Herz, 1985 for details).

A two-cue, unbiased procedure was chosen because of its high sensitivity and because it embodies internal control conditions that are needed for classical conditioning, such as for sensitization and pseudoconditioning (Rescorla, 1967). This is not the case with one-cue procedures, which require the running of separate control groups (Mucha and Iversen, 1984; Section 4.5.). Two-cue procedures also are suited for situations where both positive and negative motivational effects may be produced by the drug reinforcer (Mucha and Herz, 1985). This was a concern even for the study of precipitated withdrawal, since opioid receptor antagonists have been reported to produce positive and negative preference conditioning (Bechara and van der Kooy, 1985), and the relevance of these data for the present phenomenon has not been clarified. In the area of opiate agonists, for example, the negative consequences of not using tests sensitive to both positive and negative motivational effects required almost two decades to rectify (Bechara and van der Kooy, 1985; Lett and Grant, 1989; Mucha and Herz, 1985,1986).

In Section 4.1., the precise methods will be described. In Section 4.2., the purpose will be to demonstrate that there are stable conditions of training and testing that yield reliable, replicable, and efficiently collected findings on the aversive effects of precipitated withdrawal. Information about the choice of various parameters will also be given in the event that some aspects of the procedures need to be individualized. The descriptions will involve considerable detail, in part because they cannot be found elsewhere. Also, not every parameter has been manipulated, and the procedures should still be considered in development. Finally, to promote a comparison of findings in the literature, it is advocated that as many aspects of the procedures as possible be standardized. In the final three sections, some apparent problems of the present taste preference procedure will be discussed and possible solutions presented.

4.1. General Experimental Procedures

4.1.1. Subjects and Surgical Preparation

Drug-naive, male Sprague-Dawley rats weighing 200–265 g were housed in groups of five or six in Plexiglas™ boxes with hardwood bedding material. The colony room was maintained on a 12/12-h light/dark cycle (lights on at approx 7:00 AM). The rats were given at least 1 wk to habituate to the colony room, during which time food and water were continuously available. In preparation for subsequent taste preference conditioning, rats were first habituated on one occasion to drinking in their test cages: They were deprived of water overnight (about 15 h) and then placed for 30 min into their test cages, where a waterspout was made available.

Surgery, when necessary, was carried out under halothane aesthesia. Pellets were implanted sc in the middle of the back through a single incision that was closed with wound clips. Multiple pellets were placed individually into separate pockets. For central drug administration, a single 23-gage, stainless-steel guide cannula that terminated just dorsal to the lateral ventricle was chronically implanted using stereotaxic procedures. Five to 7 d were allowed for convalescence. After testing, the rats were killed, and an injection of dye was used to verify the position of the placement.

The rats were trained and tested in standard rat cages that were located in a dimly lit room provided with white masking noise. The cages were fitted with filled food hoppers. Richter-type drinking tubes were used to present test solutions. In the past, dripless drinking spouts have also been used (Mucha and Herz, 1985).

4.1.2. Drugs, Injections, and Flavored Solutions

The experimental drugs employed were naloxone HCl, naltrexone HCl (kindly donated by duPont deNemours Ltd., Glenolden, PA), Mr 2266, methylnaltrexone bromide, *N,N*-diallylnormorphinium bromide (kindly donated by H. Merz of Boehringer Ingelheim KG, Ingelheim, Germany), and ICI 174,864. Where possible, the concentrations of the drug solutions were prepared so that they could be administered during a 0.5–1.0 mL sc injection or a 4 μL icv infusion. The sc-administered drugs were prepared in physiological saline and icv-administered drugs usually in distilled water; Mr 2266 was mixed with an equivalent of HCl, and ICI 174,864 was prepared as a 2.5 mg/ mL solution with 0.5% sodium bicarbonate as the solvent. The vehicle was distilled water or saline, except in the case of ICI 174,864, where it was the corresponding dilution of the original bicarbonate solution. The drugs and vehicle solutions were stored until immediately before use at –70°C. The sc injections were given in the neck skin with a sharp syringe needle, whereas the icv administrations were 45-constant infusions with a 30-gage injector calibrated to pierce the ventricular ependyma.

Unless stated otherwise, the pellets were provided by C. Kim of the Addiction Research Foundation (ARF), Toronto. They were made with morphine base (British Drug Houses-BDH, Toronto, Ontario, Canada) according to the recipe of Gibson and Tingstad (1970). The final product consisted of 13 mm diameter, 1 mm thick disks produced with a ring press under 5 t of pressure. When appropriate, these contained 75 mg morphine. Also used were pellets available through the United States National Institutes of Health (NIH), Institute on Drug Abuse. These were also prepared according the formulation of Gibson and Tingstad, contained 75 mg or no morphine, and were donated by R. Hawks (Research Technology Branch, NIDA, Rockville, MD); however, they were harder and smaller than those from Toronto. Details concerning these were described by Yoburn et al. (1985).

Unless otherwise stated, the solutions for the two-flavor, unbiased method comprised (a) 9.36 m*M* monosodium glutamate (MSG) and 92.7 m*M* NaCl, and (b) 1.13 m*M* citric acid and

0.71 m*M* saccharin (all chemicals were from Sigma Chemical Co., St. Louis, MO) prepared with fresh Toronto tap water. The flavored solutions were refrigerated until used. The solutions will be referred to hereafter as MSG/NaCl and Cit/ Sacc, respectively.

4.1.3. Training and Testing Procedures

Training started by limiting water access in the home cage to 30 min/d between 6:00–7:00 PM. During a conditioning session, an individual, water-deprived rat was placed into its test cage, and 5 min later, a tube containing 3 or 4 mL of flavored solution was positioned onto the cage (3 mL of fluid were given only in the initial time-course study). Ten minutes later, the subject received an icv injection, an sc injection, or both (as required, e.g., Mucha, 1989) and then was returned to its home cage. In the standard protocol involving one cycle of training, subjects were given a training session on two consecutive days, one of which involved the pairing of drug with one of two tastes and the other pairing of vehicle with the other taste. The pairing sessions were in the mornings. In the standard three-cycle protocol, rats were given one cycle of training on each of three consecutive days. One session was in early morning, and one in the early afternoon. There were no less than 4 h between sessions, and the interval between the afternoon session and the evening watering was at least 4 h. The assignment of drug condition with a particular flavor and the order of presentation of the drug to individual subjects were balanced across the various groups; the assignment of a rat to a training condition was random.

Testing was carried out 1 d following training. The rats were water-satiated overnight and placed for 24 h into their test cages, which were fitted with two tubes containing 100 mL of each of the test solutions. The data comprised the difference between the volume consumed of the drug-paired taste and vehicle-paired taste on the test day, expressed as percent of total fluid intake, and presented as mean ± SEM. These scores were analyzed parametrically following an arcsin transformation, which is necessary for percent data (Kirk, 1968). Whether an individual dose produced conditioning was usually evaluated using the sign test or the Wilcox paired-ranks test (Siegel, 1956), since some animals responded with very high scores on one of the choices of the test and low on the other.

4.2. General Findings and Considerations

4.2.1. Preliminary Reflections

The starting point of any work concerns the composition of the two flavors. A main assumption of a two-cue, unbiased model is that, under control conditions, animals will explore and sample the two cues, and treat them as neutral relative to one another. Initially, testing should be done to determine whether there are any preferences for one flavor over the other and whether the animals are, indeed, drinking a substantial amount of both fluids. Otherwise, the sensitivity of the test will be reduced.

We have noted with our flavors that individual rats can vary in their overall amount of total fluid consumption. The mean total fluid consumed in control animals tends to be similar for a given batch of animals, but between batches, it can range from 40 to 70 mL. It should be noted that there is usually a drop in total fluid intake in groups that show a strong conditioned aversion to one of the flavors. These differences in total fluid intake should not affect comparisons of data from different groups if the test data are converted to a preference score, as done here.

A clear flavor bias will require adjustment of the flavor compositions. The need for this was first noted at the Department of Experimental Psychology, Cambridge University, in collaboration with S. Iversen, when the flavors of Stolerman et al. (1978) were tested in naive rats. It was found that every rat consumed four to five times more Cit/Sacc than MSG/NaCl. The compositions were then manipulated until the average consumption of the two flavors was similar. The mean consumptions of the MSG/NaCl and Cit/Sacc solutions made with the appropriate flavor recipes were 32.8 ± 2.3 and 27.1 ± 3.1 mL ($n = 4$), respectively. The concentrations are given in Table 1. This same process was also necessary when the work was further pursued in Munich at the Max-Planck Institute for Psychiatry in collabo-

Table 1
Examples of Different Compositions of the Flavors Used
in the Two-Flavor, Unbiased, Taste Preference Procedure

	Water	MSG/NaCl		Cit/Sacc	
		NaCl, mM	MSG, mM	Sacc, mM	Citric acid, mM
A[a]	Distilled	128	12.5	2.0	1.0
B[b]	Distilled	128	12.5	0.5	1.5
C[c]	Deionized	128	12.5	0.25	1.5
D[d]	Tap	92.7	9.36	0.71	1.13

[a]From Stolerman et al. (1978).
[b]From unpublished work (details on rats in Mucha and Iversen, 1984).
[c]From Mucha and Herz (1985).
[d]From Mucha (1989).

ration with A. Herz and then later in Toronto. The underlying factors for these variations are not clear. The laboratory environment may be important, and water sources could be critical. It is unlikely that this is owing to strain, since we have always used Sprague-Dawley rats. An additional recipe for these flavors can be found in D'Mello et al. (1977).

A further initial consideration is the minimum size of groups needed. These are estimated using power analysis (Cohen, 1987). This requires a value for the test score dispersion and the expected effect size. These can be obtained from preliminary studies or from the literature.

4.2.2. Time-Course of the Model's Sensitivity

The use of an implanted pellet to produce morphine dependence was chosen because of economy and availability. However, a pellet produces a constantly changing level of morphine. The most appropriate time after an implant for the study of the naloxone-precipitated aversion, therefore, was determined empirically (1) by measuring conditioning with different low training doses of naloxone at various times after implantation of a single pellet of morphine and (2) from the morphine levels produced by a morphine implant.

Presented in Fig. 1 are test data from a single cycle of taste training started on different days after pellet implantation. The high and low naloxone doses used produced near maximal and

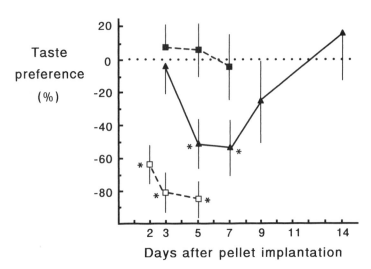

Fig. 1. Taste preference conditioning scores (ordinate) of 11 different groups of rats implanted with a single morphine pellet on day 0 and given one cycle of pairing commencing on different days after implantation (abscissa). The naloxone doses (sc) were 0.005 mg/kg (closed squares), 0.015 mg/kg (closed triangles), or 0.05 mg/kg (open squares). There were 8–12 rats/group. The dotted line depicts no flavor preference. Asterisks indicate significant conditioning.

no taste aversion, respectively, but there was a clear variation over time in the preference scores seen at 0.015 mg/kg naloxone [$F(5,53) = 2.73, p < .03$, with significant conditioning produced in this group on days 5 and 7].

Run in parallel with these subjects was a different group of similarly prepared rats; tail vein blood samples were taken repeatedly after implantation according to methods described previously (Mucha et al., 1987). Serum-free morphine levels were then kindly analyzed by C. Kim (Kim and Katz, 1984), and indicated peak levels early after pellet implantation and an absence of detectable release of morphine in any subject until after day 9 (Fig. 2). More importantly, the times of the potent responses to naloxone in the conditioning tests corresponded with a relatively slow drop in magnitude of the morphine levels.

Accordingly, the optimal period of stability of the morphine levels and of high sensitivity to naloxone was found with the

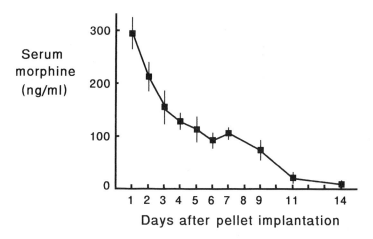

Fig. 2. Levels of free morphine in serum obtained from tail vein blood of rats implanted with one morphine pellet on day 0 ($n = 6$).

present procedure to be about 5–7 d after pellet implantation. The data collected from this training period were then replotted as shown in Fig. 3 (closed triangles) and contrasted with data collected in similar subjects implanted with a placebo pellet (open triangles). Thus, the effects measured in the morphine pellet-implanted rats were clear and the result of the morphine treatment rather than the surgical procedures.

4.2.3. Role of the Pellet Preparation

Widespread application of the present model requires that consideration be given to the fact that the methods of morphine treatment can influence the magnitude of the dependence scores (Meyer and Sparber, 1976; Cochin et al., 1979). Even the precise method of manufacture of the pellets seems to be critical (Meyer and Sparber, 1976), with results differing depending on the surface area and hardness of the pellets (Meyer and Sparber, 1976). Since the release of morphine from our pellets appeared to be complete much earlier than suggested by the literature (Manning and Jackson, 1977; Young and Thompson, 1979), another preparation of pellet was tested.

In Fig. 3 is a DR curve collected with morphine pellets available to all investigators from the NIH (closed squares). The curve

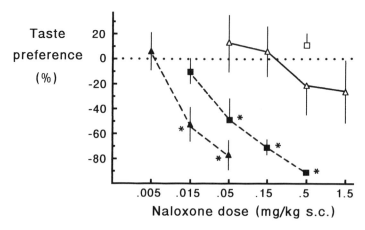

Fig. 3. Taste preference conditioning scores as a function of naloxone train-ing dose in rats implanted with a single morphine pellet (closed symbols) or placebo (open symbols) supplied by the Addiction Research Foundation in Toronto (ARF, triangles) or by the National Institutes of Health (NIH, squares). Eight to 12 rats were used in each group.

for the taste aversion in rats implanted with an NIH pellet was shifted to the right, as compared to that of the Alcohol Research Foundation (ARF) pellet, as indicated from the doses common to the two curves. [$F(1,34) = 11.6$, $p < .002$]. Serum morphine levels measured from six rats implanted with one NIH pellet were lower on day 1 (154.5 ± 21.9 ng/mL) to day 3 (130.8 ± 19.4), but higher on to days 9 (86.5 ± 20.1) and 14 (43.2 ± 10.9) than those obtained in rats treated with our usual pellet (*see* Fig. 2).

The magnitude of the taste aversion appeared to be increased with a second NIH pellet. Plotted in Fig. 4 are the results of one cycle of conditioning with 0.15 mg/kg naloxone sc beginning on day 5 in rats implanted with one-half, one, two, or three NIH pellets. Whereas all but three rats, two in the one-half and one in the three pellet treatment groups, avoided the flavor paired with naloxone, there was also evidence for a sig-nificant group effect [$F(3,37) = 4.27$, $p < .01$]. This would suggest that two NIH pellets may be required to yield results compa-rable to those collected here with the standard procedure using one ARF pellet.

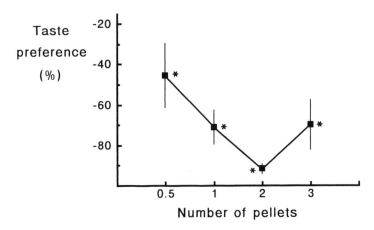

Fig. 4. Taste preference conditioning scores measured in rats given one cycle of training with 0.15 mg/kg sc naloxone starting 5 d after implantation of one-half to three morphine pellets (from NIH). Eight to 10 rats were in each group. For other details, *see* Fig. 1.

4.2.4. Specificity of Test and Treatment Drug Effects

As a control, it was also established that morphine implantation does not effect major changes in the taste aversion produced by any and all systemically administered aversive agents. To demonstrate this, lithium chloride was chosen as the appropriate test drug. Lithium chloride does not bind to opioid receptors, but is sensitive to some of the same lesion effects as naloxone (Shippenberg et al., 1988; Mucha et al., 1985). Depicted in Fig. 5 is a study of one cycle of conditioning with lithium chloride (sc) beginning 5 d after implantation. There was neither a significant effect of pellet condition. $[F(1,48) = 0.74]$ nor of pellet condition by dose interaction $[F(2,48) = 1]$. There was, however, a clear effect of dose $[F(2,48) = 12.5, p < .001]$.

The question of specificity has also been directed at the question of the treatment drug. Stereospecificity refers to the fact that only the active levo-rotatory isomer of the morphine-like agonists acts on the opiate receptor (Mucha and Herz, 1985). To confirm that the effect of the morphine treatment studied here was likely the result of chronic receptor activation, as is known to occur for the somatic signs of withdrawal (Collier et al., 1981), a

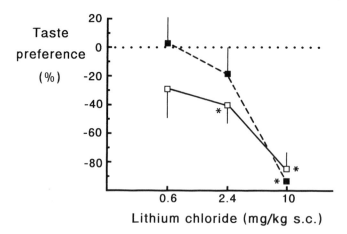

Fig. 5. Taste preference conditioning scores measured in rats implanted with a morphine or a placebo pellet and trained starting 5 d later with different doses of lithium chloride. There were 8–10 rats in each group. For details, *see* Fig. 1.

study was done using osmotic minipumps to treat rats chronically with levorphanol or dextrorphan, its isomer, which is inactive at the opioid receptor.

Rats were given a single conditioning cycle with 0.1 mg/kg naloxone sc, commencing 12 d after implantation of an Alzet 2002 minipump. Details of the pumps and drug preparations are given elsewhere (Mucha and Herz, 1986; Mucha et al., 1982, 1990). The pumps were filled with 20 mg/mL of levor-phanol- or dextrorphan-tartrate (Hoffman-LaRoche, Etobicoke, Ontario, Canada). The animals implanted with the levor-phanol pumps were also injected ip four times with 5 mg/kg levorphanol tartrate: once in the afternoon and evening of the day of surgery, with a repeat of the same the day after surgery. The dextrorphan-treated animals were similarly treated with 5 mg/kg dextrorphan tartrate. On the taste test, the levorphanol group showed a significant aversion to the flavor paired with the naloxone (–63.3 ± 18 s, n = 10), and the dextrorphan group did not (–12 ± 21 s, n = 10).

4.2.5. Number of Conditioning Trials

Careful consideration should be given to this parameter. Our experiments suggested that it may be wrong to expect (1) that a degree of standardization of the methodology can be

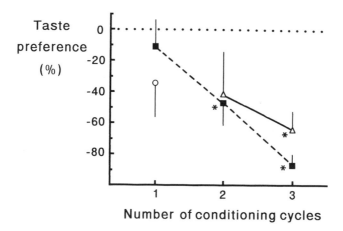

Fig. 6. Taste preference conditioning scores of rats implanted with a single morphine pellet and then given one to three cycles of training with icv administered naloxone. The doses per training session were 30 nmol (open circle), 3 nmol (closed squares), and 1 nmol (open triangles). There were eight rats in each group, except for six in the group trained for two cycles with 1 nmol and 16 and 32 in the groups trained for one cycle with 30 and 3 nmol, respectively. For other details, *see* Fig. 1.

attained by holding the trials constant in different experimental situations, and (2) that the sensitivity of the procedure increases with the number of trials. Concerning point one, the route of administration was seen to be important. Systemically administered naloxone, for example, produced a clear taste aversion in dependent animals, but not in controls, after only one cycle of training (*see* Fig. 3). However, when naloxone was administered icv, two or three cycles could be required. In Fig. 6 are data on taste preference conditioning measured after one, two, or three cycles of pairing with icv naloxone over days 4–6 following implantation of one morphine pellet. The three-cycle group was given two training sessions per day. The training sessions for the one- and two-cycle groups were distributed in a nonsystematic fashion over the training days so as to minimize the bias in comparison with the three-cycle group with regard to (a) duration since implantation and (b) last training before testing.

The nature of the antagonist used may also be important for point one. The absence of taste aversion learning, noted above with icv, as opposed to sc naloxone after one cycle of training in dependent rats, was not apparent with the antagonists methyl-naltrexone, diallylnalorphinium, and naltrexone (Mucha, 1989). In addition, it was found that naltrexone and diallynalorphinium both showed a weak aversive effect in placebo-implanted rats after one pairing (Mucha, 1989).

Indeed, the relation between number of trials and the sensitivity of the procedure is troublesome (point two above) because of aversions produced by antagonists in drug-naive animals (Stolerman et al., 1978; Mucha et al., 1982). Thus, with naloxone sc, for example, after one cycle there appears to be a qualitative effect of the morphine pellet: clear aversions with the pellet and none with the placebo. However, after three trials, the effect of the morphine pellet becomes a quantitative difference, with the dose–response curve of the pellet-implanted animals shifted to the left of that of the placebo (Mucha et al., 1982; Mucha, 1989).

Accordingly, with respect to the number of groups that are needed to establish an effect of the morphine treatment, a study may be more difficult with an increased number of trials. Problems with interpretation of the literature can also arise, if this point is not considered closely. An example of this problem may already be found in the place conditioning literature: Hand et al. (1988) reported a morphine-produced potentiation of a place aversion produced by three trials of icv methylnaloxone that was considerably less than that seen previously with a single trial with methylnaltrexone (Mucha, 1987). Whereas this may be in part owing to the different drug used, the overall sensitivity of the Hand et al. paradigm was changed through the use of multiple trials (Mucha, 1989).

4.2.6. Retention and Extinction

A study of the interval between training and testing and the effect of repeated testing on the magnitude of the conditioned aversion is also presented here to confirm further that the present model reflects learning and to indicate further the different test situations that are possible. In rats implanted with a single morphine pellet, a single cycle of training with 0.9 mg/kg naloxone was carried out starting 5 d after surgery. The rats were randomly divided into three groups: One group was tested 24 h after training, a second group was tested at weekly intervals starting 1 wk after training, and the third group was tested only after 4 wk, but received all earlier experimental manipulations, including placement in the cages, handling, and so on, as Group 2. It should be noted that, 1 d after training, the residual pellets of Groups 1 and 2 were removed to prevent any pairing of spontaneous withdrawal with the flavor presentation during the testing.

As seen in Fig. 7, the taste aversion was retained even for 4 wk after training. There was, however, a gradual loss of this effect as the interval to test increased, as confirmed by a significant group effect $[F(2,27) = 6.9, p < .004]$. It was also seen that, with repeated testing, the loss was even greater, such that, at 4 wk, there was a significant difference between the aversion score measured in the group with and without the repeated testing $[t(18) = 2.5, p < .02]$.

4.3. Choice of Antagonists and Routes of Administration

As noted in Section 4.2.5., under standard conditions of training and testing, the effects are not the same for different antagonists and routes of administration. To help overcome this problem and to facilitate the decisions regarding the choice of drug and the different advantages that they offer, Table 2 summarizes conditioning produced by four related antagonists given sc or icv to pellet- and placebo-implanted animals.

Whereas this may seem problematic for the standardization of the procedure, these differences may reflect factors that are relevant for understanding the mechanisms of the behavior studied here. In this case, central sites may be necessary for producing the precipitated withdrawal (Mucha, 1989). However, a practical advantage is that the test system may be optimized to particular needs. It can be noted that methylnaltrexone is particularly useful in discriminating the morphine treatment conditions when an icv drug is used. Naloxone and naltrexone seem to serve this purpose for the sc route.

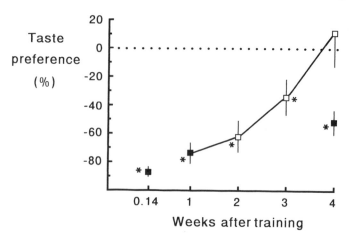

Fig. 7. Taste preference conditioning scores measured in three groups of rats (*n* = 10 group) tested at different intervals after the completion of a single cycle of training with 0.09 mg/kg sc naloxone (closed symbols) and in the 1-wk group again on three repeat tests taken at weekly intervals (open symbols). For other details, *see* Fig. 1.

Table 2
Lowest Doses of Different Intracerebral
and Systemic Opioid Antagonist Treatments
That Produced Significant Taste Aversions
in Placebo and Morphine Pellet-Implanted Rats

Drug	Morphine pellet icv, nmol	sc, nmol	Placebo pellet icv, nmol	sc, nmol
Methylnaltrexone[a]	0.1	50,000	>10[b]	50,000
Diallylnormorphinium[a]	4	10,000	20	2000[c]
Naloxone[d]	1.0	10	30	100
Naltrexone[a]	10	10	100	1000
Naltexone[d]	1	10	–	–

[a]One training cycle, from Mucha (1989).
[b]The highest dose used was 10 nmol.
[c]DR curve was not strictly linear.
[d]Three training cycles.

It can be further noted that the effect of the morphine pellet may only appear with antagonists that are selective for the μ opioid receptor. This is illustrated here in Fig. 8, where data are summarized from a study that compared conditioning in pellet

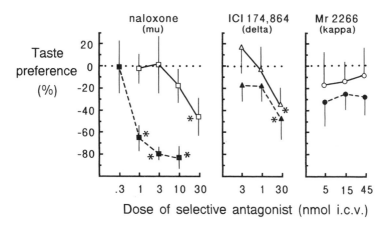

Fig. 8. Taste preference conditioning scores of 20 different groups of rats (8–13/group) implanted with a single morphine (closed symbols) or placebo pellet (open symbols) and then given three cycles of training with intracerebral dosing of the one of three different opioid receptor selective antagonists (*see* figure). For other details, *see* Fig. 1.

and placebo animals trained with icv dosings with three antagonists selective for the μ, δ, and K opioid receptors, namely naloxone, ICI 174,864, and Mr 2266, respectively. Three cycles of training were used, since preliminary experiments indicated that conditioning was produced by Mr 2266 and the ICI compound in the dependent rat after three trials.

In contrast to the effects seen with naloxone, neither of the other two antagonists showed any significant effect of the morphine treatment. The largest and smallest F values were $F(1,61) = 2.1$ and $F(2,56) = 0.1$ for the treatment effect in the peptide and the treatment dose interaction in the Mr 2266 animals, respectively (all $p > 0.15$). There was evidence that these drugs were nevertheless effective in producing an aversion in the present test. For ICI 174,864, there was a significant dose effect [$F(2,61) = 3.2$, p < .05], and this was associated with significant avoidance of the drug-paired fluid in both groups trained with 15 nmol. With Mr 2266, there was no significant dose effect, but inspection of the data indicated that, of the 61 rats run, 40 showed an avoidance of the flavor paired with the drug, suggesting that a small aversion was produced (binomial test, $p < .03$).

4.4. Problems of Motor Artifacts

A major difficulty in the measurement of learning is the existence of performance problems. This has been shown to be important in the case of a number of simple models of learning and motivation (Rossi and Reid, 1976; Pinel and Mucha, 1973). With the present taste conditioning procedure, which involves a choice test, the problem of motor artifacts is minimized, but not necessarily eliminated.

A possible interpretation involving motor artifact of the present effects could follow from the work of Wikler (1948,1980), demonstrating that the somatic signs of withdrawal can act as unconditioned responses (UCRs) for Pavlovian conditioning. Presentation of a withdrawal conditioned stimulus (CS) outside of the withdrawal situation elicits various indications of somatic signs of withdrawal in several species (O'Brien et al., 1977). Since drinking involves approach behavior, it can be suggested that failure to continue drinking after the flavor is tasted may be because a motor response, such as jumping or vigorous shaking, is reflexively elicited; these would compete with the drinking. To test whether this was the case, the defensive burying test for avoidance learning was used. In this test, there is an approach component in the way a subject responds to an aversive CS (Pinel and Wilkie, 1983). In order to eliminate the CS, the subject must approach it and engage in a complex motor behavior—burying. Therefore, it would not be possible for the conditioned withdrawal mechanism offered above to produce defensive burying. A protocol similar to that of the taste conditioning was used, except that the CSs were two objects (a 2.5-cm sphere and a similarly sized cube). Rats implanted with a morphine pellet 4 d earlier were injected on three occasions with naloxone (0.1 mg/kg, sc) and placed for 30 min into a clear Plexiglas™ box with either the cube or the sphere on its grid floor and, on a similar number of occasions, were injected with saline, and placed into the same box with the alternate object. Rats were tested for 15 min in a box (43 × 43 × 60 cm) with the two test objects placed at opposite corners on a layer of sawdust. The rats showed a clear defensive response to the object paired with the withdrawal by selectively burying it, as opposed to the control object. Measures

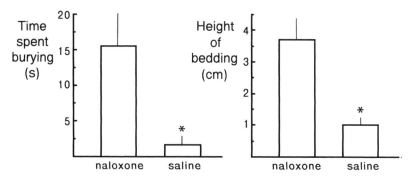

Pairing condition of test object

Fig. 9. Results of a burying test in 12 morphine pellet-implanted rats that were given three pairings of one of two distinctive, small objects (sphere or cube) with 0.1 mg/kg sc naloxone and three pairings of the alternate object with saline. Testing was 1 d later in a chamber with a sawdust floor and both objects present. *Left panel*: Mean time spent burying the object paired with naloxone and the one paired with saline. *Right panel*: Mean height of bedding material piled agonist the naloxone and the saline objects. The asterisks indicate a significant difference between the activity directed at the two objects (data from Mucha and Massaro, 1991).

of the sawdust around the objects and the time spent burying confirmed this (Fig. 9). Since the burying response underwent extinction, was proportional to the number of training trials, was retained for over 40 d after training, and was not seen in placebo-pell-eted or naive rats, it was concluded that it reflected conditioned aversive effects of the withdrawal precipitated by naloxone (Mucha and Massaro, 1991). This confirmed that preference conditioning produced by opiate withdrawal likely reflects the motivational property of the withdrawal.

4.5. Complexities of Quantifying the Aversive Effect

Certain studies of biological mechanisms may be hindered by limitations of the present techniques for analysis of DR curves; psychopharmacological manipulations generally require these. The two-cue, unbiased taste conditioning methods give rats the choice of two solutions whose palatabilities are difficult to match perfectly for every rat. This, coupled with the fact that the experimental subjects are free to choose all, none, or some of each of the two flavors, makes the variability of the data of individual points higher than desired for analytical uses of the resulting DR curves. Even for the simple Litchfield and Wilcoxon (1949) analysis, there may be problems. In this method, the individual subjects of a group are partitioned into responders and nonresponders. Under control conditions, half of the subjects show a preference for one of the flavors. This means that the DR will range from about 50% preference to 100%, making one-half of the data useless.

This is mentioned here, since it is likely that additional developmental work on the taste model may be required if its full potential is to be applied to DR curve analyses. One possible solution is to develop better criteria of a conditioned preference in order to apply the Litchfield and Wilcoxon method to the data from the present method. An alternative approach that M. Manna introduced into our laboratory uses the well-known palatability of saccharin-flavored solutions to rats as a baseline. It can be expected that, since all rats show a very high preference for the flavor, some of the variability of the two-cue, unbiased method will be overcome. This was tested in subjects prepared exactly as described previously. On the day prior to conditioning, the water bottles were removed from each of the cages at about 6:00 PM. Approximately 18 h later, each subject was placed into its test cage and 5 min later given a tube with 4.0 mL of the flavored solution. This comprised 0.1% saccharin and 0.025% citric acid in tap water. Ten minutes later, each subject was injected sc with drug or saline, and returned to its home cage. At the end of the day, each rat was again given free access to water in its home cage. Testing was carried out the next day with one tube containing ordinary tap water, and the other the flavor. The data comprised the ratio of the volume of the flavor consumed divided by the total fluid intake.

In Fig. 10, it is seen that preference for the saccharin is a rather long, slowly declining, and fairly linear function of the dose of naloxone used for conditioning. Also, the clustering of the control data (Fig. 10, points marked saline) makes the establishment of criteria of change for the Litchfield and Wilcoxon method more accessible. However, what must be realized is that,

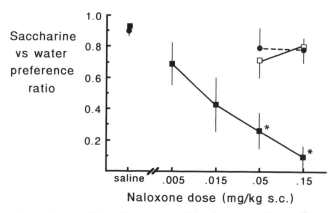

Fig. 10. Data from a biased taste conditioning paradigm. Groups of rats (n = 6/group) were trained by pairing the drinking of a saccharin solution with a single injection of saline or naloxone 5 d after implantation of a morphine pellet (closed squares). Serving as control were groups (*n* = 8/group) implanted with a placebo pellet (circles) and groups implanted with a morphine pellet that underwent pseudoconditioning (open squares): These latter rats were given naloxone *unpaired* with the flavor; half the rats were given their naloxone 17 h before and the other half 8 h after the flavor; the data for the two training conditions were then combined. The asterisks indicate a significant difference relative to the two control data.

to draw any conclusions about the presence of conditioning, more control conditions must be run than are required by the two-cue, unbiased method. For example, simply noting whether injecting with saline would alter the saccharin consumption is not sufficient. In addition, the drugs used had to be tested for pseudoconditioning. That is, they were injected in a fashion that was not correlated with the flavor (Fig. 10, open squares). There was a consistent reduction in the flavor intake in these animals compared to the saline controls, but it was not sufficient to account for the magnitude of effects in the groups experiencing pairing of drug and flavor.

5. Summary

Existing animal models appear to be making little actual impact on the understanding of withdrawal motivation. This may follow from notions advanced in the early clinical literature that somatic signs of withdrawal are equivalent to an aversive state, which has permitted study of these to be thought of as contributory to an understanding of withdrawal motivation. It was argued that this study provides little clear information on withdrawal motivation and that other conventional measures of withdrawal motivation are too complex to break this stagnation in the immediate future. Therefore, preference conditioning using opioid antagonists in opiate-dependent rats was presented as an efficient, simple research model of withdrawal motivation.

A procedure involving taste aversion produced by naloxone in morphine pellet-implanted rats was detailed. Data were presented suggesting that the aversions are reliable, sensitive, and specific to opiate antagonists and opiate treatment. They are robust, long lasting, and confirmed to be a clear indication of motivation produced by the precipitated withdrawal. Information relevant to applying the present or related techniques effectively to microinjection, opioid receptor subtyping, and other pharmacological investigations of withdrawal motivation was emphasized.

Acknowledgments

Most of the experimentation described here was carried out in the Biobehavioral Research Department of the Addiction Research Foundation, Toronto, but the work began at the University of Cambridge, England, and continued at the Max-Planck Institute for Psychiatry, Munich. The final preparation was done at the University of Cologne. The work was supported by a Toronto University grant, MRC of Canada grant MA-9552, and NRC of Canada grant A-2000 awarded to the author. Individuals gratefully acknowledged are L. Adamson, L. Currin, V. Fisher, M. Massaro, R. Konop, C. Kim, D. Lesperance, G. Mutz, S. Samir, and E. Werk for their technical assistance, and J. Alpert, M. Massaro, M. Linseman, M. Mana, C. Pilcher, K. Walker, and D. van der Kooy for useful discussions. Special gratitude is extended to Professors Harold Kalant of the University of Toronto and Egon Stephan of the University of Cologne for their support and encouragement.

References

Acquas E., Carboni E., Leone P., and DiChiara G. (1989) SCH23390 blocks drug-induced conditioned place preference and place aversion: anhedonia (lack of reward) or apathy (lack of motivation) after dopamine-receptor blockade? *Psychopharmacol.* **99**, 151–155.

Alexander B. K., Peele S., Hadaway P. F., Morse S. J., Brodsky A., and Beyerstein B. L. (1985) Adult, infant, and animal addiction, in *The Meaning of Addiction: Compulsive Experience and Its Interpretation* (Peele S., ed.), D. Heath Co., Lexington/Toronto, pp. 73–96.

Azrin N. H. and Hake D. F. (1969) Positive conditioned suppression: conditioned suppression using positive reinforcers as the unconditioned stimuli. *J. Exptl. Anal. Behav.* **12**, 167–173.

Bechara A. and van der Kooy D. (1985) Opposite motivational effects of endogenous opioid in brain and periphery. *Nature* **314**, 533,534.

Bergman J. and Schuster C. R. (1985) Behavioral effects of naloxone and nalorphine preceding and following morphine maintenance in the rhesus monkey. *Psychopharmacol.* **86**, 324–327.

Blackburn J. R. and Pfaus J. G. (1988) Is motivation really modulation? A comment on Wise. *Psychobiol.* **16**, 303,304.

Bläsig J., Herz A., Reinhold K., and Ziegelgänsberger S. (1973) Development of physical dependence on morphine in respect to time and dosage and quantification of the precipitated withdrawal syndrome. *Psychopharmacol.* **33**, 181–191.

Bozarth M. A. (1987) Conditioned place preference: a parametric analysis using systemic heroin injections, in *Methods of Assessing the Reinforcing Properties of Abused Drugs* (Bozarth M.A., ed.), Springer-Verlag, New York, pp. 241–273.

Bozarth M. A. and Wise R. A. (1984) Anatomically distinct opiate receptor fields mediate reward and physical dependence. *Science* **224**, 516,517.

Brown D. R. and Goldberg L. I. (1985) The use of quaternary narcotic antagonists in opiate research. *Neuropharmacol.* **24**, 181–191.

Cannon W. B. (1929) Organization for physiological homeostasis. *Physiol. Rev.* **9**, 399–431.

Cappell H. and LeBlanc A. E. (1975) Conditioned aversion by psychoactive drugs: Does it have significance for an understanding of drug dependence? *Addict. Behav.* **1**, 55–64.

Cappell H. and LeBlanc A. E. (1981) Tolerance and physical dependence. *Res. Adv. Alc. Drug Probl.* **6**, 159–196.

Carr G. D., Fibiger H. C., and Phillips A. G. (1988) Conditioned place preference as a measure of drug reward, in *Monographs in Psychopharmacology. vol. 1: The Neuropharacological Basis of Reward* (Cooper S. J. and Liebman J., eds.), Oxford University Press, Oxford, UK, pp. 264–319.

Carroll M. E. and Lac S. T. (1987) Cocaine withdrawal produces behavioral disruptions in rats. *Life Sci.* **40**, 2183–2190.

Cochin J., Miller J. M., Rosow C. E., Grell R., and Poulsen J. L. (1979) The influence of the mode of morphine administration on tolerance and dependence. *N.I.D.A. Res. Monogr.* **27**, 36–47.

Cohen J. (1987) *Statistical Power Analysis for the Behavioral Sciences.* Lawrence Erlbaum Associates Publishers, Hillsdale, NJ.

Collier H. O. J., Cuthbert N. J., and Francis D. L. (1981) Model of opiate dependence in the guinea-pig isolated ileum. *Brit. J. Pharmacol.* **73**, 921–932.

Colpaert F. C. (1978) Theoretical review: discriminative stimulus properties of narcotic analgesic drugs. *Pharmacol. Biochem. Behav.* **9**, 863–887.

D'Mello G. D., Stolerman I. P., Booth D. A., and Pilcher C. W. T. (1977) Factors influencing flavour aversions conditioned with amphetamine in rats. *Pharmacol. Biochem. Behav.* **7**, 185–190.

Downs D. A. and Woods J. A. (1976) Naloxone as a negative reinforcer in rhesus monkeys: effect of dose, schedule, and narcotic regimen. *Pharmacol. Rev.* **27**, 397–436.

Dymshitz J. and Lieblich I. (1987) Opiate reinforcement and naloxone aversion, as revealed by place preference paradigm, in two strains of rats. *Psychopharmacol.* **92**, 473–477.

Eddy N. B., Halbach H., Isbell H., and Seevers M. H. (1964) Drug dependence: its significance and characteristics. *Bull. Wld. Hlth. Org.* **32**, 721–733.

Emmett-Oglesby M. W., Harris C. M., Lane J. D., and Lal H. (1984) Withdrawal from morphine generalizes to a pentylenetetrazol stimulus. *Neuropeptides* **5**, 37–40.

Emmett-Oglesby M. W., Spencer D. G., Elmesallamy F., and Lal H. (1983) The pentylenetetrazol model of anxiety detects withdrawal from diazepam. *Life Sci.* **33**, 161–168.

Ericsson K. A. and Simon H. A. (1980) Verbal reports as data. *Psychol. Rev.* **87**, 215–250.

Erlenmeyer A. (1887) *Die Morphiumsucht und ihre Behandlung,* 3rd ed., Heuser's Verlag, Berlin/Leipzig/Neuwied.

Estes W. K. and Skinner B. F. (1941) Some quantitative properties of anxiety. *J. Exp. Psychol.* **29**, 390–400.

Finlay J. M., Jakubovic A., Phillips A. G., and Fibiger H. C. (1988) Fentanyl-induced conditional place preference: lack of associated conditional neurochemical events. *Psychopharmacol.,* **96**, 534–540.

France C. P. and Woods J. H. (1985) Effects of morphine, naloxone, and dextrorphan in untreated and morphine-treated pigeons. *Psychopharmacol.* **85**, 377–382.

Frumkin K. (1976) Differential potency of taste and audiovisual stimuli in the conditioning of morphine withdrawal in rats. *Psychopharmacol.* **46**, 245–248.

Fry J. P., Herz A., and Zieglgänsberger W. (1980) A demonstration of naloxone-precipitated opiate withdrawal on single neurones in the morphine-tolerant/dependent rat brain. *Brit. J Pharmacol.* **68**, 585–592.

Garcia J., Hankins W. G., and Rusniak K. W. (1974) Behavioral regulation of the milieu interne in man and rat. *Science* **185,** 824–831.

Gellert V. F. and Holtzman S. G. (1979) Discriminative stimulus effects of naltrexone in the morphine-dependent rat. *J. Pharmacol. Exp. Ther.* **211,** 596–605.

Gibson R. D. and Tingstad J. E. (1970) Formulation of a morphine implantation pellet suitable for tolerance-physical dependence studies in mice. *J. Pharmacol. Sci.* **59,** 426,427.

Goldberg S. (1976) Conditioned behavioral and physiological changes associated with injections of a narcotic antagonist in morphine-dependent monkeys. *Pav. J. Biol. Sci.* **11,** 203–221.

Goldberg S. R. and Schuster C. R. (1967) Conditioned suppression by a stimulus associated with nalorphine in morphine-dependent monkeys. *J. Exptl. Anal. Behav.* **10,** 235–242.

Goudie A. J. (1991) Animal models of drug abuse and dependence, in *Behavioural Models in Psychopharmacology: Theoretical, Industrial and Clinical Perspectives* (Willner P., ed.), Cambridge University Press, Cambridge, UK, pp. 453–485.

Grinker R. G. (1938) *Neurology,* C. C. Thomas, Springfield/Baltimore, pp. 890–906.

Hand T. H., Koob G. F., Stinus L., and LeMoal M. (1988) Aversive properties of opiate receptor blockade: evidence for exclusively central mediation in naive and morphine-dependent rats. *Brain Res.* **474,** 364–368.

Hebb D. O. (1949) *The Organisation of Behavior,* Wiley, New York, pp. 171–206.

Herz A. and Shippenberg T. S. (1989) Neurochemical aspects of addiction: opioids and other drugs of abuse, in *Molecular and Cellular Aspects of the Drug Addictions* (Goldstein A., ed.), Springer-Verlag, Heidelberg, pp. 111–140.

Himmelsbach C.K. (1943) IV. With reference to physical dependence. *Fed. Proc.* **2,** 201–203.

Hoffmeister F. (1986) Negative reinforcing properties of naloxone in the nondependent rhesus monkey: influence on reinforcing properties of codeine, tilidine, buprenorphine, and pentazocine. *Psychopharmacol.* **90,** 441–450.

Jaeger T. V. and Mucha R. F. (1990) A taste aversion model of drug discrimination learning: Training drug and condition influence rate of learning, sensitivity and drug specificity. *Psychopharmacol.* **100,** 145–150.

Katz J. L. and Valentino R. J. (1986) Pharmacological and behavioral factors in opioid dependence in animals, in *Behavioral Analysis of Drug Dependence* (Goldberg S. L. and Stolerman I. P., eds.) Academic, London, pp. 287–327.

Kehoe P. and Blass E. (1986) Behaviorally functional opioid systems in infant rats: I. Evidence for olfactory and gustatory classical conditioning. *Behav. Neurosci.* **100,** 359–367.

Kim C. and Katz T. (1984) Rapid and sensitive analysis of morphine in serum by reversed phase high performance liquid chromatography with electrochemical detection. *J. Anal. Toxicol.* **8**, 135–137.

Kirk R. E. (1968) *Experimental Design: Procedures for the Behavioral Sciences.* Wadsworth, Belmont.

Kolb L. (1925) Pleasure and deterioration from narcotic addiction. *Ment. Hygiene* **9**, 699–724.

Kolb L. and Himmelsbach C. K. (1938) Clinical studies of drug addiction, III. A critical review of the withdrawal treatments with method of evaluating abstinence syndromes. *Am. J. Psychiat.* **94**, 759–799.

Koob G. F., Wall T. L., and Bloom F. E. (1989) Nucleus accumbens as a substrate for the aversive stimulus effects of opiate withdrawal. *Psychopharmacol.* **98**, 520–534.

Kornetsky C. and Esposito R. U. (1979) Euphorigenic drugs: effects on the reward pathways of the brain. *Fed. Proc.* **38**, 2473–2476.

Lal H. and Emmett-Oglesby M. E. (1983) Behavioral analogues of anxiety. *Neuropharmacol.* **22**, 1423–1441.

Lett B. T. and Grant V. L. (1989) Conditioned taste preference produced by pairing a taste with a low dose of morphine or sufentanil. *Psychopharmacol.* **98**, 236–239.

Litchfield J. T. and Wilcoxon F. (1949) A simplified method of evaluating dose–effect experiments. *J. Pharmacol. Exp. Ther.* **96**, 99–113.

Lorens S. A. and Mitchell C. L. (1973) Influence of morphine on lateral hypothalamus self-stimulation in the rat. *Psychopharmacol.* **32**, 271–277.

Mackintosh N. J. (1974) *The Psychology of Animal Learning*, Academic, New York, pp. 4,5.

Manning F. J. and Jackson M. C. (1977) Enduring effects of morphine pellets revealed by conditioned taste aversion. *Psychopharmacol.* **51**, 279–283.

Martin G. M., Bechara A., and van der Kooy D. (1988) Morphine preexposure attenuates the aversive properties of opiates without preexposure to the aversive properties. *Pharmacol. Biochem. Behav.* **30**, 687–692.

McGregor D. (1986) Ethics of animal experimentation. *Drug Metab. Rev.* **17**, 349–361.

Meggendorfer F. (1928) Intoxikationspsychosen, in *Handbuch der Geisteskrankheiten: vol. 7: Die Exogenen Reaktionsformen und die Organischen Psychosen* (Bumke O., ed.), Springer Verlag, Berlin, pp. 151–400.

Mello N. K., Mendelson J. H., and Bree M. P. (1981) Naltrexone effects on morphine and food self-administration in morphine-dependent rhesus monkeys. *J. Pharmacol. Exp. Ther.* **218**, 550–557.

Meyer D. R. and Sparber S. B. (1976) A comparison of withdrawal in rats implanted with different types of morphine pellets. *Pharmacol. Biochem. Behav.* **5**, 603–607.

Montgomery M. F. (1931) The role of the salivary glands in the thirst mechanism. *Am. J. Physiol.* **46**, 221–227.

Mucha R. F. (1987) Is the motivational effect of opiate withdrawal reflected by common indices of precipitated withdrawal? A place conditioning study in the rat. *Brain Res.* 418, 214–220.

Mucha R. F. (1989) Taste aversion involving central opioid antagonism is potentiated in morphine-dependent rats. *Life Sci.* 45, 671–678.

Mucha R. F. (1991) What is learned during opiate withdrawal conditioning? Evidence for a cue avoidance model. *Psychopharmacology* 104, 391–396.

Mucha R. F. and A. Herz. (1985) Motivational properties of kappa and mu opioid agonists studied with place and place preference conditioning. *Psychopharmacol.* 86, 274–280.

Mucha R. F. and Herz A. (1986) Preference conditioning produced by opioid active and inactive isomers of levorphanol and morphine in rat. *Life Sci.* 38, 241–249.

Mucha R. F. and Iversen S. D. (1984) Reinforcing properties of morphine and naloxone revealed by conditioned place preferences: a procedural examination. *Psychopharmacol.* 82, 241–247.

Mucha R. F. and Walker M. J. K. (1988) Aversive property of opioid receptor blockade in drug-naive mice. *Psychopharmacol.* 93, 483–488.

Mucha R. F., Gritti M. D., and Kim C. (1986) Aversive properties of opiate withdrawal in rats. *N.I.D.A. Res. Monogr.* 75, 567–570.

Mucha R. F., Kalant H., and Kim C. (1987) Tolerance to hyperthermia produced by morphine in rat. *Psychopharmacol.* 92, 452–458.

Mucha R. F., Kalant H., and Linseman M. A. (1979) Quantitative relationships among measures of morphine tolerance and physical dependence in the rat. *Pharmacol. Biochem. Behav.* 10, 387–405.

Mucha R. F., Millan M. J., and Herz A. (1985) Aversive properties of naloxone in non-dependent (naive) rats may involve blockade of central beta-endorphin. *Psychopharmacol.* 86, 281–285.

Mucha R. F., Walker M. J. K., and Fassos F. F. (1990) Parker and Radow test of drug withdrawal aversion: opposite effect in rats chronically infused with sufentanil or amphetamine. *Pharmacol. Biochem. Behav.* 35, 219–224.

Mucha R. F., van der Kooy D., O'Shaughnessy M., and Bucenieks P. (1982) Drug reinforcement studied by the use of place conditioning in rat. *Brain Res.* 243, 91–105.

North R. A. and Vitek L. V. (1980) The effect of chronic morphine treatment on excitatory junction potentials in the mouse vas deferens. *Br. J. Pharmacol.* 68, 399–405.

O'Brien C. P., Testa T., O'Brien T. J., Brady J. P., and Wells B. (1977) Conditioned narcotic withdrawal in humans. *Science* 195, 1000–1002.

Olds M. E. (1976) Effectiveness of morphine and ineffectiveness of diazepam and phenobarbital on the motivational properties of hypothalamic selfstimulation behaviors. *Neuropharmacol.* 15, 117–131.

Olds J. and Travis R. P. (1960) Effects of chloropromazine, meprobamate, pentobarbital, and morphine on self-stimulation. *J. Pharmacol. Exp. Ther.* **128**, 389–404.

Overton D. A. (1979) Influence of shaping procedures and schedules of reinforcement on performance in the two-bar drug discrimination task: A methodological report. *Psychopharmacol.* **65**, 291–298.

Overton D. A. (1982) Comparison of the degree of discriminability of various drugs using the T-maze drug discrimination paradigm. *Psychopharmacol.* **76**, 385–395.

Parker L. A. and Radow B. L. (1974) Morphine-like physical dependence: a pharmacologic method for drug assessment using the rat. *Pharmacol. Biochem. Behav.* **2**, 613–618.

Parker L., Failor A., and Weidman K. (1973) Conditioned preferences in the rat with an unnatural need state: morphine withdrawal. *J. Comp. Physiol. Psychol.* **82**, 294–300.

Paterson S. J. (1988) The *in vitro* pharmacology of selective opioid ligands, in *The Psychopharmacology of Addiction* (Lader M., ed.), Oxford University Press, Oxford, pp. 97–114.

Peachy J. E. and Lei H. (1989) Assessment of opioid dependence with naloxone. *Br. J. Addict.* **83**, 193–201.

Pellow S. and File S. E. (1984) Multiple sites of action for anxiogenic drugs: behavioural, electrophysiological and biochemical correlations. *Psychopharmacol.* **83**, 304–315.

Pfeiffer A., Brantl V., Herz A., and Emrich H. M. (1986) Psychomimesis mediated by kappa opiate receptors. *Science* **233**, 774–776.

Pilcher C. W. T. and Stolerman I. P. (1976a) Recent approaches to assessing opiate dependence in rats, in *Opiates and Endogenous Opioid Peptides* (Kosterlitz H. W., ed.), Elsevier, Amsterdam, pp. 327–334.

Pilcher C. W. T. and Stolerman I. P. (1976b) Conditioned flavor aversions for assessing precipitated withdrawal morphine abstinence in rats. *Pharmacol. Biochem. Behav.* **4**, 159–163.

Pinel J. P. J. and Mucha R. F. (1973) Incubation and Kamin effects in the rat: changes in activity and reactivity after footshock. *J. Comp. Physiol. Psychol.* **84**, 661–668.

Pinel J. P. J. and Wilkie D. M. (1983) Conditioned defensive burying: Biological and cognitive approach to avoidance learning, in *Animal Cognition and Behavior* (Mellgren R. L., ed.), North-Holland Publishing, Amsterdam, pp. 285–318.

Redmond D. E. and Krystal J. H. (1984) Multiple mechanisms of withdrawal from opioid drugs. *Ann. Rev. Neurosci.* **7**, 443–478.

Reid L. D. (1985) Endogenous opioid peptides and regulation of drinking and feeding. *Am. J. Clin. Nutr.* **42**, 1099–1132.

Rescorla R. A. (1967) Pavlovian conditioning and its proper control procedures. *Psychol. Rev.* **74**, 71–80.

Rossi N. A. and Reid L. D. (1976) Affective states associated with morphine injections. *Physiol. Psychol.* **4**, 269–274.

Rozin P. and Kalat J. W. (1971) Specific hungers and poison avoidance as adaptive specializations of learning. *Psychol. Rev.* **78,** 459–486.

Sawynok J., Pinsky C., and Labella F. S. (1979) *Life Sci.* **25,** 1621–1632.

Schaefer G. J. and Michael R. P. (1983) Morphine withdrawal produces differential effects on the rate of lever-pressing for brain self-stimulation in the hypothalamus and midbrain in rats. *Pharmacol. Biochem. Behav.* **18,** 571–577.

Schenk S., Ellison F., Hunt T., and Amit Z. (1985) An examination of heroin conditioning in preferred and nonpreferred environments and in differentially housed mature and immature rats. *Pharmacol. Biochem. Behav.* **22,** 215–220.

Schulz R. (1988) Dependence and cross-dependence in the guinea-pig myenteric plexus. *Naunyn-Schmiedeberg's Arch. Pharmacol.* **337,** 644–648.

Schulz R. and Herz A. (1984) Opioid tolerance and dependence in light of the multiplicity of opioid receptors. *N.I.D.A. Res. Monogr.* **54,** 70–80.

Schuster C. R. (1976) Discussion of symposium on conditioning and addiction. *Pav. J. Biol. Sci.* **11,** 263–266.

Shippenberg T. S. and Herz A. (1988) Motivational effects of opioids: influence of D-1 versus D-2 receptor antagonists. *Eur. J. Pharmacol.* **151,** 233–242.

Shippenberg T. S., Millan M. J., Mucha R. F., and Herz A. (1988) Involvement of β-endorphin and K-opioid receptors in mediating the aversive effects of lithium in the rat. *Eur. J. Pharmacol.* **154,** 135–144.

Siegal S. (1956) *Nonparametric Statistics for the Behavioral Sciences.* McGraw Hill, New York.

Siegel R. K., Gusewelle B. E., and Jarvik M. E. (1975) Naloxone-induced jumping in morphine-dependent mice: stimulus control and motivation. *Int. Pharmacopsychiat.* **10,** 17–23.

Spyraki C. (1987) Drug reward studied by the use of place conditioning in rats, in *The Psychopharmacology of Addiction* (Lader M., ed.), Oxford University Press, Oxford, pp. 97–114.

Spyraki C., Nomikos G. G., Galanoulou P., and Daifotis Z. (1988) Drug-induced place preference in rats with 5,7-dihydroxtryptamine lesions of the nucleus accumbens. *Behav. Brain Res.* **29,** 127–134.

Stolerman I. P. and D'Mello G. D. (1981) Oral self-administration and the relevance of conditioned taste aversions. *Adv. Behav. Pharmacol.* **3,** 169–214.

Stolerman I. P., Pilcher C. W. T., and D'Mello G. D. (1978) Stereospecific aversive property of narcotic antagonists in morphine-free rats. *Life Sci.* **22,** 1755–1762.

Swerdlow N. R., Gilbert D., and Koob G. F. (1988) Conditioned drug effects on spatial preference: critical evaluation, in *Neuromethods, vol. 13: Psychopharmacology* (Boulton A. A., Baker G. B., and Greenshaw A. J., eds.), Humana Press, Clifton, NJ.

Ternes J. (1975) Naloxone-induced aversion to sucrose in morphine-dependent rats. *Bull. Psychonom. Soc.* **5,** 311,312.

Turkington D. and Drummond D. C. (1989) How should opiate withdrawal be measured? *Drug Alc. Depend.* **24,** 151–153.

Undeutsch U. (1967) Die Beurteilung der Glaubhaftig von Aussagen, in *Handbuch der Psychologie. vol. 11: Forensische Psychologie* (Undeutsch U., ed), Hogrefe, Göttingen, pp. 26–181.

van der Kooy D. (1987) Place conditioning: A simple and effective method for assessing the motivational properties of drugs, in *Methods of Assessing the Reinforcing Properties of Abused Drugs* (Bozarth M. A., ed.), Springer-Verlag, New York, pp. 229–240.

van der Kooy D., Mucha R. F., O'Shaughnessy M., and Bucenieks P. (1982) Reinforcing effects of brain microinjections of morphine revealed by conditioned place preference. *Brain Res.* **243,** 107–117.

van Ree J. M., Slangen J. L., and DeWied D. 1978) Intravenous self-administration in rats. *J. Pharmacol. Exptl. Therap.* **204,** 547–557.

Vezina P. and Stewart J. (1987) Morphine conditioned place preference and locomotion: the effect of confinement during training. *Psychopharmacol.* **93,** 257–260.

Vitello M. V. and Woods S. C. (1977) Evidence for withdrawal from caffeine by rats. *Pharmacol. Biochem. Behav.* **6,** 553–555.

Wangensteen O. H. and Carson A. J. (1931) Hunger sensations in a patient after total gastrectomy. *Proc. Soc. Exptl. Biol. Med.* **28,** 545–547.

Wei E. (1973) Assessment of precipitated abstinence in morphine-dependent rats. *Psychopharmacol.* **28,** 35–44.

Wikler A. (1948) Recent progress in research on the neurophysiological basis of morphine addiction. *Am. J. Psychiat.* **67,** 672–684.

Wikler A. (1980) *Opioid Dependence: Mechanisms and Treatment.* Plenum, New York.

Wise R. A. (1987a) The role of reward pathways in the development of drug dependence. *Pharmacol. Therap.* **35,** 227–263.

Wise R. A. (1987b) Sensorimotor modulation and the variable action pattern (VAP): Toward a noncircular definition of drive. *Psychobiol.* **15,** 7–20.

Yoburn V. C., Chen J., Huang T., and Inturrisi C. E. (1985) Pharmacokinetics and pharmacodynamics of subcutaneous morphine pellets in the rat. *J. Pharmacol. Exp. Ther.* **235,** 282–286.

Young A. and Thompson T. (1979) Naloxone effects on schedule-controlled behavior in morphine-pelleted rats. *Psychopharmacol.* **62,** 307–314.

Yuille J. C., ed. (1989) *Nato ASI Series. Series D. vol. 47: Credibility Assessment.* Kluwer Academic Publishers, Dordrecht/Boston/London.

Zellner D. A., Dacanay R. J., and Riley A. L. (1984) Opiate withdrawal: the result of conditioning or physiological mechanisms? *Pharmacol. Biochem. Behav.* **20,** 175–180.

Zilm D. and Sellers E.M. (1978) The quantitative assessment of physical dependence on opiates. *Drug Alc. Depend.* **3,** 419–428.

Zito K. A., Bechara A., Greenwood C., and van der Kooy D. (1988) The dopamine innervation of the visceral cortex mediates the aversive effects of opiates. *Pharmacol. Biochem. Behav.* **30,** 693–699.

A Rodent Model for Nicotine Self-Administration

William A. Corrigall

1. Concepts of Drug Addiction

An obvious preliminary to the discussion of models of "drug addiction" is to consider what we mean by that term. Over the past several decades, a number of definitions of drug addiction have been used. Although some of the historically earliest definitions of addiction or drug dependence relied in part on the consequences of protracted drug exposure as defining characteristics, contemporary definitions generally recognize that use of a psychoactive drug itself is the central element in the process of addiction (*see* review in Clarke et al., 1989). For example, the recent report of the US Surgeon General on Nicotine Addiction used three primary criteria in defining addiction (US DHHS, 1988):

1. Drug-seeking and drug-taking behavior is driven by strong, often irresistible urges and can persist despite a desire to quit or even repeated attempts to quit;
2. The drug has psychoactive or mood-altering effects in the brain; and
3. The drug is capable of functioning as a reinforcing agent that directly strengthens behavior leading to further drug-taking.

Other consequences of drug administration may be important in modifying the extent or pattern of use. For example, aversive consequences of drug administration may limit intake,

From: *Neuromethods, Vol. 24: Animal Models of Drug Addiction*
Eds: A. Boulton, G. Baker, and P. H. Wu ©1992 The Humana Press Inc.

whereas development of tolerance may be an important factor in the escalation of drug use, either because tolerance to the aversive properties reduces barriers to increased intake, or because development of tolerance to the positive properties of a drug causes the user to need more drug to attain the desired effect. Physical dependence may contribute to a sustained pattern of drug use when avoidance of the unpleasant consequences of withdrawal becomes important to the user.

For most drugs of abuse, animal models already exist to study both the reinforcing properties of the substance and the consequences of its administration. With respect to nicotine specifically, methods are available to study the development of tolerance (e.g., Corrigall et al., 1988b; Hendry and Rosecrans, 1982), the production of behavioral arousal and stimulation (e.g., Clarke and Kumar, 1983a,b; Clarke et al., 1988; Corrigall et al., 1988a), the nicotine discriminative stimulus (e.g., Chance et al., 1977, Pratt et al., 1983; Stolerman et al., 1984,1987), and some of the consequences of chronic treatment with nicotine (Corrigall and Coen, 1988b; Levin et al., 1987). For most drugs of abuse, it has been possible to study their reinforcing properties by means of self-administration techniques (e.g., Collins et al., 1984; Johanson and Schuster, 1981), which generally have been found to provide a reliable measure of addictive liability (Griffiths et al., 1980). In this respect, however, nicotine has been an exception; prior attempts to develop an animal model of nicotine self-administration have had variable results. Whereas some studies have shown that nicotine serves as a reinforcer in primates in some conditions, studies with rodents have typically concluded that (1) the conditions under which nicotine is self-administered are more limited than those for other drugs, and/or (2) that low rates of self-administration behavior are maintained by nicotine (Henningfield and Goldberg, 1983a). This chapter will discuss the development of a rodent model for nicotine self-administration in which response rates are comparable to those maintained by other drugs under the same conditions of access. As will become evident later in this chapter, although the dose regulation of nicotine intake may be different than that of other addictive drugs, there is no question that nicotine can reinforce behavior leading to its delivery when a rodent model of self-administration is used.

2. Previous Studies of Nicotine Self-Administration

A few studies have reported attempts to establish oral self-administration of nicotine. These will not be discussed here, because the slow onset of effect after nicotine ingestion orally does not mimic the rapid delivery of nicotine to the brain that occurs during smoking. Since most cigaret smokers inhale (Health and Welfare Canada, 1981), there is very rapid absorption of nicotine into the blood and subsequent rapid delivery of the drug to the brain. Thus, nicotine is delivered to the brain in essentially the same time as it would if it were injected intravenously, making iv delivery of nicotine a valid approach to use in self-administration studies in animals. A number of attempts have been made to establish a model of iv nicotine self-administration with a range of species of laboratory animals (Henningfield and Goldberg, 1983a,1988), as summarized in the following section.

2.1. Unlimited Access Models

With the benefit of hindsight, it is convenient to divide previous reports into studies that used continuous access schedules and studies that used limited access models. One of the first studies of nicotine reinforcement in rats used a continuous reinforcement schedule (CRF, in which each lever press results in the delivery of the reinforcer). In this research, animals that were of normal body wt were compared to ones that were reduced to 80% body wt by restricted feeding (Lang et al., 1977). Significant nicotine self-administration was evident only in the food-deprived group. When, in addition, the animals experienced presentation of a 45-mg food pellet every 60 s, there was a marked increase in nicotine self-administration. Although schedule-induced enhancement of drug self-administration by means of concurrent presentation of another reinforcer is well known today, nicotine was one of the first drugs for which this was demonstrated (Slifer, 1983). Once self-administration was established with this model, infusions of nicotine continued to exceed saline infusions when body weights were returned to free-feeding values. However, the number of nicotine infusions obtained by free-feeding animals was decreased compared to the number received by animals at 80% body wt (Singer et al., 1978). In addition, self-administration was sensitive to the dose of nicotine that was available only during a brief acquisition period; thereafter, changes in responding did not occur when the dose of nicotine was altered, although extinction did occur when saline was substituted for nicotine (Latiff et al., 1980).

The excretion of nicotine can be changed by altering the urinary pH; there is greater excretion of ionized nicotine in acidic urine. With the above model, acidification of the urine with ammonium chloride caused the expected change in nicotine self-administration during acquisition—greater drug intake to compensate for the greater excretion (Latiff et al., 1980). Similarly, there was a decrease in self-administration when the urine was made more alkaline than normal with sodium bicarbonate. However, if the initial exposure to nicotine occurred at normal pH, alterations in urine pH did not affect nicotine intake. Since excretion of unmetabolized nicotine, relative to its metabolism to cotinine, represents a relatively minor pathway for nicotine removal, the small effects of urinary pH manipulations on self-administration may not be surprising (Benowitz and Jacob, 1987).

Schedule induction may operate because the controlling or inducing schedule provides periodic reinforcement, which increases the reinforcing value of other stimuli such that they maintain behavior to a greater degree than they do alone. Although the act of taking a drug in such a model can be termed "voluntary," this approach to self-administration leads to ambiguity about the ability of the drug under investigation to serve as a reinforcer in itself and the conditions under which reinforcement of drug self-administration will occur.

However, subsequent work by Smith and Lang (1980) showed that schedule induction was not necessary. First, when nicotine self-administration was induced with a food-delivery schedule for 14 d, animals continued to respond for nicotine if the schedule was removed. This was not simply sustained responding resulting from the previous presence of the schedule, because operant behavior extinguished if saline was substituted for nicotine. Second, without schedule-induction, animals at 80% body wt gradually acquired self-administration behavior for nicotine, such that by the end of 3–4 wk their rate of self-administration was comparable to that of animals that had acquired self-administration under the influence of a concurrent schedule of food presentation. These observations showed, therefore, that neither acquisition nor maintenance of nicotine self-administration in this model was dependent upon the fixed-time schedule. Rather, the food presentation schedule appeared only to hasten acquisition. However, food deprivation was a necessary condition; animals not food-deprived did not acquire nicotine self-administration.

In summary, then, this group of experiments showed that nicotine self-administration could be established in food-deprived animals with the drug available on a CRF schedule. It was not necessary to induce adjunctive behavior by means of the concurrent presentation of another reinforcer, but this approach would speed acquisition. Response rates for nicotine in these studies were necessarily low, since only one lever press was required for each infusion of drug; response rates ranged typically between 10–20 presses/h.

Hanson and his colleagues (1979) used a different approach to obtain nicotine self-administration. In their experiments, rats were treated intravenously with nicotine every 30 min for a period of 48 h, with doses ranging from 0 to 0.06 mg/kg/infusion. Animals were then allowed to self-administer nicotine at the same dose that they had received chronically, in 24-h sessions with a CRF schedule. Pretreatment with nicotine did increase the rate of self-administration. Nicotine-maintained responding was greater than saline-maintained responding and was dose-dependent, with a maximum occurring at 0.03 mg/kg/infusion. Responding extinguished slowly following saline substitution and was only transiently sensitive to treatment with the nicotinic antagonist mecamylamine. This study suggests that nicotine can serve as a reinforcer after chronic treatment, but does not address whether the drug can serve as a reinforcer in naive subjects. In addition, with this model, as with the one discussed above, response rates were low, probably because of the use of a CRF schedule and 24-h access conditions.

A report by Cox et al. (1984), showed that the conditions under which nicotine self-administration would occur were not as restricted as had been previously thought. In this study, animals had access to nicotine on a CRF schedule for 24 h each day, conditions similar to those used by Hanson et al. (1979), but without chronic pretreatment. Nicotine supported self-administration at several doses, particularly at 0.03 mg/kg/infusion. There was some question of the specificity of operant responding, since the number of responses made on a second, activity-control lever was high, and increased with dose and time of exposure to the drug. However, compensatory responding occurred when the nicotine unit dose was decreased, suggesting that animals were responding to maintain a certain level of nicotine. Unfortunately, the amount of nicotine actually taken in a given time was low (about 1 mg/kg in a 12-h period), once again probably because of the 24-h access to the drug. Nonetheless, this study showed that nicotine could act as a reinforcer in naive animals without the use of weight reduction, schedule induction, or chronic pretreatment with nicotine itself.

In all of these studies, in which rats were used as experimental subjects, response rates were low. However, the low rates of nicotine self-administration may be owing to the choice of schedule (CRF) rather than to the species (Clarke, 1987; Dougherty et al., 1981; Henningfield and Goldberg, 1983a). Even with primates, nicotine has been found to maintain low-rate responding; for example, Ator and Griffiths (1983) obtained low rates of self-administration with a FR2 schedule of reinforcement.

2.2. Intermittent Schedules of Access

In contrast, higher rates of responding generally occur when intermittent schedules of access are used. For example, squirrel monkeys responded at a high rate under a fixed-interval or second-order fixed-interval schedule when nicotine was available (Goldberg et al., 1981,1983). Nicotine-maintained responding was decreased by mecamylamine treatment and by saline substitution, and was sensitive to the dose of nicotine available. In some subjects, response rates for nicotine were comparable to those during cocaine self-administration on the same schedule. Risner and Goldberg (1983) also found that nicotine would maintain substantial responding in dogs on either a FR15 or a progressive ratio schedule.

However, high rates of responding have not been consistently found with intermittent schedules of access to nicotine. Ator and Griffiths (1983) reported that rates of nicotine self-administration by baboons on a FI-5-min schedule were well below those obtained for cocaine or food, and did not increase as the interval duration decreased. Reasons for the differences between this study and the reports of Goldberg and colleagues are not obvious. There has, however, been general agreement that scheduling of access is important in establishing a model of nicotine self-administration, and that to obtain optimum rates of drug-maintained behavior for nicotine, limited access schedules of reinforcement should be used (Goldberg et al., 1983; Henningfield and Goldberg, 1983a).

Until recently, use of limited access schedules had not been explored with rodents (Corrigall and Coen, 1989b). The remainder of this chapter discusses our approach to nicotine self-administration with rodents as experimental subjects.

3. Specific Methods

Many of the details of our self-administration techniques have been described in previous research reports (e.g., Corrigall, 1987; Corrigall and Coen, 1989b; Corrigall and Vaccarino, 1988), and in principle are similar to those used in other laboratories (e.g., Collins et al., 1984; Roberts and Goeders, 1987). In common with other researchers, for example, we use dual-lever operant chambers to control for the specificity of responding on the drug-reinforced lever. For drug delivery, we use a pneumatic pump (Weeks, 1981) that allows rapid administration of a small-volume bolus of drug. However, our techniques do differ from others in the way in which we prepare and install iv catheters, and these methods are therefore described in detail. In addition, several factors are particularly important in establishing iv nicotine self-administration, and these are also discussed in this section.

3.1. Animal Care and Training

For all studies of drug self-administration that are carried out in our laboratory, animals are housed under conditions of a reversed light-dark cycle, with the lights on between 7:00 PM and 7:00 AM. When received from the supplier, animals are first habituated to the colony room for a 1-wk period before training procedures are begun. When habituated, animals are deprived of food for a period of approx 36 h, and trained to press a lever for 45-mg food pellets on a schedule of continuous reinforcement. Training can take from one to several days depending upon the particular subject. During this time, food supplements are provided to animals that are slow to acquire operant responding. Once trained, animals are no longer maintained in a food-deprived condition; they are fed approx 20 g of chow each day, as a single meal, for the duration of the experiment.

Since depriving animals of food is known to increase drug self-administration, it is important to point out that subjects in our self-administration experiments, although not allowed *ad libitum* access to food, are not food-deprived. The food ration (20 g) constitutes the daily nutritional requirement for the rat (CCAC, 1980). Subjects in our studies gain body wt during the months of an experiment with this feeding schedule (Dai et al., 1989), but are smaller than animals with free access to food. We have not examined whether increasing the amount of food that the animals receive affects self-administration of nicotine or of other drugs, but it is likely that it would. The objective, however, in establishing an animal model for drug self-administration is to have a reasonable degree of drug-seeking behavior generated by animals that are otherwise "normal." It seems likely that feeding an animal its daily nutritional requirement would produce a more normal subject than allowing unrestricted feeding, and therefore, we have adopted the former approach in our studies.

3.2. Construction of IV Catheters

Our method for constructing iv catheters has been modified from that of Weeks, and many of the basic techniques used are described in the bulletins he has prepared (Weeks, 1983). The catheter consists of three separate pieces of tubing, a 37 mm length of silicon rubber tubing (Silastic Medical-Grade Tubing, Dow-Corning Corp., Medical Products, Midland, MI), a 65 mm length of polyethylene tubing with an internal diameter of 0.28 mm (Intramedic, PE10, Clay Adams, Division of Becton Dickinson and Co., Parsippany, NJ), and a 170 mm length of polyethylene tubing with a similar internal diameter, but a much thicker wall (PE20, Clay Adams). It is the Silastic tubing that enters the circulatory system, and the PE20 tubing, because of its wall thickness, that comprises part of the external end of the catheter; the PE10 tubing forms the junction between these two. Figure 1 shows a completed catheter.

The catheter is assembled in the following way. First the Silastic-to-PE10 joint is made. To do this, the Silastic tubing is soaked in xylene for approx 1 min; this makes the silicone rubber much easier to stretch. The Silastic tubing is pushed over the end of the PE10 tubing so that two overlap for about 7 mm. The excess xylene is removed immediately from the lumen of the tubing by flushing with water. A length of stainless-steel surgical wire (the largest size that will fit) is passed through the lumen of the Silastic-PE10 assembly. A piece of polyolefin heat shrink tubing (initial diameter 3/64 in.) approx 6 mm long is then shrunk over the Silastic-to-polyethylene joint by using the heat from a fine-tipped soldering iron. In doing so, it is necessary to make sure that the heat shrink tubing does not extend beyond the area of overlap with the PE10 tubing, since there is the risk that the lumen of the Silastic tubing may be occluded by the shrink tubing.

The next step is construction of the joint between the Silastic-PE10 assembly and the PE20 tubing. To effect this joint, the ends of the PE10 and PE20 polyethylene segments are melted in a fine jet of hot air and held together until cool. Again a piece of

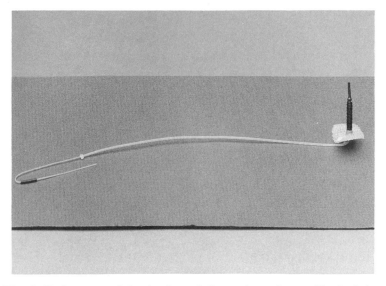

Fig. 1. Catheter used for iv drug delivery in rodent self-administration research.

surgical wire is used as a mandrel, being inserted through the lumen of both pieces of tubing before heat is applied so that the lumen is not closed.

Our catheters differ from those used in other laboratories mainly in the design of the external end. We have adapted the external end of the catheter to allow a reliable connection to be made to the drug delivery system, and also to allow rapid connection and disconnection of the subject and the drug delivery line in the operant chamber. To permit this, the end of the catheter is constructed as follows. A nylon bolt (6–32 × 3/4-in.) is prepared with an axially drilled hole sufficiently large to allow passage of the PE20 tubing. The end of the bolt opposite the head is countersunk with a larger hole, sufficient to accommodate PE20 tubing with a sleeve of 3/64-in heat shrink tubing. The PE20 end of the catheter is passed through the hole in the nylon bolt, starting at the head of the bolt; a piece of 3/64-in heat shrink tubing, approx 15 mm long, is passed over the PE20 tubing, to be used later. A 33-gage wire is then inserted into the end of the PE20 tubing, and the end of the tubing is heated in a jet of hot air in order to raise a small bump on it. The length of heat shrink tubing is then positioned over the end of the PE20 tubing such that it extends about 8 mm beyond the bump at the terminal end of the polyethylene.

A mandrel must be prepared before the terminal end of the catheter can be finished. This consists of a a 23-gage blunt-cut hypodermic needle with a piece of 33-gage suture wire extending beyond its end. The mandrel is inserted into the PE20 tubing such that the end of the 23-gage tubing rests against the bump at the end of the polyethylene, and the shrink tubing is heated sufficiently to cause it to wrap snugly around the polyethylene tubing and needle. The heat shrink tubing must form a fluid-tight junction with the polyethylene. The mandrel is removed when the tubing is cool. A small drop of cyanoacrylate adhesive is placed on this last piece of heat shrink in the area over the PE20 tubing, and the heat shrink-tubing assembly is pushed gently into the countersunk hole in the bolt, so that only the bump and the heat shrink sleeve extend above it. The heat shrink provides a more resilient end at which connections can be made to the drug-delivery line, since polyolefin is harder than polyethylene and can be reshrunk to its previous size if it stretches during regular use. (As described here, the heat shrink end of the catheter can be connected to the drug-delivery line with a length of 22-gage needle tubing.)

After installation of the nylon bolt, two permanent bends are made in the polyethylene tubing components of the catheter. This is done by bending the tubing to the desired orientation, and quickly dipping the tubing into hot water (about 90–100°C) and then into cold. One bend, consisting of a 180° turn, is made in the PE10 tubing just past the heat shrink joint. The other is an approximate 90° bend made in the PE20 tubing just beyond the head of the bolt. The 180° bend allows the catheter to be inserted into the jugular vein pointing toward the heart, while keeping the remainder of the catheter tubing pointing rostrally. The 90° bend ensures that the bolt can be positioned approximately perpendicularly to the animal's back with the PE20 tubing lying flat subcutaneously.

Surgical mesh provides the means through which the bolt assembly is anchored to the animal when the catheter is implanted. A piece of surgical mesh (Marlex Mesh, Bard Cardiosurgery Division, Billerica, MA) is cut to be ellipsoid in shape, with the long axis about 30 mm. A small hole is made in the center of the piece of mesh, and the shaft of the bolt is pushed through so that the head of the bolt rests against the mesh. The mesh is oriented so that its long axis is perpendicular to the PE20 tubing, and fastened to the head of the bolt with a layer of dental acrylic (Dentsply International, York, PA). It is important to ensure that there are no sharp points on the dental acrylic layer, since these can act as irritants. It is also important that the dental acrylic be localized to the section around the head of the bolt, leaving most of the mesh free to be encapsulated by connective tissue.

The catheter can be pressure-tested at various stages of construction to check for leaks, but certainly it should be tested at least when complete. To do so, a 1-cc syringe is filled with water or saline, equipped with a blunt-cut 22-gage hypodermic needle, and connected to the heat shrink termination at the external end of the catheter; fluid is then flushed to the end of the Silastic tubing. The end of the Silastic tubing can then be pinched off between the fingers, while pressure is applied to the catheter via the hypodermic syringe to check for leaks in the joints. (Too much pressure will cause the Silastic tubing to swell—this should be avoided.)

The catheter is now complete. At any time prior to its implantation, the tip of the Silastic tubing is cut to a bevel to facilitate inserting it into the vein.

3.3. Surgical Techniques

Catheters are implanted into the jugular vein under surgical anesthesia induced by a combination of acepromazine maleate (10 mg/kg, administered ip) and ketamine hydrochloride (100 mg/kg, administered im). We find that this is a more reliable anesthetic than pentobarbital, since there are fewer respiratory complications. The first step is to expose the right jugular vein and position two lengths of 4 – 0 silk sutures under it. A second incision is then made on the dorsal surface between the scapulae, exactly at midline. A small sc "pocket" is made by blunt dissection; this "pocket" will eventually contain the surgical mesh on the bolt end of the catheter. It is therefore important to ensure that the dorsal incision is indeed at midline; positioning it in this way makes it difficult for the animal to grasp the external end of the catheter when grooming and, therefore, minimizes potential damage to it. Again by means of blunt dissection, the tips of a pair of forceps are passed sc from the ventral incision to the dorsal incision, traveling caudal to the animal's right front leg. To facilitate passage of the catheter, a piece of large-diameter polyethylene tubing (PE380) is drawn through the channel so made, and in turn, the catheter is passed through the PE380 tubing from dorsal to ventral. The PE380 tubing is then removed via the ventral incision, leaving the catheter in place subcutaneously.

The next step is to install the tip of the catheter in the jugular vein, so that the tip lies just outside the heart. The catheter should be filled with saline prior to implantation. The jugular is tied off distal to the area of implant, and a small V-shaped cut is made in the wall of the vein. This is best done with the vein carefully raised from the body cavity. The tip of the catheter is inserted into the vein to the level of the heat shrink tubing at the Silastic-to-PE10 joint. The heat shrink tubing on the joint is then sutured at either end to the outer surface of the vein and also glued to the vein with cyanoacrylate adhesive. The PE10 tubing of the catheter is sutured to deep muscle, and the muscle tissue is then drawn back and sutured together; overlaying skin is next sutured together.

At the dorsal surface, the excess PE20 tubing and the head of the bolt are placed in the subcutaneous pocket, with the tubing distributed in a smooth curve and the surgical mesh flat against the animal's back. The PE20 tubing should point caudally as it exits the bolt assembly. The skin overlying the mesh is then sutured together. It is not necessary to suture the mesh to tissue, and it is risky to do so, since it is possible to puncture the tubing of the catheter. A nylon nut is threaded onto the bolt until it is just above the skin and glued into place with a small droplet of cyanoacrylate adhesive on the threads.

Following catheter implantation, the animal is given a single dose of penicillin (30,000 U, im). For the first week after surgery, the catheter is flushed with sterile saline containing heparin (0.1 mL daily, 5 U USP/mL) and thereafter with sterile saline alone. When not in use, each catheter is capped with a Silastic plug to prevent backup of blood in it and the entry of unwanted foreign material.

Our system of externalizing the catheter has several advantages. First, it allows the animals to be disconnected quickly and easily from the operant chambers when the self-administration session is over. Second, animals need not be habituated to wearing a saddle. Third, unlike methods that rely on fastening the catheter to the head, this method allows experimental manipulations of the brain to be carried out at any time. Fourth, and most important, the surgical mesh provides a solid anchor point for the catheter; tissue grows through the mesh and encapsulates it to yield a reliable attachment of the external end of the catheter to the animal.

The patency of catheters can be determined in several ways. An obvious fault in a catheter is mechanical blockage of flow. This is apparent when the catheter is flushed with solution. For example, in our laboratory, catheters are flushed twice each day, once before the animal begins its self-administration session, and once when the session is completed. If flow in the catheter is becoming restricted, the increased resistance is obvious. Since we flush approx 0.1 mL of sterile saline into the catheter at these times (several times the dead volume of the catheter), this serves the additional purpose of replacing the residual drug solution remaining in the catheter after the self-administration session with sterile saline solution.

Catheter failure can also occur because of externalization of the tip from the circulatory system. We have determined that, in some cases when this happens, the tip of the catheter may deliver solution to the region of the brachial plexus. The resulting pressure on nerve fibers in that area when the catheter is flushed can cause the animal to lift its right paw, thereby providing an indication of catheter failure.

Catheter failure may not always be so apparent, however. A catheter may exhibit normal resistance to saline infusion, yet be unusable because it is no longer delivering drug into the venous circulation. This may happen because of leakage at the tip, failure of the joints in the catheter, or a combination of these. A simple way to test the catheter is to infuse the short-acting barbiturate methohexital at a dose of approx 3 mg/kg (we test with 0.1 mL of a 10 mg/kg solution). If the catheter is delivering to the venous circulation, the animal will show an immediate effect of the barbiturate—within a few s, the animal will become ataxic or lose righting reflex. If the catheter is not delivering solution into the circulatory system, there will be no effect of methohexital.

Although it is important to be cautious of catheter patency, the stability of the self-administration data itself provides an indication of catheter condition. Following acquisition, drug-maintained responding usually remains very constant as long as the dose available to the animals is also constant. A change in the rate of drug self-administration in the absence of experimental manipulations is a sign that should suggest testing the patency of the catheter with methohexital.

3.4. Factors Particularly Important to Nicotine

Previous studies with primates have suggested that scheduling of access to nicotine is an important variable in achieving successful self-administration. The difficulty in attempting to use intermittent access schedules in rodent studies is the general lack of studies showing what parameters might be successful. Although occasional studies have reported use of progressive ratio and fixed-interval schedules for cocaine or heroin self-administration (Corrigall and Coen, 1989a; Dougherty and Pickens, 1973; Roberts et al., 1989), the bulk of rodent studies of drug self-administration have been carried out with CRF schedules or low fixed-ratio schedules. Therefore, we opted to use a fixed-ratio schedule with a modest requirement of five for studies with rats.

The aversive properties of nicotine could figure prominently in determining the extent of self-administration

(Goldberg and Henningfield, 1988). We therefore limit the extent of nicotine self-administration temporally within a given session by imposing a minimum time delay of 1 min between infusions (i.e., a 1-min time-out). Also, daily access to nicotine is limited to a single 1-h session.

Aversive properties also depend upon dose. In monkeys, doses as low as 0.01 mg/kg/infusion have been shown to act as a punisher (Goldberg et al., 1983). However, in the same study, nicotine self-administration was best maintained at doses of 0.03 mg/kg/infusion, and in some cases higher, although vomiting occurred after doses of 0.1 mg/kg/infusion. In the Cox et al. (1984) study, maximal response rates occurred at the 0.03 mg/kg unit dose. Based on these studies, we decided to use a dose of nicotine of 0.03 mg/kg/infusion for acquisition (all doses are delivered as a vol of 0.1 mL/kg/infusion). We use the [–]-nicotine bitartrate salt. It is important to point out that doses reported here refer to the base. Some of the early studies of nicotine self-administration do not specify whether reported concentrations refer to the salt or the base, and for nicotine bitartrate, there is almost a threefold difference between these values.

Solutions of nicotine bitartrate are highly acidic. Indeed, as has been noted by Clarke (1987), use of nicotine solutions with the pH value not adjusted was probably why human research subjects reported discomfort at the site of iv injection. In our studies of nicotine self-administration, the pH of nicotine solutions is adjusted to 7.0 ± 0.1 with sodium hydroxide.

Sterility of solutions is an important factor when agents are administered intravenously. Nicotine solutions are prepared fresh weekly and, like all solutions used in self-administration studies in this laboratory, are sterilized by passing them through a 0.22-μm filter prior to use. This is particularly important for nicotine solutions since they appear to sustain bacterial growth readily. Other than these specifics, the same methodology was used to obtain nicotine self-administration as has been used in this laboratory to establish self-administration of other drugs.

4. Nicotine Self-Administration

4.1. Acquisition and Maintenance of Self-Administration

Acquisition of drug self-administration is generally begun 1 wk after catheter implantation. Although the final schedule is an FR5, for the first week of self-administration nicotine is made available to the animals on a schedule of continuous reinforcement in order to increase the association of lever pressing with nicotine delivery (this schedule still includes a 1-min time-out after each drug delivery). During the second week, schedule requirements are increased to FR2 and then to the final value of FR5. Responding is usually stable by the end of the third week of self-administration and remains stable as long as catheters are patent.

Figure 2 shows data for the first 3 wk of acquisition of nicotine self-administration for a group of animals. Each time the ratio requiremement was increased, the number of drug infusions obtained by the animals decreased somewhat, and then gradually returned to its previous value over successive sessions. Notice that by the end of the third week of self-administration, substantial drug-maintained responding had already been established and was beginning to stabilize.

Figure 3 shows the temporal pattern of responding in a 1-h session in a typical subject at the end of the third week of self-administration at 0.03 mg/kg/infusion. Nicotine maintains a high level of responding on the lever that results in drug delivery, whereas responding on the inactive or control lever is virtually absent. The number of nicotine infusions tends to be slightly greater at the beginning of the session than at the end.

4.2. Regulation of Nicotine Intake

It is important to establish the relationship between the dose of nicotine available and the extent of self-administration. This can be done by altering the nicotine unit dose and permitting responding to stabilize at the new dose level. As shown in Fig. 4,

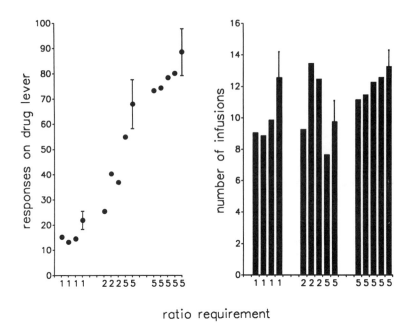

ratio requirement

Fig. 2. Acquisition of nicotine self-administration over the first 3 wk of drug availability. The left-hand graph shows the average total number of responses on the drug-reinforced lever (including responses during the time-out period), whereas the right graph shows the mean number of infusions received by the animals. Each data point or column is the mean of 14 animals on consecutive weekday sessions; error bars in this and subsequent figures represent standard errors (SEMs) from the mean. Note that, as the ratio requirement increases, the amount of responding increases; by contrast, increases in the ratio requirement result in a transient decrease in the number of infusions obtained by the animals, followed by a progressive increase in infusions to the previous value. By the end of the third week of self-administration, both number of responses and number of infusions have begun to stabilize.

the curve of responding vs unit dose has an inverted-U shape, with peak responding at the 0.01–0.03 mg/kg/infusion values. Doses both lower and higher than these maintain less responding.

To understand the regulation of nicotine self-administration, it is useful to compare it to another drug. Figure 5 (A) shows data from the same group of animals as Fig. 4, but plotted as the number of infusions vs unit dose. Also shown is the dose–effect curve for cocaine, obtained in a different group of animals, but

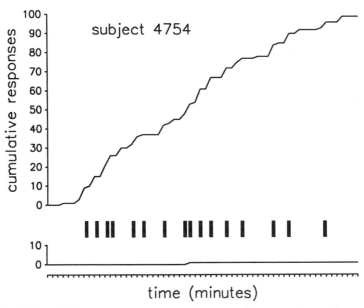

Fig. 3. Cumulative response record for one subject within a 1-h session at the end of the third week of self-administration. The top graph shows cumulative responding of the drug-reinforced lever, and the bottom tracing shows responding on the inactive lever. The vertical bars between the two cumulative tracings show the times at which drug infusions were obtained by the subject.

with the identical operant schedule and training history (Fig. 5 [B]). For nicotine, as predicted by the curve in Fig. 4, the infusions vs dose curve has an inverted-U shape. Similarly, for cocaine, the dose–effect curve also has an inverted-U shape, and this qualitative similarity might at first glance suggest that nicotine and cocaine are regulated similarly in a self-administration paradigm. Closer examination reveals that this is not so.

Self-administration of cocaine follows a particularly orderly dose relationship; as the unit dose of cocaine decreases from 1.0 to 0.1 mg/kg, the number of infusions received increases, presumably because of the subject's attempt to compensate for the decreased drug dose. Eventually, the dose of cocaine becomes so small (0.03 mg/kg/infusion) that responding extinguishes to values similar to those obtained when saline is substituted for drug.

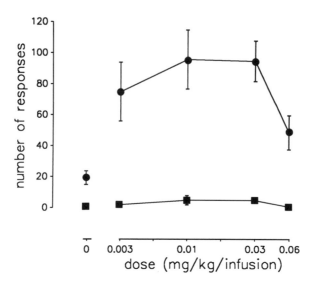

Fig. 4. Dose–effect curve for nicotine self-administration. Points plotted are the means for eight rats at each dose, taken when responding had stabilized at the respective dose. The filled circles represent responding on the nicotine-reinforced lever, and the filled squares are the means of responding on the inactive lever. Where not shown, the error bar falls within the plotted point. From Corrigall and Coen (1989b), with permission.

For nicotine, there is less obvious compensation in responding as the unit dose is changed downward. For example, there is no increase in responding at a unit dose of 0.01 as compared to the acquisition dose of 0.03 mg/kg/infusion; rather the curve is flat compared to cocaine. At a unit dose one-half logarithmic step lower than 0.01 mg/kg infusion, responding and the number of infusions received begin to decline.

In addition, several facts suggest that the regulation of intake of cocaine and nicotine differs at both the high-dose and low-dose ends of the curve. First, for cocaine, the fewer number of infusions that occur as the unit dose increases corresponds nonetheless to an increasing total drug intake in the session (bar graphs in Fig. 5B). For nicotine, however, although fewer infusions are obtained at the 0.06 mg/kg as compared to the 0.03 mg/kg unit dose, the total session intake of nicotine actually remains very similar at the two doses (Fig. 5[B]). This suggests

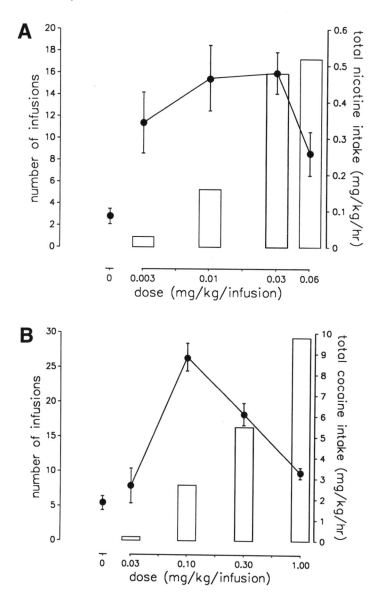

Figure 5. (A). Nicotine dose–effect curve for nicotine plotted as the number of infusions self-administered (line graph) in the 1-h session vs the unit dose of drug available. The bar graph shows the total session intake of nicotine at each unit dose. From Corrigall and Coen (1989b), with permission. (B). Dose–effect curve for cocaine self-administration on an FR5 schedule showing the number of infusions obtained (line graph) and the total cocaine intake (bar graph).

the possibility of an upper limit on the amount of nicotine that the subjects will administer in a given time.

In addition to the total dose of nicotine obtained in the session, it may be that the size of the unit dose itself has important consequences in determining nicotine self-administration. In pilot studies that we have carried out, we observed that an iv dose of nicotine of 0.1 mg/kg/infusion could cause seizures in rats; similar findings were reported by Hanson et al. (1979). This, too, could account in part for the decrease in self-administration at the high end of the dose–effect curve. For drugs, such as cocaine, the unit dose itself may not have aversive properties that are as prominent in the dose range that supports self-administration; certainly doses higher than even 1 mg/kg/infusion have been used in self-administration studies of cocaine in rodents (e.g., Roberts et al., 1989). The existence of an upper limit to nicotine self-administration is in keeping with observations in research on cigaret smoking by humans (*see* Section 5).

Another piece of evidence supports the idea that there may be an upper limit to nicotine intake. If the duration of the time-out period is decreased at the 0.03 mg/kg unit dose, there is no significant increase observed in the number of infusions obtained (Corrigall and Coen, 1989b). Therefore, although the animals are provided more time within the session to obtain nicotine, there is no increase in the number of infusions obtained.

Nicotine self-administration appears to differ from cocaine self-administration at the low end of the dose–effect curve as well. Whereas low doses of cocaine maintain substantially less responding than do midrange doses, for nicotine there is still considerable responding maintained at the lowest unit dose that we tested, even though the nicotine intake was small. Nicotine, therefore, may present a somewhat unique pattern of self-administration as compared to other drugs. Regulation may be less dependent upon the magnitude of the dose, up to a maximum, than is the case for drugs like cocaine.

An additional comment that should be made about the regulation of nicotine self-administration concerns the transient effects of changing the dose on response output (that is, what happens immediately after the nicotine dose is altered). In the case of other drugs, such as psychomotor stimulants, there is a compensatory increase in responding when the drug dose is decreased. For nicotine, although some subjects do show a compensatory increase in responding when the dose of nicotine is decreased, other subjects simply produce fewer responses over successive sessions (i.e., the compensatory increase in responding does not occur). Similarly, pretreatment with the nicotinic antagonist mecamylamine prior to self-administration sessions results in dose-dependent decreases in nicotine self-administration with minimal evidence for compensatory increases in responding (Corrigall and Coen, 1989b). Therefore, these data also suggest that nicotine self-administration is regulated in a different way than psychomotor stimulant drugs, which are often used as general examples of addictive substances.

5. Comparison to Nicotine Use by Humans

Smokers alter their smoking behavior apparently in response to the amount of nicotine that they receive (i.e., there is a dose–effect relationship) (Henningfield et al., 1987; Russell, 1987). This has been demonstrated in several ways. If cigarets with lower nicotine yield are substituted for smokers' normal brands, at least some individuals will compensate by increasing the number of cigarets that are smoked; if higher-yield cigarets are substituted for the usual brand, the number of cigarets smoked will decrease. Upregulation of smoking behavior also occurs when human subjects are treated with the nicotinic antagonist mecamylamine (Stolerman et al., 1973). However, it has been noted that upregulation of smoking generally results in only partial compensation for the decreased effect of nicotine. In other words, although it appears that regulation of smoking behavior does occur as a consequence of changes in nicotine availability, there is only partial compensation made by cigaret smokers for the decreased availablity of nicotine.

If, on the other hand, nicotine levels are increased through administration of the drug by an alternate route to smoking, by switching to higher-yield cigarets, or by altering urinary pH to change the rate of nicotine excretion, downregulation of cigaret smoking and nicotine intake can also be demonstrated (Benowitz and Jacob, 1987; Henningfield et al., 1987; Russell, 1987; Schacter et al., 1977). In these instances, the downregulation that occurs is much more precise than upregulation. These observations suggest that the compensation that occurs in smoking behavior functions more to keep nicotine levels from increasing beyond a certain level that may be aversive than it does to maintain a given level of intake (Russell, 1987). Our animal data are in keeping with these observations.

The upper aversive limit for nicotine intake is likely related to the drug's ability to act as a punisher in some situations in both humans (Henningfield and Goldberg, 1983b) and animals (Goldberg et al., 1983; Goldberg and Spealman, 1982). Nicotine self-administration in animals is supported at some doses that have effects that are presumably aversive—for example, monkeys self-administered nicotine at doses that also produced emesis (Goldberg and Spealman, 1982; Spealman and Goldberg, 1982)—showing the close link between positive and negative effects for this drug. Nicotine was also self-administered iv by cigaret smokers in a laboratory setting (Henningfield et al., 1983; Henningfield and Goldberg, 1983b), and in these studies, subjects reported experiencing nausea after receiving nicotine.

Subjects given nicotine, as opposed to those self-administering it to themselves, also report both positive and negative effects (Henningfield et al., 1983; Jasinski et al., 1984). Intravenous nicotine increased scores of "drug-liking" and was frequently identified as cocaine by cigaret smokers with histories of drug abuse. Both iv administered and inhaled nicotine increased scores on the morphine-benzedrine (MBG) scale, corresponding to subjective reports of a "rush" or "high" following nicotine. However, subjects given nicotine iv also reported feelings of fear and discomfort, and correspondingly their scores on the lysergic acid diethylamide (LSD) scale of the Addiction Research Center Inventory were increased (Henningfield et al., 1985). Scores were also increased, although to a lesser extent, after inhalation of nicotine in cigaret smoke. Aversive factors may therefore comprise the upper boundary of smoking by humans, or iv nicotine self-administration by animals or humans.

The consistency in the quantitative dependence on dose in this nicotine self-administration paradigm compared to what we know about smoking behavior by humans (Kozlowski and Herman, 1984; Henningfield et al., 1987; Russell, 1987) speaks to the validity of this model for studies of nicotine reinforcement.

6. Conclusions

This model of self-administration satisfies a number of criteria necessary to show that nicotine functions as a reinforcer in rodents (Goldberg and Henningfield, 1988). First, the amount of responding maintained by nicotine is similar to the amount of responding maintained by a known reinforcer, such as cocaine, at least at some doses. Second, responding for nicotine occurs in a temporally distributed fashion across the session, with drug infusions being obtained in a relatively regular fashion.

Nicotine self-administration with this model is dependent on the dose of the drug that is available. In comparing nicotine to a stimulant like cocaine, one finds that the dose regulation is not the same for the two drugs. An upper limit appears to exist for nicotine self-administration, governed perhaps by several factors that do not exist for cocaine self-administration. It may be that for nicotine, more so than for other drugs, processes other that the positive properties of the drug are important in regulation.

In this model, lever-pressing behavior is directed specifically toward obtaining nicotine. Several facts support this conclusion. First, the centrally acting nicotinic antagonist mecamylamine reduces nicotine self-administration. In addition, responding for nicotine is significantly greater than responding for saline vehicle, and responding on the inactive or nonreinforced manipulandum virtually does not occur. Taken together, these data provide conclusive evidence that nicotine can serve as a reinforcer in rodents.

What uses can this model serve? First, along with studies in other animals, it demonstrates that nicotine can act as a reinforcer in a range of species. Second, it seems clear that one of the directions of neuroscience research over the next decade will be investigation of the basic central nervous system mechanisms of drug reinforcement. It is likely that models such as this one will be used to further our understanding of the processes of addiction, addressing questions about the commonality or separateness of brain reinforcement pathways underlying drug-taking behavior, and the mechanisms that are operative in these pathways. For example, we have recently found that nicotine self-administration is reduced following acute treatment with specific dopamine antagonists (Corrigall and Coen, 1991b) and that 6-hydroxydopamine lesions of the mesolimbic system at the level of the nucleus accumbens reduce nicotine self-administration (Corrigall, Franklin, Clarke, and Coen, 1992), just as identical lesions attenuate cocaine self-administration (Roberts et al., 1980). Recent data have also shown that opioid antagonists reduce cocaine self-administration (Carroll et al., 1986; De Vry et al., 1989); we have confirmed this observation, and demonstrated that these antagonists do not alter nicotine reinforcement (Corrigall and Coen, 1991a). Although on one hand the former findings suggest a commonality of function for the mesolimbic dopamine system in each of the cocaine and nicotine reinforcements, the latter data suggest that the two drugs may be under different influences of a endogenous opioid system. Moreover, it is clear from clinical studies that nicotine differs from other drugs in its rewarding or pleasurable affective properties on one hand and its ability to maintain drug-taking behavior on the other (Kozlowski et al., 1989). It will therefore be important to determine what other brain pathways, either separate to the mesolimbic system or modifying its action, are participants in nicotine reinforcement. Finally, at a more applied end, models of nicotine self-administration can provide a preclinical system with which to test pharmacological interventions that might be useful adjuncts to smoking cessation programs.

References

Ator N. A. and Griffiths R. R. (1983) Nicotine self-administration in baboons. *Pharmacol. Biochem. Behav.* **19,** 993–1003.
Benowitz N. L. and Jacob P. III (1987) Metabolism, pharmacokinetics, and pharmacodynamics of nicotine in man, in *Advances in Behavioral Biol-*

ogy, vol. 31, Tobacco Smoking and Nicotine (Martin W. R., Van Loon G.R., Iwamoto E.T., and Davis L., eds), Plenum, New York, pp. 81–99.

Carroll M. E., Lac S. T., Walker M. J., Kragh R., and Newman T. (1986) Effects of naltrexone on intravenous cocaine self-administration in rats during food satiation and deprivation. *J. Pharmacol. Exp. Ther.* **238,** 1–7.

CCAC (1980) *Guide to the Care and Use of Experimental Animals.* Canadian Council on Animal Care, Ottawa, Ontario.

Chance W. T., Murfin D., Krynock G. M., and Rosecrans J. A. (1977) A description of the nicotine stimulus and tests of its generalization to amphetamine. *Psychopharmacology* **55,** 19–26.

Clarke P. B. S. (1987) Nicotine and smoking: a perspective from animal studies. *Psychopharmacology* **92,** 135–143.

Clarke P. B. S. and Kumar R. (1983a) The effects of nicotine on locomotor activity in non-tolerant and tolerant rats. *Br. J. Pharmacol.* **78,** 329–337.

Clarke P. B. S. and Kumar R. (1983b) Characterization of the locomotor stimulant action of nicotine in tolerant rats. *Br. J. Pharmacol.* **80,** 587–594.

Clarke P. B. S., Fu D. S., Jakubovic A., and Fibiger H. C. (1988) Evidence that mesolimbic dopaminergic activation underlies the locomotor stimulant action of nicotine in rats. *J. Pharmacol. Exp. Ther.* **246,** 701–708.

Clarke P. B. S., Corrigall W. A., Ferrence R. G., Friedland M. L., Kalant H., and Kozlowski L. T. (1989) Tobacco, Nicotine and Addiction, A Committe Report prepared at the request of the Royal Society of Canada, Health and Welfare Canada, Ottawa, Ontario.

Collins R. J., Weeks J. R., Cooper M. M., Good P. I., and Russell R. R. (1984) Prediction of abuse liability of drugs using IV self-administration by rats. *Psychopharmacology* **82,** 6–13.

Corrigall W. A. (1987) Heroin self-administration: effects of antagonist treatment in lateral hypothalamus. *Pharmacol. Biochem. Behav.* **27,** 693–700.

Corrigall W. A. and Coen K. M. (1989a) Fixed-interval schedules for drug self-administration in the rat. *Psychopharmacology* **99,** 136–139.

Corrigall W. A. and Coen K. M. (1989b) Nicotine maintains robust self-administration in rats on a limited-access schedule. *Psychopharmacology* **99,** 473–478.

Corrigall W. A. and Coen K. M. (1991a) Opiate antagonists reduce cocaine but not nicotine self-administration. *Psychopharmacology* **104,** 167–170.

Corrigall W. A. and Coen K. M. (1991b) Selective dopamine antagonists reduce cocaine but not nicotine self-administration. *Psychopharmacology* **104,** 171–176.

Corrigall W. A. and Vaccarino F. J. (1988) Antagonist treatment in nucleus accumbens or periaqueductal grey affects heroin self-administration. *Pharmacol. Biochem. Behav.* **30,** 443–450.

Corrigall W. A., Franklin K. B. J., Coen K. M., and Clarke P. B. S. (1992) The mesolimbic dopaminergic system is implicated in the reinforcing effects of nicotine. *Psychopharmacology* **107,** 285–289.

Corrigall W. A., Herling S., and Coen K. M. (1988a) Evidence for opioid mechanisms in the behavioral effects of nicotine. *Psychopharmacology* **96,** 29–35.

Corrigall W. A., Herling S., and Coen K. M. (1988b) Evidence for a behavioral deficit during withdrawal from chronic nicotine treatment. *Pharmacol. Biochem. Behav.* **33,** 559–562.

Cox B. M., Goldstein A., and Nelson W. T. (1984) Nicotine self-administration in rats. *Br. J. Pharmacol.* **83,** 49–55.

Dai S., Corrigall W. A., Coen K. M., and Kalant H. (1989) Heroin self-administration by rats: influence of dose and physical dependence. *Pharmacol. Biochem. Behav.* **32,** 1009–1015.

De Vry J., Donselaar I., and Van Ree J. M. (1989) Food deprivation and acquisition of intravenous cocaine self-administration in rats: effect of naltrexone and haloperidol. *J. Pharmacol. Exp. Ther.* **251,** 735–740.

Dougherty J. and Pickens R. (1973) Fixed-interval schedules of intravenous cocaine presentation in rats. *J. Exp. Anal. Behav.* **20,** 111–118.

Dougherty J., Miller D., Todd G., and Kostenbauder H. B. (1981) Reinforcing and other behavioral effects of nicotine. *Neurosci. & Biobehav. Rev.* **5,** 487–495.

Goldberg S. R. and Henningfield J. E. (1988) Reinforcing effects of nicotine in humans and experimental animals responding under intermittent schedules of iv drug injection. *Pharmacol. Biochem. Behav.* **30,** 227–234.

Goldberg S. R. and Spealman R. D. (1982) Maintenance and suppression of behavior by intravenous nicotine injections in squirrel monkeys. *Fed. Proc.* **41,** 216–220.

Goldberg S. R., Spealman R. D., and Goldberg D. M. (1981) Persistent behavior at high rates maintained by intravenous selfadministration of nicotine. *Science* **214,** 573–575.

Goldberg S. R., Spealman R. D., Risner M. E., and Henningfield J. E. (1983) Control of behavior by intravenous nicotine injections in laboratory animals. *Pharmacol. Biochem. Behav.* **19,** 1011–1020.

Griffiths R. R., Bigelow G. E., and Henningfield J. E. (1980) Similarities in animal and human drug-taking behavior, in *Advances in Substance Abuse,* vol. 1 (Mello N. K., ed.), JAI Press, Greenwich CT, pp. 1–90.

Hanson H. M., Ivester C. A., and Morton B. R. (1979) Nicotine self-administration in rats, in NIDA Research Monograph 23, *Cigarette Smoking as a Dependence Process* (Krasnegor N. A., ed.), U.S. Department of Health and Human Services, Rockville, MD, pp. 70–90.

Health and Welfare Canada, Statistics Canada, Canada Health Survey, Ottawa, Supply and Services Canada, 1981.

Hendry J. S. and Rosecrans J. A. (1982) The development of pharmacological tolerance to the effect of nicotine on schedule-controlled responding in mice. *Psychopharmacology* **77,** 339–343.

Henningfield J. E. and Goldberg S. R. (1983a) Nicotine as a reinforcer in human subjects and laboratory animals. *Phamacol Biochem. Behav.* **19,** 989–992.

Henningfield J. E. and Goldberg S. R. (1983b) Control of behavior by intravenous nicotine injections in human subjects. *Pharmacol. Biochem. Behav.* **19,** 1021–1026.

Henningfield J. E. and Goldberg S. R. (1988) Pharmacologic determinants of tobacco self-administration by humans. *Pharmacol. Biochem. Behav.* **30,** 221–226.

Henningfield J. E., Goldberg S. R., and Jasinski D. R. (1987) Nicotine: Abuse liability, dependence potential and pharmacologic treatment of dependence, in *Advances in Behavioral Biology, vol. 31: Tobacco Smoking and Nicotine* (Martin W. R., Van Loon G. R., Iwamoto E. T., and Davis L., eds.), Plenum, New York, pp. 81–99.

Henningfield J. E., Miyasato K., and Jasinski D. R. (1983) Cigarette smokers self-administer intravenous nicotine. *Pharmacol. Biochem. Behav.* **19,** 887–890.

Henningfield J. E., Miyasato K., and Jasinski D. R. (1985) Abuse liability and pharmacodynamic characteristics of intravenous and inhaled nicotine. *J. Pharmacol. Exp. Ther.* **234,** 1–12.

Jasinski D. R., Johnson R. E., and Henningfield J. E. (1984) Abuse liability assessment in human subjects. *Trends Pharmacol. Sci.* **5,** 196–200.

Johanson C. E. and Schuster C. R. (1981) Animal models of drug self-administration, in *Advances in Substance Abuse,* vol. 2 (Mello N. K., ed.), JAI Press, Greenwich, CT, pp. 219–297.

Kozlowski L. T. and Herman C. P. (1984) The interaction of psychosocial and biological determinants of tobacco use: more on the boundary model. *J. App. Soc. Psychol.* **14,** 244–256.

Kozlowski L. T., Wilkinson D. A., Skinner W., Kent C., Franklin T., and Pope M. (1989) Comparing tobacco cigarette dependence with other drug dependencies: Greater or equal "difficulty quitting" and "urges to use," but less pleasure from cigarettes. *J. Am. Med. Assoc.* **261,** 898–901.

Lang W. J., Latiff A. A., McQueen A., and Singer G. (1977) Self administration of nicotine with and without a food delivery schedule. *Pharmacol. Biochem. Behav.* **7,** 65–70.

Latiff A. A., Smith L. A., and Lang W. J. (1980) Effects of changing dosage and urinary pH in rats self-administering nicotine on a food delivery schedule. *Pharmacol. Biochem. Behav.* **13,** 209–213.

Levin E. D., Morgan M. M., Galvez C., and Ellison G. D. (1987) Chronic nicotine and withdrawal effects on body weight and food and water consumption in female rats. *Physiol. Behav.* **39,** 441–444.

Pratt J. A., Stolerman I. P., Garcha H. S., Giardini V., and Feyerabend C. (1983) Discriminative stimulus properties of nicotine: further evidence for mediation at a cholinergic receptor. *Psychopharmacology* **81,** 54–60.

Risner M. E. and Goldberg S. R. (1983) A comparison of nicotine and cocaine self-administration in the dog: Fixed-ratio and progressive-ratio schedules of intravenous drug infusion. *J. Pharmacol. Exp. Ther.* **224,** 319–326.

Roberts D. C. S. and Goeders N. (1987) Drug self-administration: experimental methods and determinants, in *Neuromethods*, vol. 13 (Boulton A. A., Baker G. B., and Greenshaw A. J., eds.), Humana Press, Clifton, NJ, pp. 349–398.

Roberts D. C. S., Loh E. A., and Vickers G. (1989) Self-administration of cocaine on a progressive ratio schedule in rats: Dose–response relationships and effect of haloperidol pretreatment. *Psychopharmacology* **97**, 535–538.

Roberts D. C. S., Koob G. F., Klonoff P., and Fibiger H. C. (1980) Extinction and recovery of cocaine self-administration following 6-hydroxy-dopamine lesions of the nucleus accumbens. *Pharmacol. Biochem. Behav.* **12**, 781–787.

Russell M. A. H. (1987) Nicotine intake and its regulation by smokers, in *Advances in Behavioral Biology, vol. 31, Tobacco Smoking and Nicotine* (Martin W. R., Van Loon G. R., Iwamoto E. T., and Davis L., eds.), Plenum, New York, pp. 25–50.

Schacter S., Kozlowski L. T., and Silverstein B. (1977) Effects of urinary pH on cigarette smoking. *J. Exp. Psychol. (Gen.)* **106**, 13–19.

Singer G., Simpson F., and Lang W. J. (1978) Schedule induced self injections of nicotine with recovered body weight. *Pharmacol. Biochem. Behav.* **9**, 387–389.

Slifer B. L. (1983) Schedule-induction of nicotine self-administration. *Pharmacol. Biochem. Behav.* **19**, 1005–1009.

Smith L. A. and Lang W. J. (1980) Changes occurring in self administration of nicotine by rats over a 28-day period. *Pharmacol. Biochem. Behav.* **13**, 215–220.

Spealman R. D. and Goldberg S. R. (1982) Maintenance of schedule-controlled behavior by intravenous injections of nicotine in squirrel monkeys. *J. Pharmacol. Exp. Ther.* **223**, 402–408.

Stolerman I. P., Garcha H. S., Pratt J. A., and Kumar R. (1984) Role of training dose in discrimination of nicotine and related compounds by rats. *Psychopharmacology* **84**, 413–419.

Stolerman I. P., Goldfarb T., Fink R., and Jarvik M. E. (1973) Influencing cigarette smoking with nicotine antagonists. *Psychopharmacologia* **28**, 247–259.

Stolerman I. P., Kumar R. K., Pratt J. A., and Reavill C. (1987) Discriminative stimulus effects of nicotine: correlation with binding studies, in *Advances in Behavioral Biology, vol. 31, Tobacco Smoking and Nicotine* (Martin W. R., Van Loon G. R., Iwamoto E. T., and Davis L., eds.), Plenum, New York, pp. 113–124.

US DHHS (1988) The Health Consequences of Smoking: Nicotine Addiction. A Report of the Surgeon General, US Department of Health and Human Services, Rockville, MD.

Weeks J. R. (1981) An improved pneumatic syringe for self-administration of drugs by rats. *Pharmacol. Biochem. Behav.* **14**, 573,574.

Weeks J. R. (1983) Cardiovascular techniques using unanesthetized and freely moving rats, unpublished documentation available from J. R. Weeks, Kalamazoo, MI.

Animal Models for Assessing Hallucinogenic Agents

Richard A. Glennon

1. Introduction

1.1. Animal Models and Classical Hallucinogens

Hallucinogenic activity, by definition, can be measured only in human subjects. Nevertheless, there may be an animal counterpart, or a behavioral or physiological response that parallels hallucinogenic activity/potency in humans and will be a useful predictor of hallucinogenic potential or abuse liability. Over the years, a variety of tests or models have been explored for the purpose of identifying such agents. Although some of these models have contributed enormously to our understanding of hallucinogenic agents, their mechanism of action, and their structure–activity relationships, a single, all-purpose, reliable animal model remains elusive. The situation is rooted in problems that only now are becoming understood—for example, the confusion associated with the classification of hallucinogenic substances, and the lack of a "pure" prototypic hallucinogen. Is it appropriate to lump together in a single class all agents capable of producing a psychotomimetic response? Exactly what constitutes a hallucinogenic/psychotomimetic response in humans? Humans can differentiate the effects of different hallucinogenic agents; some of these differentiating effects may be of a central origin, whereas others may be peripheral in nature. Some hallucinogens produce their effects upon administration of a single

From: *Neuromethods, Vol. 24: Models of Drug Addiction*
Eds: A. Boulton, G. Baker, and P. H. Wu ©1992 The Humana Press Inc.

effective dose, whereas other agents (such as amphetamine and cocaine) produce psychotomimetic effects only after chronic administration of large doses. Other agents, not generally considered as being hallucinogenic, may produce a toxic delirium that occasionally involves hallucinatory episodes. In addition to their behavioral subtleties, hallucinogenic agents may often be distinguished from one another on the basis of their somatic effects (e.g., nausea and other side effects). Indeed, Naranjo (1973) has commented that a given hallucinogen may produce different effects in the same individual on different occasions of administration. Thus, it is no wonder that identification of an animal model of this activity poses a significant problem. With regard to a prototype hallucinogenic agent, (+)lysergic acid diethylamide (LSD) immediately comes to mind. Yet, the effects of LSD in humans are clearly distinguishable from those of, for example, tetrahydrocannabinol (THC) or phencyclidine (PCP). LSD has been, and continues to be, the best investigated of all the hallucinogenic agents. The actions of this agent have served as the basis for a variety of animal models, but now that certain agents are known to produce effects that differ dramatically from those of LSD, it is realized that LSD may not be *the* prototypic hallucinogen.

It is becoming increasingly clear that hallucinogenic/psychotomimetic agents do not constitute a single class of drugs and, even if some of these agents are capable of producing similar effects, that many of them produce effects that readily allow for their pharmacological differentiation. We, and others, have frequently employed the term "classical hallucinogens" to describe certain hallucinogenic agents. Although a membership roster has never been published, classical hallucinogens generally refer to those that possess an indolealkylamine or phenylalkylamine backbone. Membership is not exclusive, nor is it inclusive. Not all indolealkylamine and phenylalkylamine derivatives are hallucinogenic. In addition to the classical hallucinogens, other major classes of agents that produce psychotomimetic effects include THC and related cannabinoids, phencyclidine-related agents, certain opiates, and atropine-like drugs. Each of these classes of agents produces distinguishable effects. They most likely act via different mechanisms of action and are perhaps best studied individually on a class-by-class basis. No single animal model is capable of identifying members of every one of these classes. The present chapter will deal only with the classical hallucinogens.

There are numerous examples of indolealkylamine hallucinogens; these include the ergolines (e.g., LSD), β-carbolines (e.g., harmaline), and simple indolealkylamines, such as *N,N*-dialkyltryptamines and α-methyltryptamines. Hallucinogenic phenylalkylamines include phenethylamines, such as mescaline, and phenylisopropylamines, such as 1-(2,5-dimethoxy-4-methylphenyl)-2-aminopropane (DOM) and 1-(2,5-dimethoxy-4-bromophenyl)-2-aminopropane (DOB). Certain di- and tri-methoxy analogs of phenylisopropylamine (amphetamine) (i.e., DMAs and TMAs, respectively) are also hallucinogenic. The 3,4-methylenedioxy analog of phenylisopropylamine (MDA) is both a hallucinogenic agent and a central stimulant. LSD is among the most potent of the hallucinogens and is generally considered an indolealkylamine prototype. Mescaline is one of the oldest known hallucinogens and is a prototypical phenalkylamine hallucinogen. Many animal models have been developed using one or both of these agents as a frame of reference. This is probably unfortunate, because LSD is known to produce a wide variety of effects (some of which are not shared by certain other classical hallucinogenic agents), and mescaline, although of historical significance, is one of the least potent of the hallucinogens. Nevertheless, early attempts to develop animal models were based on the observation and categorization of the various behavioral and physiological effects produced by these two agents.

1.2. Classification of Models

There are two types of behavioral animal models that can be used to study hallucinogenic agents: analog models and assay models (Stoff et al., 1978). These models have also been referred to as isomorphic models and parallel models, respectively (Jacobs and Trulson, 1978). An analog or isomorphic model is an animal model that is faithful to drug-induced hallucinosis in humans; that is, the model rests on some intrinsic similarity in behavior in animals and humans. Analog behaviors consist of excitatory syndromes with sympathomimetic stimulation, "hallucinatory-like" behaviors, and disturbances in attention (Stoff et al., 1978). It has been argued that this type of model runs the risk of anthropomorphic interpretation (Stoff et al., 1978). The assay or parallel model represents behaviors that are independent of the behavior associated with hallucinations and have no apparent human counterpart. The model only requires that there be at least one formal shared characteristic (e.g., rank order of potency for a series of agents) and that other pharmacological interventions result in parallel responses. It has been acknowledged that not all animal behaviors fit neatly into one of these two categories. Furthermore, this classification ignores in vitro models, which, nonetheless, might also be considered as assay models. The assay model, though more stochastic than the analog model, may or may not be useful for mechanistic interpretations, but does have value in its predictive potential.

In addition to behavioral models and in vitro (e.g., isolated tissue) pharmacological models, there are also nonanimal models. The activity/potency of a series of hallucinogenic agents might be correlated with a physicochemical property, a quantum chemical parameter, or some other descriptor of the agents' structure. These types of models can be useful from a predictive standpoint, and they can also assist in understanding underlying mechanisms of action at the molecular or receptor level. However, these types of models are usually developed using animal or human data as input. Any predictions made on the basis of such models require verification in animal models and, ultimately, in human subjects. Nonanimal models will not be discussed here.

The ideal animal model should:

1. Selectively identify only those agents capable of producing hallucinogenic effects in humans;
2. Account for the activity and potency of agents from different chemical classes;
3. Provide results in a dose-dependent manner;

4. Be reproducible from laboratory to laboratory; and
5. Be convenient to use (i.e., relatively inexpensive, uncomplicated, and easy to quantify).

Stoff and coworkers (1978) have developed a list of ten criteria that should be met by an animal model of hallucinations and have scored several behavioral models on how well they meet these criteria.

1.3. Problem Agents

Certain agents are recurrently detected by some animal models as being potential hallucinogens. Among the more frequently cited examples are amphetamine, lisuride, quipazine, and certain serotonin (5-hydroxytryptamine, 5-HT) antagonists, such as methysergide and 2-bromo-LSD (BOL). In human subjects, single doses of amphetamine are not hallucinogenic. The situation is a bit more complex with the other agents mentioned. Generally, lisuride, quipazine, methysergide, and 2-bromo-LSD are considered as nonhallucinogens. However, lisuride has been reported to produce hallucinatory episodes (e.g., Todes, 1986). In one study, treatment of Parkinsonian patients with lisuride produced vivid hallucinations in six of seven subjects, whereas in another study, 13 of 26 patients reported "vivid hallucinations and/or paranoid ideation in a clouded sensorium" (*see* Calne et al., 1983 for other examples). Likewise, there is some evidence that methysergide (Hale and Reed, 1962) and 2-bromo-LSD (Bertino et al., 1959; Isbell et al., 1959) may be weakly hallucinogenic. Quipazine is probably the most problematic of agents and is repeatedly identified by various animal models as a potential hallucinogen. In the one reported clinical study on the endocrinological effects of quipazine, nothing was mentioned about hallucinogenic side effects (Parati et al., 1980). There is, however, a personal communication suggesting that quipazine produces "low-dose mescaline-like effects" in normal human subjects (Winter, 1979). On the one hand, an animal model should perhaps not be discredited for detecting one of these latter agents; on the other hand, it must be remembered that these agents do not *consistently* produce hallucinogenic effects in humans.

2. Methods for Assessing Hallucinogenic Agents

What follows is a discussion of some of the more common or useful models that have been employed over the past 20 years. Each model has its unique advantage(s) or disadvantage(s), and those interested in using a particular model must carefully weigh not only these factors, but also the goals of their own individual studies. Every attempt has been made to indicate whether a particular model is useful for identifying examples of the different families of classical hallucinogens, and whether or not the "problem agents" give false positives. Space limitations do not allow a compilation of all agents evaluated in each of the different models, and readers are urged to consult the cited literature. An attempt has also been made to identify, where possible, which neurotransmitter mechanisms underlie each of the described behaviors.

2.1. Mouse Ear-Scratch Reflex

Deegan and Cook (1958) noticed that oral doses of mescaline produced a curious behavior in mice—scratching of the head and ear area with a hind limb. This behavior has been referred to as the ear-scratch response, the scratch reflex, and scratch-reflex stereotypy. Mescaline induces this effect via the oral, ip, and im routes of administration. Although detailed investigations have not been reported, the response is apparently specific for the mouse and has not been observed in other animal species (Deegan and Cook, 1958). The scratch reflex is induced by various hallucinogenic phenylisopropylamines, such as DOM, DOET, and DOIP, but not by the inactive DOTB (Kulkarni, 1973). The effect is stereoselective, and the R(–)isomer of DOM is twice as potent as racemic DOM, whereas the S(+)isomer is inactive (Nichols et al., 1978; Yim et al., 1979). DOI also elicits the scratch reflex in a stereoselective manner, and the response produced by R(–)DOI can be antagonized by spiperone and the 5-HT$_2$ antagonist ketanserin (Glennon et al., 1991).

Indolealkylamine hallucinogens do not elicit the scratch reflex; psilocybin, for example, is ineffective, and LSD is essentially inactive at doses ranging from 0.1 to 125 µg/kg (Yim et al.,

1979). In fact, LSD is capable of antagonizing the mescaline-induced scratch reflex (Chen and Bohner, 1960; Borsy et al., 1964). Although this method may not be generally useful for the identification of hallucinogenic agents, Yim and coworkers (1979) suggest that it may have utility for studying phenalkylamine-type hallucinogens.

The scratch reflex induced by mescaline can be antagonized by a wide variety of agents, including LSD, 2-bromo-LSD, dihydroergotamine, promazine, chlorpromazine, prochlorperazine, serotonin, reserpine, hydroxyzine, mephenesin, pipradrol, amphetamine, picrotoxin, pentylenetetrazol, morphine, methadone, meperidine, codeine, chloral hydrate, zoxazolamine, diphenhydramine, atropine, benactyzine, and hyoscine (Deegan and Cook, 1958; Chen and Bohner, 1960; Borsy et al., 1964). Neither quipazine nor l-tryptophan produce the response (Yim et al., 1979), and the mescaline-induced reflex is enhanced by coadministration of sodium barbital (Chen and Bohner, 1960). This method, then, is perhaps of questionable value for investigating the mechanism of action of hallucinogenic agents in that it provides little insight in this regard.

In summary, phenalkylamine, but not indolealkylamine, hallucinogens elicit the ear-scratch phenomenon in mice. It may be a useful model for examining phenalkylamine hallucinogens, but is obviously not universally applicable. The mechanism underlying this response is unknown at this time.

2.2. Serotonin (5-HT) Syndrome

Agents that increase synaptic levels of 5-HT (e.g., 5-HT agonists, 5-HT precursors, 5-HT-releasing agents) produce effects in rodents that are collectively termed the "serotonin syndrome" (Green and Heal, 1985; Tricklebank, 1985; Glennon and Lucki, 1988). The syndrome consists of fine body tremor, lateral head weaving, reciprocal forepaw treading, hindlimb abduction, Straub tail, hyperactivity, hyperreactivity, and backward walking. In the past, some investigators considered head-twitch as part of the serotonin syndrome; this will be discussed as a separate topic. Although the serotonin syndrome is generally studied in rodents, all or part of the syndrome also occurs in other animal species, including cats, dogs, hamsters, pigeons, and rabbits (Jacobs, 1976).

Many classical hallucinogens, particularly the indolealkylamine derivatives, such as LSD and 5-methoxy-N,N-dimethyltryptamine, produce all or portions of the serotonin syndrome (Gessner, 1970; Sloviter et al., 1978,1980). Individual symptoms of the syndrome, such as fine body tremor (Gessner, 1970) and backward locomotion (Yamamoto and Ueki, 1975), have also been examined separately. Nevertheless, nonhallucinogenic agents are capable of producing certain aspects of the syndrome, and there is also some evidence for the involvement of neurotransmitters other than serotonin (Jacobs, 1976; Silbergeld and Hruska, 1979). In general, however, the syndrome appears to be mediated primarily by 5-HT$_1$ receptors and is induced by 5-HT$_1$-selective, as well as by nonselective, serotonergic agonists (e.g., *see* Glennon and Lucki, 1988 for a review). Those aspects of the syndrome produced by agents, such as DOI, can be readily differentiated from those produced by 5-HT$_1$-selective agonists (Pranzatelli, 1988). As such, and because there is no evidence that 5-HT$_1$-selective agents are hallucinogenic, this behavioral model does not appear to be suitable as a reliable model for hallucinogenic activity.

2.3. Head-Twitch Response

Examples of every family of classical hallucinogens have been demonstrated to produce a head-twitch response in rodents. This behavior consists of a tic-like head movement in mice, whereas in rats, the response may involve the animal's shoulders as well as the head and is commonly referred to as "wet-dog shake." In the rat, the response seems to be slower in onset, longer in duration, and somewhat more difficult to quantify (Green, 1984).

The head-twitch response, as described by Corne and coworkers (1963), was initially proposed as an animal test for evaluating the presumptive hallucinogenic potential of novel psychotherapeutic agents (Corne and Pickering, 1967). In their first paper on hallucinogens, they described the effects both of indolealkylamines, such as LSD, psilocin, and α-methyltryptamine, and phenalkylamines, such as mescaline and 1-(3, 4, 5-trimethoxyphenyl)-2-aminopropane (3, 4, 5-TMA) (Corne and Pickering, 1967). They also showed that examples of other classes of hallucinogens (e.g., phencyclidine, JB329) can elicit the response, as do such agents as atropine, hyoscine, yohimbine, and 5-hydroxytryptophan (5-HTP). The head-twitch response has subsequently been shown to be one of the most reliable models for identifying hallucinogenic agents (Silva and Calil, 1975; Calil, 1978). It is not, however, without serious drawbacks. For example, certain nonhallucinogens (e.g., 5-HTP) are very potent in producing head-twitch. Quipazine (Malick et al., 1977), 5-HT (e.g., Mawson and Wittington, 1970), amphetamine (Taylor and Sulser, 1973), and intraventricularly administered *para*-hydroxyamphetamine (Tadano et al., 1986) also elicit head-twitch. Some agents (e.g., 5-HTP, *para*-hydroxyamphetamine, LSD) seem to produce an autoinhibitory effect (i.e., the dose–response curves are of an inverted-U shape) (Corne and Pickering, 1967; Vetulani et al., 1980; Colpaert and Janssen, 1983; Tadano et al., 1986). Additionally, a single dose of hallucinogen may antagonize the effect of another; for example, single doses of psilocybin, DMT, and mescaline can antagonize the effect of LSD (Corne and Pickering, 1967). Some agents act as agonists at low doses and as antagonists at higher doses; LSD, for example, produces head-twitch at doses of about 0.02–0.2 mg/kg (Corne and Pickering, 1967; Vetulani et al., 1980), but antagonizes the effect of 300 mg/kg of 5-hydroxytryptophan at a dose of 8.1 mg/kg (Corne et al., 1963). Both LSD and lisuride can antagonize the effect of 5-HTP-induced head-twitch in the rat (Gerber et al., 1985).

Agents from different pharmacological classes can produce head-twitch behavior, but they may not all involve a common mechanism (Handley and Singh, 1986). Although there is evidence that adrenergic, GABAergic, opiate, and other mechanisms may be involved in head-twitch (particularly when nonserotonergic agents are employed to induce the effect), 5-HT also plays a substantial role. The response, when produced by serotonergic agents, appears to involve activation of central 5-HT receptors (Lucki et al., 1984) and cannot be antagonized by the peripherally acting 5-HT antagonist xylamadine (Matthews and

Smith, 1980). 5-HTP-induced head-twitch in rodents can be antagonized by a wide variety of 5-HT antagonists, including cinanserin, cyproheptadine, metergoline, methysergide, and mianserin (Corne et al., 1963; Bedard and Pycock, 1977; Ortmann et al., 1982; Colpaert and Janssen, 1983; Green, 1984; Lucki et al., 1984; Gerber et al., 1985). In the rat, cyproheptadine also antagonizes the head-twitch induced by LSD (Vetulani et al., 1980; Lucki et al., 1984), quipazine (Vetulani et al., 1980), and 5-OMe-DMT (Matthews and Smith, 1980). Peroutka and coworkers (1981) were the first to recognize that head-twitch might be specifically related to a particular population of 5-HT receptors—5-HT$_2$ receptors. Subsequently, Ortmann et al. (1982) demonstrated a significant correlation ($r = 0.80$) between antagonism of 5-HTP-induced head-twitch in mice and 5-HT$_2$ receptor affinity for a series of 22 antagonists. Lucki and coworkers (1984) provided similar evidence for head-twitch in the rat. Antagonism of mescaline-induced head-twitch in the rat also correlates ($r = 0.88$) with the 5-HT$_2$ affinity of 19 antagonists (Leysen et al., 1982). It should be noted, however, that although 5-HT$_2$ binding studies are typically conducted using rat frontal cortex homogenates, the head-twitch response does not seem to involve 5-HT$_2$ receptors located in this region of the brain (Lucki and Minugh-Purvis, 1987).

Chronic administration of certain antidepressants decreases the number of 5-HT$_2$ receptors in frontal cortex. Under these conditions, 5-HTP-induced head-twitch behavior is decreased (Green et al., 1983; Goodwin et al., 1984). More direct evidence for 5-HT$_2$ involvement comes from antagonism studies using 5-HT$_2$-selective antagonists; ketanserin, pirenperone, and ritanserin can attenuate 5-HTP- and/or mescaline-induced head-twitch in rodents (Leysen et al., 1982; Green et al., 1983; Colpaert and Janssen, 1983; Lucki et al., 1984; Green, 1984). Examining eight 5-HT antagonists (2-bromo LSD, cinanserin, cyproheptadine, methiothepin, methysergide, metergoline, mianserin, pizotifen), Colpaert and Janssen (1983) found that all blocked 5-HTP-induced head-twitch in the rat, but did so in a biphasic manner. There is an initial sharp decrease in response frequency followed by a further, but shallower, rate of decrease. The first phase is attributed to a 5-HT$_2$ mechanism, whereas the second phase is speculated to involve a paradoxical agonist effect (possibly involving 5-HT$_1$ receptors). It is noteworthy in this regard that the 5-HT$_2$ antagonist pirenperone exhibits only a monophasic inhibition curve (Colpaert and Janssen, 1983). DOI, now thought to be a 5-HT$_2$ agonist, stereoselectively induces head-twitch in mice with R(–)-DOI being the more potent isomer; this effect can be antagonized by the 5-HT$_2$ antagonist ketanserin (Darmani et al., 1990). Interestingly, it has been demonstrated that 8-OH DPAT, a 5-HT$_{1A}$ agonist, can also antagonize the head-twitch produced by DOI (Glennon et al., 1991). These results suggest that a functional relationship exists between 5-HT$_{1A}$- and 5-HT$_2$-mediated behaviors and may explain why nonselective agonists, such as LSD, can antagonize the effects produced by 5-HT$_2$ agonists.

In summary, the head-twitch response in rodents seems to be a very useful model for the investigation of classical hallucinogens. Because of the high subjectivity of response measurement, it may be difficult to obtain objective quantification (Silva and Calil, 1975). There may also be high variability between subjects; however, the response seems to be stable for individual animals (Silva and Calil, 1975; Vetulani et al., 1980). Nonhallucinogens produce the response, and it is not unusual to obtain false positives. Nevertheless, the convenience of the procedure and the lack of expense make it quite an attractive model. Classical hallucinogens appear to produce the head-twitch response via activation of central 5-HT$_2$ receptors; 5-HT$_2$ agonists induce the effect, and 5-HT$_2$ antagonists attenuate it. Even with these agents, though, there is evidence that an intact adrenergic system is necessary in order for the behavior to be observed (Handley and Singh, 1986).

2.4. Rabbit Hyperthermia

In a now classic paper, Jacob and Lafille (1963) demonstrated that classical hallucinogens produce hyperthermia in the rabbit. The effect is dose-dependent and provides a convenient measure of potency. Indolealkylamines and phenalkylamines are capable of eliciting this response. Although only a few examples have been investigated, other classes of psychotomimetic agents can also produce this hyperthermic effect. However, hallucinogens have been reported to produce little effect, an increase, or a decrease in the body temperatures of other species of animals (*see* Otis et al., 1978 and Myers and Waller, 1978 for discussion). Another problem associated with this method is that certain stimulants (e.g., amphetamine) and pyretic agents (e.g., 2,4-dinitrophenol) also produce a pronounced hyperthermia (Jacob and Lafille, 1963). Aldous and coworkers (1974) believe that rabbit hyperthermia is a reliable quantitative measure of hallucinogenic potency once an agent has been shown by other tests to be mainly LSD-like rather than merely amphetamine-like. Otis and coworkers (1978) have concluded that: central stimulants generally increase body temperature of rats and rabbits, hallucinogenic agents tend to decrease body temperature in rats, but increase body temperature in rabbits, and agents that decrease body temperature in both species are most probably neither stimulants nor hallucinogens. *See* Clark and Lipton (1986) for an extensive compilation of the effects of drugs on body temperature.

LSD is among the most potent of the hallucinogens in producing this hyperthermic effect and is generally employed as a standard agent for purposes of comparison. Aldous and coworkers (1974) examined the hyperthermic activity of approx 50 phenalkylamines and demonstrated that hyperthermia is produced in a stereoselective manner (the R[–]-isomers generally being approximately an order of magnitude more potent than the corresponding S[+]-enantiomers), and that there is a significant parallelism between hyperthermic potency and human psychotomimetic potency. Among the most potent phenalkylamines are DOM and DOB; nevertheless, most agents are considerably less potent than LSD (Aldous et al., 1974; Standridge et al., 1976).

Serotonin has long been known to be involved in thermoregulation (e.g., *see* Myers and Waller, 1978). Thus, it is not unusual that various serotonergic agents can alter body temperature in animals. Curiously, however, Gessner (1970) noticed that the hallucinogenic agent 5-methoxy-N,N-dimethyltryptamine (5-OMe-DMT) produces a biphasic effect in mice; that is, this agent produces a hyperthermic effect followed, after about 15–20 min by a hypothermic response. It is now known that 5-HT$_{1A}$ agonists produce hypothermia in rodents whereas 5-HT$_2$ agonists produce hyperthermia (*see* Glennon and Lucki, 1988 and Glennon, 1990a for reviews). 5-OMe-DMT has since been shown to be a nonselective 5-HT agonist, and Gudelsky and coworkers (1988) have demonstrated that the hypothermic effect of this agent in rats is most likely 5-HT$_{1A}$-mediated, whereas the hyperthermic effect is 5-HT$_2$-mediated. Interestingly, such agents as DOM and DOB are now considered to be fairly selective 5-HT$_2$ agonists, and it has been demonstrated that a significant correlation ($r = 0.978$) exists between the hyperthermic potency and 5-HT$_2$ receptor affinity of a series of 11 such agents (Glennon, 1990a). It should be noted, however, that for this same series of agents, there is also a correlation ($r = 0.944$) between hyperthermic potency and 5-HT$_{1C}$ affinity (Glennon, 1990a).

In summary, classical hallucinogens produce a reliable increase in the body temperature of rabbits. Nonhallucinogens can produce a similar effect. There is evidence that the effect produced by hallucinogens involves a serotonergic mechanism and, in particular, a 5-HT$_2$ (or 5-HT$_{1C}$) mechanism. Although this is a useful model, false positives are frequently encountered.

2.5. Startle Reflex

Startle reflexes induced by sound (i.e., acoustic startle) or by, for example, air-puffs (i.e., tactile startle) and measured using a stabilimeter can be increased or decreased by a variety of drugs and neurochemical interventions (Davis, 1980). On the basis that hallucinogens produce profound sensory alterations in human subjects, it was originally anticipated that they might also have an influence on the startle reflex in animals.

The hallucinogens mescaline, DOM, DOET, and DOPR produce reliable increases in tactile startle in rats, whereas nonhallucinogenic analogs of DOM, such as DOTB and DOAM, are inactive (Geyer et al., 1978). Mescaline, however, has no effect on acoustic startle (Bridger and Mandel, 1967). Although LSD increases both acoustic and tactile startle reflexes Davis and Sheard, 1974a; Davis et al., 1977), high doses depress the reflex. Simple indolealkylamine hallucinogens, such as DMT, 5-OMe-

DMT, and psilocin, also elicit this biphasic response (Davis and Bear, 1972; Davis and Sheard, 1974b; Davis and Walters, 1977; Davis et al., 1988). Interestingly, simple indolealkylamines do not alter tactile startle (Geyer and Mandell, 1978). Nonhallucinogens can enhance the startle reflex (Davis, 1980); for example, apomorphine, but not amphetamine, increases startle. The neurochemical mechanisms mediating the startle response have yet to be elucidated; nevertheless, the neurotransmitter 5-HT is thought to play a substantial role (Davis et al., 1986). 5-HT_{1A} as well as 5-HT_2 agonists enhance startle (Davis et al., 1988). With regard to the application of this technique as a general model for the identification of hallucinogenic agents, Geyer and Mandell (1978) have concluded that the startle response neither reliably predicts the human potency of hallucinogens nor clearly discriminates between hallucinogens and other drugs.

2.6. Limb-Flick

Jacobs and Trulson (1976,1978) have identified several behaviors in cats that appear upon administration of hallucinogenic agents: limb-flicking, abortive grooming, and hallucinatory-like behavior. These behaviors are essentially nonexistent in control animals. The first two behaviors are the most reliable and the easiest to score. In some studies, abortive grooming does not occur with significant frequency (Rusterholz et al., 1977). The limb-flick response is similar to the response a cat may make to remove a foreign substance from its paw; it can be quantitated by a simple count of limb-flicks per unit time.

In a group of cats, where a total of two limb-flicks were registered over a period of 1 h after administration of saline, administration of 50 µg/kg of LSD elicited 49.5 responses/h. The effect is dose-related, and low doses (e.g., 2.5 µg/kg) resulted in a low, albeit significant, number of limb-flicks. However, the effect is biphasic, and fewer responses were recorded for 200 µg/kg than for 50 µg/kg (Jacobs and Trulson, 1978). Other hallucinogenic agents produce similar effects: Examples are DMT, 5-OMe-DMT, psilocin, mescaline, DOM, DOET, and DOB (Jacobs et al., 1977; Trulson and Jacobs, 1977,1979; Rusterholz et al., 1977). The response produced by DOB is stereoselective; at

0.1 mg/kg, R(–)-DOB produces approximately five times the number of limb-flicks as 0.1 mg/kg of S (+)-DOB (Rusterholz et al., 1977). Hallucinogens, such as LSD and mescaline, also induce limb-flick in rats (Nielsen et al., 1980).

Certain nonhallucinogenic agents, (atropine, caffeine, chlorpheniramine [Jacobs and Trulson, 1978]); do not elicit limb-flick, whereas other nonhallucinogens (apomorphine [White et al., 1981], 2-bromo-LSD [Jacobs and Trulson, 1978], lisuride [Marini et al., 1981], methysergide [Jacobs and Trulson, 1978], and quipazine [Trulson et al., 1981a; White et al., 1981]) do. Amphetamine does not elicit limb-flick behavior in cats (Jacobs and Trulson, 1978); however, chronic administration of amphetamine elicits the behavior in rats (Nielsen et al., 1980). On the basis of their investigations, White and coworkers (1981) have concluded that this model lacks specificity.

It is likely that limb-flick behavior is mediated by more than one neurotransmitter system; nevertheless, the limb-flick response produced by LSD and DOM is consistently antagonized by 5-HT antagonists, such as cinanserin, cyproheptadine, methysergide, mianserin, and pizotyline (Marini and Sheard, 1981; White et al., 1981,1983; Trulson et al., 1981b) and probably involves a 5-HT$_2$ mechanism (Heym and Jacobs, 1988).

Schlemmer and colleagues (e.g., *see* Schlemmer and Davis, 1983) have investigated the effects of hallucinogenic agents on monkeys and have proposed the limb-jerk response as a model for studying hallucinogens. The limb-jerk differs from the limb-flick in cats in that the former is a sudden myoclonic spasm that involves the entire limb. The limb-jerk is seen infrequently in untreated animals. The response is produced by members of the different subtypes of classical hallucinogens, including LSD, DMT, 5-OMe-DMT, psilocin, DET, mescaline, and DOM. Agents not producing the effect include amphetamine, apomorphine, 2-bromo-LSD, fluoxetine, lisuride, and phencyclidine. Quipazine elicits the limb-jerk response (Schlemmer and Davis, 1986). The limb-jerk response can be antagonized by the 5-HT antagonists cinanserin, cyproheptadine, methysergide, and methiothepin; haloperidol and trifluperazine antagonized the response produced by 5-OMe-DMT, but did not return them to baseline levels.

In summary, the limb-flick response in cats is a proposed model of hallucinogenic activity that, like many of the other models, seems to involve a serotonergic mechanism. False positives have been obtained. Although members of different subtypes of classical hallucinogens have been examined, relatively few examples have been reported. Methodologically, the limb-flick response is fairly easy to quantitate; however, there may be considerable interanimal variation (Rusterholz et al., 1977). The limb-jerk response in monkeys may be related to the limb-flick in cats; additional studies are warranted.

2.7. Investigatory Behavior

In rodents, LSD and other hallucinogens are known to influence what might be termed exploratory activity, which consists of, for example, ambulation, rearing, and investigatory behavior. These effects, which have been investigated individually and in combinations, can be reduced or enhanced depending upon the particular drug and dose examined, the test apparatus, and the test conditions (e.g., *see* Woolley and Shaw, 1957; Brimblecombe, 1963, Schneider, 1968; Dandiya et al., 1969; Kabes et al., 1972; Hughes, 1973; Silva and Calil, 1975).

One measure of investigatory activity is "holepoke" behavior. Using a hole-board, or a chamber with holes in the walls and/or floor of the apparatus, the frequency and/or duration of holepokes made by an animal with its nose appears to be related to investigatory tendency (File and Wardill, 1975). LSD initially decreases and subsequently increases holepokes over time (Geyer and Light, 1979). Similar effects are produced by members of different families of the classical hallucinogens, including DMT, mescaline, and DOM (File, 1977; Geyer et al., 1979). This behavior is also observed with apomorphine; amphetamine, methysergide, and 2-bromo-LSD are inactive (Geyer et al., 1979).

These types of studies evolved into an analysis of patterns of locomotion and exploratory behavior using an apparatus with hole-boards that is equipped with an array of photocells to monitor the animal's activity (Adams and Geyer, 1985b; Gold et al., 1988). Using this behavioral pattern monitor, Adams and Geyer (1985a) found that LSD elicits three distinguishable effects:

1. Increased avoidance of central areas of the chamber;
2. Suppression of rearing; and
3. Disruption of the spatial patterning of locomotion.

This overall suppression of exploration (or enhancement of neophobia) was proposed as an animal model for investigating hallucinogenic agents. The amount of exploratory activity seen in LSD-treated animals is influenced by the degree of threat in the environment; thus, it was suggested that this might be an analog model of hallucinations in view of the analogy to the enhanced responsiveness to threatening situations produced by hallucinogens in humans (Adams and Geyer, 1985a). The model was subsequently modified on the basis that reduction of rearing was not a suitable measure of hallucinogenic activity (Adams and Geyer, 1985b).

The potentiating effect of hallucinogens on normal neophobia exhibited by rats confronted with a normal environment probably involves a 5-HT_2 mechanism. To further document this, the effects of several 5-HT_{1A} agonists, including 8-hydroxy-2-(di-*n*-propylamino) tetralin (8-OH DPAT), were examined using the above behavioral pattern monitor (Mittman and Geyer, 1989). 5-HT_{1A} agonists generally decreased locomotor activity; holepokes, rearing, and the duration of time spent in the center of the apparatus were also reduced, but they were reduced in proportion to the decrease in locomotor activity. It was concluded that the 5-HT_{1A} agonist actions of LSD contribute to the behavioral effects of LSD on locomotor activity. Also, although the effects of 5-HT_{1A} agonists were similar in some respects to those of the hallucinogens, the behaviorally suppressant effect of the 5-HT_{1A} agonists were relatively independent of the environmental situation in which the animals are tested (Mittman and Geyer, 1989).

2.8. Disruption of Fixed-Ratio Responding

Using rats trained to lever-press for food reward on a fixed-ratio schedule of reinforcement, administration of hallucinogenic agents decreases response rates (Freedman et al., 1964). More accurately, hallucinogenic agents produce disruptions of oper-

ant behavior characterized by periods of nonresponding or pausing. A counter can be incorporated to monitor the interresponse
times; the pausing behavior is referred to as the "hallucinogenic
pause" (e.g., *see* Rech and Commissaris, 1982). The pausing
behavior is produced both by indolealkylamine (e.g., LSD, DMT)
and phenalkylamine (e.g., mescaline, DOM) hallucinogens
(Commissaris et al., 1980) and appears to involve a serotonergic
mechanism (Commissaris et al., 1981; Rech et al., 1988). Amphetamine decreases rates of responding, but it does so in a manner
that is distinguishable from that of hallucinogenic agents
(Commissaris et al., 1980). Lisuride and quipazine produce
"hallucinogenic pausing" (Commissaris et al., 1980; Rech et al.,
1988). The effect of hallucinogens, lisuride, and quipazine can
be antagonized by 5-HT antagonists. Hallucinogens and
quipazine, but not lisuride, are effectively antagonized by the 5-
HT_2 antagonist pirenperone (Rech and Mokler, 1986). Based on
these and other findings, Rech and coworkers have concluded
that the behavior may involve both a $5\text{-}HT_1$ and $5\text{-}HT_2$ component; *see* Rech and Mokler (1986) and Rech et al. (1988) for more
detailed discussion.

2.9. Drug Discrimination

The drug discrimination paradigm has never been claimed
to serve as a model of hallucinogenic activity in animals. However, it does provide results that are worthy of discussion. The
general premise behind this procedure is that animals (typically
rats, pigeons, or monkeys) can be trained to distinguish or discriminate the effects of one drug vs those of another drug (or
vehicle). A variety of procedures and techniques have been
developed, and operant responding using a two-lever test chamber is perhaps the most commonly used apparatus. At one time,
it was thought that the ease with which a drug was discrimi-
nated might serve as an indicator of its abuse liability, but clearly,
this is not the case (Overton and Batta, 1977). Neither is there a
relationship between the discriminability of a drug and its reinforcing properties (Overton and Batta, 1977). In fact, hallucinogenic agents are not normally self-administered by animals
(Johanson and Balster, 1978). This procedure can, nevertheless,

be used to study drugs of abuse. Once animals have been trained to discriminate a "training drug" reliably from, for example, saline vehicle, tests of stimulus generalization can be conducted. Graded doses of the training drug can be administered to the animals, and a dose–response curve can be obtained; from these data, a potency measure (e.g., ED_{50} value) can be calculated. Perhaps more important from the perspective of this current chapter is that tests of stimulus generalization can also be conducted with agents other than nontraining doses of the training drug. If administration of a novel agent to animals trained to discriminate a particular training drug from saline results in stimulus generalization, this is evidence that the training drug and the novel drug are producing similar stimulus effects in the animals. Routinely, agents known to produce similar effects in humans result in stimulus generalization in animals. Potency measures can also be examined for the novel drug (or "challenge drug"). Thus, the drug discrimination paradigm is essentially a drug detection technique; animals can be queried as to whether a challenge drug produces stimulus effects similar to those of a training drug, and if it does, potency comparisons can be made. For example, *see* Glennon et al. (1982,1983a), Overton (1984), and Jarbe (1986) for reviews.

Theoretically, it should be possible to train animals to discriminate a hallucinogenic agent from saline and, through tests of stimulus generalization, examine novel agents in order to determine if they are hallucinogens. This has not been found to be the case. In fact, early on it was concluded that drug discrimination was of limited value for screening hallucinogenic agents (Silva and Calil, 1975). It is now recognized that stimulus generalization does not ordinarily occur between classes; that is, animals trained to discriminate LSD from saline, for example, will not recognize phencyclidine or THC. The process of stimulus generalization appears to be mechanism based. Animals trained to discriminate a classical hallucinogen from saline will recognize other classical hallucinogens (Glennon et al., 1983a), animals trained to discriminate THC will recognize other active canabinoids (Jarbe, 1986), and animals trained to discriminate phencyclidine will recognize other phencyclidine-like agents

(Balster and Willetts, 1988). Thus, the technique is very drug-class-, but not chemical-class-specific. Stimulus generalization will occur with agents from different chemical classes, as long as they produce similar pharmacological effects.

Animals have been trained to discriminate examples of each type of classical hallucinogen: LSD (Hirschhorn and Winter, 1971) 5-OMe-DMT (Glennon et al., 1979), mescaline (Hirschhorn and Winter, 1971), and DOM (Young et al., 1981).

Stimulus generalization occurs among these agents regardless of which agent is used as the training drug. There is evidence, then, that each of these agents can produce similar (although not necessarily identical!) stimulus effects in animals. Although the hallucinogen that has seen the greatest use as a training drug is LSD (Stolerman, 1989), the greatest number and variety of agents have been examined using DOM as the training drug. For a series of 14 phenalkylamines, there is a significant correlation ($r = 0.96$) between ED_{50} values for stimulus generalization and human hallucinogenic potencies (Glennon et al., 1983a). Indeed, for an extended series of agents ($n = 33$), including examples of each type of classical hallucinogen, the correlation is still significant ($r = 0.94$) (Glennon et al., 1982).

With DOM as the training drug, structure–activity relationships have been formulated (Glennon et al., 1983a; Glennon, 1989), and the mechanism of action has been investigated in detail. Stimulus generalization is stereoselective; for example, the R(−)isomers of the phenylisopropylamine hallucinogens are generally five to ten times more potent than their S(+) enantiomers. There has long been evidence that 5-HT might be involved in the mechanism of action of hallucinogenic agents, but it was not until the discovery that pirenperone was able to antagonize the stimulus effects of LSD (Colpaert et al., 1982) and that ketanserin and pirenperone were able to antagonize the stimulus effects of DOM (Glennon et al., 1983b) that it was postulated that hallucinogens might be acting as $5-HT_2$ agonists (Glennon et al., 1983b). Subsequent studies with DOM (Glennon, 1988,1989) and LSD (Appel and Cunningham, 1986) support this hypothesis. In fact, it was on the basis of drug discrimination studies that it was first postulated that certain hallucinogens (e.g., DOM and its structurally related derivatives DOB and DOI) might constitute the first class of 5-HT$_2$-selective agonists (Glennon et al., 1983b; Glennon, 1988). The stimulus effects of LSD, DOM, R(–)-DOB, and/or DOI can be attenuated by a variety of 5-HT antagonists and, in particular, by 5-HT$_2$ antagonists (*see* Glennon, 1988 for a review). Interestingly, however, although low doses of 5-HT antagonists are able to attenuate an LSD stimulus, higher doses mimic the effect of LSD (e.g., methysergide, mianserin) or result in partial generalization (e.g., 2-bromo-LSD) (Colpaert et al., 1982). The 5-HT$_2$-selective antagonist pirenperone, however, does not mimic the effect of LSD (Colpaert et al., 1982). These results suggest that certain 5-HT antagonists are in fact partial agonists or mixed agonist-antagonists.

To date, there are no reported instances (at least where DOM has been used as the training drug) of false negatives; that is, there are no instances where DOM-stimulus generalization has failed to occur with another example of a classical hallucinogen where human hallucinogenic activity has been adequately documented in the literature. Furthermore, stimulus generalization does not occur with structurally related agents that lack hallucinogenic activity. For example, DOM-stimulus generalization does not occur with the phenylisopropylamine stimulant amphetamine. On the other hand, there are a few instances of what might be false positives. For example, using animals trained to discriminate LSD or DOM from saline, stimulus generalization occurs with lisuride and quipazine. It should also be mentioned that, although lisuride and LSD produce similar stimulus effects, it is possible to train rats to discriminate the two agents from one another (White and Appel, 1982). This suggests that LSD and lisuride may produce a common effect, but that they are capable of producing effects that are sufficiently dissimilar to allow them to be differentiated. Quipazine produces LSD- and DOM-like effects regardless of which is used as the training drug; nevertheless, stimulus generalization does not occur with quipazine when the more selective DOB is used as the training drug (Glennon et al., 1988). It has been speculated that DOM may produce its stimulus effects via 5-HT$_2$ agonism. Thus, it might be anticipated, if this is the case, that DOM-stimulus generalization will occur with agents that increase synaptic 5-HT levels (5-HT itself being a nonselective serotonergic agent). Indeed, DOM-stimulus generalization occurs with the 5-HT-releasing agent fenfluramine (Glennon, 1988). The DOM-stimulus also generalizes to other nonselective 5-HT agonists (such as the hallucinogenic agent 5-OMe-DMT), but does not generalize to serotonergic agents that are selective for other populations of 5-HT receptors (e.g., 5-HT$_{1A}$ agonists) (Glennon, 1988). Both 5-OMe-DMT and LSD produce discriminative stimuli that seem to be dose-dependent (*see* Glennon, 1988 for a review); that is, results of stimulus generalization studies may vary depending upon the training dose employed. This may simply reflect the lower selectivity of 5-OMe-DMT and LSD, relative to DOM, for 5-HT$_2$ binding sites. For example, the 5-OMe-DMT stimulus (but not a DOM stimulus) generalizes to the 5-HT$_{1A}$ agonist 8-hydroxy-2-(di-*n*-propylamino)tetralin; also, in 5-OMe-DMT-trained rats, 5-OMe-DMT-stimulus generalization occurs with DOM (but not with the more selective DOB). In this regard, the drug discrimination paradigm is similar to other behavioral tests where nonhallucinogenic 5-HT agonists are active.

In summary then, drug discrimination offers a powerful technique for investigating hallucinogenic agents. Using animals trained to a particular hallucinogen, stimulus generalization occurs with other agents that are known to produce similar hallucinogenic effects in humans. Indeed, the technique is sufficiently powerful to distinguish among different classes of hallucinogen/psychotomimetic agents (e.g., classical hallucinogens vs cannabinoids). Yet, it has never been claimed that the drug discrimination paradigm is a model of hallucinogenic activity or a behavioral model of hallucinogenic drug action. On the negative side, false positives (e.g., lisuride, quipazine, fenfluramine) have been reported.

Also, there may be considerable expense involved in setting up the apparatus (depending upon the particular method employed) and maintaining the trained animals for extended periods of time; however, the daily costs of running of the procedure are no more than those associated with most other routines.

2.10. Radioligand Binding

Many of the above assays/models implicate a role for the neurotransmitter 5-HT in the mechanism of action of the classical hallucinogens. On the basis of drug discrimination studies, it was postulated that certain of these hallucinogens act as 5-HT_2 agonists and that the phenalkylamine hallucinogens might constitute the first class of 5-HT_2-selective agonists (Glennon et al., 1983b). It was subsequently demonstrated that the classical hallucinogens bind in a stereoselective manner and display significant affinity for 5-HT_2 binding sites (rat cortex) when [^{3}H]ketanserin is used as radioligand. Furthermore, phenalkylamine hallucinogens are rather selective for 5-HT_2 sites, whereas most of the indolealkylamine hallucinogens, although they may bind at 5-HT_2 sites with high affinity, also bind with high affinity at various other populations of 5-HT_1 sites (e.g., 5-HT_{1A}, 5-HT_{1B}, 5-HT_{1C}, 5-HT_{1D} sites). Because all of the classical hallucinogens share this common 5-HT_2 component, it seemed reasonable to assume, as a working hypothesis, that a 5-HT_2 receptor interaction is an important factor in mediating the hallucinogenic effects of these agents. For a series of 22 agents, there is a significant correlation ($r = 0.94$) between discrimination-derived ED_{50} values and 5-HT_2 affinities (i.e., K_i values) (Glennon et al., 1984). For 15 of these agents where human data were available, there is also a significant correlation ($r = 0.92$) between human hallucinogenic potencies and 5-HT_2 receptor affinities (Glennon et al., 1984). More recently, this study has been replicated with [^{3}H]DOB as the radioligand using rat brain ($r = 0.90$; $n = 17$) (Titeler et al., 1988) and human brain ($r = 0.97$; $n = 10$) (Sadzot et al., 1989) homogenates.

These results support the hypothesis that hallucinogens act via a 5-HT_2 mechanism. However, hallucinogens also bind with significant affinity at 5-HT_{1C} sites, and here, too, there is a significant correlation ($r = 0.78$; $n = 17$) between affinity and human potency (Titeler et al., 1988). At this time, it is not known which (or if both) of these two types of receptors is involved in hallucinogenic activity. Nevertheless, hallucinogens bind with significant affinity at these sites, and it may be worthwhile including radioligand binding in future screens for hallucinogenic potential. Agents with high affinity for these sites may behave either as agonists or as antagonists. Thus, radioligand binding studies alone are insufficient for making predictions. It is noteworthy, however, that 5-HT$_2$ antagonists display similar affinities regardless of whether [^{3}H]ketanserin or [^{3}H]DOB is used as the radioligand, whereas agonists display a higher affinity for sites labeled by the latter radioligand. This difference in affinity, then, might be exploited and might also have some predictive value. Nevertheless, functional in vivo data are required to support any radioligand binding results. Regardless, although it remains to be documented, radioligand binding may prove to be a rapid and economical primary screen.

2.11. Combination Models

As can be seen from the foregoing discussion, no single animal model, analog, or assay model reliably identifies hallucinogenic agents without some false positives and/or false negatives. From this standpoint, then, no model meets the above-mentioned criteria. A model that examines more than one response or, alternatively, a combination of several animal models might be expected to provide more reliable information.

In the chronic spinal dog (usually transected at the T-5 or T-10 level), a flexor reflex is invoked by electrical stimulation of the toe. Hallucinogens, such as LSD, psilocin, mescaline, and DOM, facilitate this flexor reflex and induce a stepping response (Martin and Eades, 1970). Amphetamine, methysergide, tryptamine, 5-HT, 5-HTP, and methoxamine produce a similar effect; however, cyproheptadine antagonizes the effect of LSD, has little effect on methoxamine, and actually enhances the action of amphetamine (Martin and Eades, 1970). The technique evolved over time, and Martin and coworkers ultimately monitored a series of responses in the chronic spinal dog: flexor reflex, stepping reflex, pupil diameter, respiration, pulse rate, effect on nictitating membrane, skin-twitch latency, rectal temperature, and occasionally, appetite (Martin et al., 1978a). In this manner, in conjunction with the use of certain neurotransmitter antagonists to block some of the effects, they were able to classify agents as being either LSD-like, primarily LSD-like with other effects, amphetamine-like, primarily amphetamine-like with other effects, and LSD- and amphetamine-like (for reviews, *see* Martin and Sloan, 1977; Martin et al., 1978b). On the basis of their studies, Martin and Sloan (1986) conclude that their model may involve a tryptaminergic mechanism distinct from a serotonergic mechanism. This procedure, although promising, has never been very popular with other groups of investigators and has not seen widespread application; this may be a direct consequence of the expense of purchasing and maintaining the animals and/or of the large number of responses being monitored.

Likewise, Otis and coworkers (1978) have used a battery of tests to identify hallucinogenic agents. They identified from the literature 56 different procedures that might be potentially useful in identifying hallucinogenic agents; they investigated 12 of these and subsequently narrowed their list to: rat body temperature, rabbit body temperature, photocell activity of the rat, a modified version of the Irwin mouse screen (Irwin, 1968), the spontaneous beating rabbit atrial preparation, and fixed-ratio lever pressing by rats. Ultimately, they focused only on the first three measures. Using this battery, they were able to classify a group of agents as being either hallucinogens (which included classical hallucinogens, phencyclidine, and PMA), stimulants, or "other." They also correctly classified all of 28 agents submitted for a double-blind evaluation (Otis et al., 1978).

Chemically, such agents as DOM and DOB are phenylisopropylamines. The central stimulant amphetamine is also a phenylisopropylamine. Thus, it has always been considered important that animal behaviors produced by phenylisopropylamine hallucinogens be distinguishable from those produced by amphetamine. Using the drug discrimination paradigm with rats trained to discriminate either DOM or (+)-amphetamine from saline, Glennon (1989) has been able to classify agents as being either DOM-like or amphetamine-like on the basis of stimulus generalization studies. As with the investigations of Martin and coworkers (1978b), drug discrimination studies correctly identified MDA as an agent that produces both DOM-like (or LSD-like) effects and amphetamine-like effects (Glennon, 1989).

2.12. Miscellaneous Methods

In addition to the procedures listed above, numerous other methods and animal models have been employed to investigate hallucinogenic agents. Some of these have seen only limited application or have not been validated by examining a sufficiently large number of classical hallucinogens and nonhallucinogens; others have long since been invalidated, and/or have been modified or combined into one of the above-mentioned procedures. A sampling of some of these methods includes:

1. Open-field behavior of rodents (e.g., Brimblecombe, 1963);
2. Bovet-Gatti profiles and shuttle avoidance in rats (e.g., Tilson et al., 1977; Beaton et al., 1978; Calil, 1978; Davis and Hatoum, 1987);
3. Disruption of swim-maze performance by rats (Uyeno and Mitoma, 1969);
4. Facilitative vs disruptive effects of drugs in animals (Bridger et al., 1978); and
5. Nest-building behavior in mice (Schneider and Chenoweth, 1970) and characteristic electroencephalographic patterns in the cat (Winters and Wallach, 1970; Wallach et al., 1972).

This list is not comprehensive and there are other models described in the literature.

3. Summary

Numerous methods have been employed to identify the classical hallucinogens, yet, to date, no single model has proven infallible. Some methods, although quite controversial, are still used. Despite the problems involved, most methods provide results that are reproducible, and all of the methods have advanced our understanding of hallucinogenic agents, neurotransmitter mechanisms, and/or behavioral pharmacology. Although there have been attempts to compare some of the different methods directly (Silva and Calil, 1975; Calil, 1978; Stoff et al., 1978), no investigator has ever compared all (or even most) of the methods in a single study. Thus, it is virtually impossible to determine which is the best of the different methods. It is likely that the use of several different methods will provide the most reliable results; in this way, false negatives and false positives may be minimized. However, even with a battery of four tests, such as, for example, radioligand binding, head-twitch, rabbit hyperthermia, and drug discrimination (using animals trained to either DOM or LSD), such agents as quipazine would still be identified as being potential hallucinogens.

Many of the animal models seem to implicate a role for the neurotransmitter 5-HT. Indeed, it has been argued that classical hallucinogens produce their psychoactive effects primarily via a 5-HT_2 agonist mechanism (Glennon et al., 1983b,1990b). However, this was before the discovery of 5-HT_{1C} sites. Because the structure–activity relationships and because the affinities of the classical hallucinogens for 5-HT_{1C} and 5-HT_2 sites seem to be parallel, a role for 5-HT_{1C} sites cannot be excluded. Nevertheless, application of radioligand binding (5-HT_2 and 5-HT_{1C}) techniques is perhaps the most convenient of the new methods for quickly screening a series of agents. To date, all of the classical hallucinogens investigated display a significant affinity for 5-HT_2 sites. However, any agent identified by this method will still require further study using an in vivo animal model.

Perhaps more important than identifying a reliable animal behavior that parallels human hallucinogenic activity is a greater availability of human data. Some new agents, although predicted to be active on the basis of animal testing, have never been examined in humans. Other agents, although examined in humans, may have been evaluated at only one dose or in only one subject; in many cases, human data are only anecdotal, yet these data are often cited as gospel and/or are propagated by those who do not read the original literature to determine the validity of the particular human study. Admittedly, hallucinogenic activity is difficult to measure. Shulgin and coworkers (Shulgin et al., 1986; Shulgin and Shulgin, 1991) have recently published a protocol that might help in this regard. More human data will be needed in order to continue the development and validation of animal models that might ultimately prove useful in the reliable identification of hallucinogens.

Acknowledgments

Work from the author's laboratory has been supported, in part, by funds from PHS grant DA 01642.

References

Adams L. M. and Geyer M. A. (1985a) A proposed animal model for hallucinogens based on LSD's effects on patterns of exploration in rats. *Behav. Neurosci.* **99**, 881–900.

Adams L. M. and Geyer M. A. (1985b) Effects of DOM and DMT in a proposed animal model of hallucinogenic activity. *Prog. Neuropsychopharmacol. Biol. Psychiat.* **9**, 121–132.

Aldous F. A. B., Barrass B. C., Brewster K., Buxton D. A., Green D. M., Pinder R. M., Skeels R. M. and Tutt K. J. (1974) Structure-activity relationships in psychotomimetic phenylalkylamines. *J. Med. Chem.* **17**, 1100–1111.

Appel J. B. and Cunningham K. A. (1986) The use of drug discrimination to characterize drug actions. *Psychopharmacol. Bull.* **22**, 959–967.

Balster R. L. and Willetts J. (1988) Receptor mediation of the discriminative stimulus properties of phencyclidine and sigmaopioid agonists, in *Transduction Mechanisms of Drug Stimuli* (Colpaert F. C. and Balster B. L., eds.), Springer-Verlag, Berlin, pp. 122–135.

Beaton J. M., Bradley R. J., and Smythies J. R. (1978) Behavioral measures of hallucinogenic behavior, in *Psychopharmacology of Hallucinogens* (Stillman R. C. and Willette R. E., eds.), Pergamon, New York, pp. 241–267.

Bedard P. and Pycock C. J. (1977) "Wet-dog" shake behavior in the rat: A possible quantitative model of central 5-hydroxytryptamine activity. *Neuropharmacology* **16**, 663–670.

Bertino J. R., Klee G. D., and Weintraub M. D. (1959) Cholinesterase, D-lysergic acid diethylamide, and 2-bromolysergic acid diethylamide. *J. Clin. Exp. Psychopathol.* **20**, 218–227.

Borsy J., Huszeti Z., and Fekete M. (1964) Antimescaline properties of some lysergic acid derivatives. *Int. J. Neuropharmacol.* **2**, 273–277.

Bridger W. H. and Mandel I. J. (1967) The effects of dimethoxy-phenethylamines and mescaline on classical conditioning in rats as measured by the potentiated startle response. *Life Sci.* **6**, 775–781.

Bridger W. H., Barr G. A., and Gorelick D. A. (1978) Dual effects of LSD, mescaline, and DMT, in *Psychopharmacology of Hallucinogens* (Stillman R. C. and Willette R. E., eds.), Pergamon, New York, pp. 150–180.

Brimblecombe R. W. (1963) Effects of psychotropic drugs on open-field behavior in rats. *Psychopharmacologia* **4**, 139–147.

Calil H. M. (1978) Screening hallucinogenic drugs. *Psychopharmacology* **56**, 87–92.

Calne D. D., McDonald R. J., Horowski R. and Wuttke W. (1983) *Lisuride and Other Dopamine Agonists*. Raven, New York.

Chen G. and Bohner B. (1960) A study of certain CNS depressants. *Arch. Int. Pharmacodyn.* **125**, 1–20.

Clark W. G. and Lipton J. M. (1986) Changes in body temperature after administration of adrenergic and serotonergic agents and related drugs including antidepressants. *Neurosci. Biobehav. Rev.* **10**, 153–220.

Colpaert F. C. and Janssen P. A. J. (1983) The head-twitch response to intraperitoneal injection of 5-hydroxytryptophan in the rat: Antagonist effects of purported 5-hydroxytryptamine antagonists and of pirenperone, an LSD antagonist. *Neuropharmacology* **22**, 993–1000.

Colpaert F. C., Niemegeers C. J. E., and Janssen P. A. J. (1982) A drug discrimination analysis of lysergic acid diethylamide (LSD): In vivo agonist and antagonist effects of purported 5-hydroxytryptamine antagonists and of pirenperone, a LSD-antagonist. *J. Pharmacol. Exp. Ther.* **221**, 206–214.

Commissaris R. L., Lyness W. H., Moore K. E., and Rech R. H. (1981) Differential antagonism by metergoline of the behavioral effects of indolealkylamine and phenethylamine hallucinogens in the rat. *J. Pharmacol. Exp. Ther.* **219**, 170–174.

Commissaris R. L., Semeyn D. R., Moore K. E., and Rech R. H. (1980) The effects of 2,5-dimethoxy-4-methylamphetamine (DOM) on operant behavior: Interactions with other neuroactive agents. *Commun. Psychopharmacol.* **4**, 393–404.

Corne S. J. and Pickering R. W. (1967) A possible correlation between drug-induced hallucinations in man and a behavioral response in mice. *Psychopharmacologia* **11**, 65–78.

Corne S. J., Pickering R. W., and Warner B. T. (1963) A method for assessing the effects of-drugs on the central actions of 5-hydroxytryptamine. *Br. J. Pharmacol.* **20**, 106–120.

Dandiya P. C., Gupta P. D., Gupta M. L., and Patni S. K. (1969) Effects of LSD on field performance in rats. *Psychopharmacology* **15**, 330–340.

Davis M. (1980) Neurochemical modulation of sensory-motor activity: Acoustic and tactile startle reflexes. *Neurosci. Biobehav. Rev.* **4**, 241–263.

Davis M. and Bear H. D. (1972) Effects of N,N-dimethyltryptamine on retention of startle habituation in the rat. *Psychopharmacologia* **27**, 29–44.

Davis W. M. and Hatoum H. T. (1987) Comparison of stimulants and hallucinogens on shuttle avoidance in rats. *Gen. Pharmacol.* **18**, 123–128.

Davis M. and Sheard M. H. (1974a) Effects of lysergic acid diethylamide (LSD) on habituation and sensitization of the startle response in the rat. *Pharmacol. Biochem. Behav.* **2**, 675–683.

Davis M. and Sheard M. H. (1974b) Biphasic dose–response effects of N,N-dimethyltryptamine on the rat startle reflex. *Pharmacol. Biochem. Behav.* **2**, 827–829.

Davis M. and Walters J. K. (1977) Psilocybin: Biphasic dose-response effects on the acoustic startle reflex in the rat. *Pharmacol Biochem. Behav.* **6,** 427–431.

Davis M., Cassella J. V., Wrean W. H., and Kehne J. H. (1986) Serotonin receptor subtype agonists: Differential effects on sensorimotor reactivity measured with acoustic startle. *Psychopharmacol. Bull.* **22,** 837–843.

Davis M., Cassella J. V., and Kehne J. H. (1988) Serotonergic modulation of sensorimotor reactivity and conditioned fear as measured with the acoustic startle reflex, in *5-HT Agonists as Psychoactive Drugs* (Rech R. H. and Gudelsky G. A., eds.), NPP, Ann Arbor, MI, pp. 163–184.

Davis M., Gallager D. W., and Aghajanian G. K. (1977) Tricyclic antidepressant drugs: Attenuation of excitatory effects of D-lysergic acid diethylamide on the acoustic startle reflex. *Life Sci.* **20,** 1249–1258.

Deegan J. F. and Cook L. (1958) A study of the anti-mescaline property of a series of CSN-active agents in mice. *J. Pharmacol. Exp. Ther.* **122,** 17.

File S. E. (1977) Effects of *N,N*-dimethyltryptamine on behavioral habituation in the rat. *Pharmacol. Biochem. Behav.* **6,** 163–168.

File S. E. and Wardill A. G. (1975) Validity of head-dipping as a measure of exploration in a modified hole-board. *Psychopharmacologia* **44,** 53–59.

Freedman D. X., Appel J. D., Hartman F. R., and Molliver M. D. (1964) Tolerance to the behavioral effects of LSD-25 in rats. *J. Pharmacol. Exp. Ther.* **143,** 309–313.

Gerber R., Barbaz B. J., Martin L. L., Neale R., Williams M., and Liebman J. M. (1985) Antagonism of 1-5-hydroxytryptophan-induced head-twitching in rats by lisuride: A mixed 5-hydroxytryptamine agonist-antagonist? *Neurosci. Lett.* **60,** 207–213.

Gessner P. K. (1970) Pharmacological studies of 5-methoxy-*N,N*-dimethyltryptamine, LSD and other hallucinogens, in *Psychotomimetic Drugs* (Efron D. H., ed.), Raven, New York, pp. 105–118.

Geyer N. A. and Light R. K. (1979) LSD-induced alterations of investigatory responding in rats. *Psychopharmacology* **65,** 41–47.

Geyer M. A. and Mandell A. J. (1978) Euphorohallucinogens—Toward a behavioral model, in *The Psychopharmacology of Hallucinogens* (Stillman R. C. and Willette R. E., eds.), Pergamon, New York, pp. 310–323.

Geyer M. A., Light R. K., Rose G., Petersen L. R., Horwitt D. D., Adams L. M., and Hawkins R. L. (1979) A characteristic effect of hallucinogens on investigatory responding of rats. *Psychopharmacology* **65,** 35–40.

Geyer M. A., Petersen L. R., Rose G. J., Horwitt D. D., Light R. K., Adams L. M., Zook J. A., Hawkins R. L., and Mandell A. J. (1978) The effects of lysergic acid diethylamide and mescaline-derived hallucinogens on sensory-integrative function: Tactile startle. *J. Pharmacol. Exp. Ther.* **207,** 837–847.

Glennon R. A. (1988) Site-selective serotonin agonists as discriminative stimuli, in *Transduction Mechanisms of Drug Stimuli* (Colpaert F. C. and Balster B. L. eds.), Springer-Verlag, Berlin, pp. 16–31.

Glennon R. A. (1989) Synthesis and evaluation of amphetamine analogues, in *Clandestinely Produced Drugs, Analogues and Precursors* (Klein M., Sapienza F., McClain H., and Khan I., eds.), US Government Printing Office, Washington, D.C., pp. 39–65.

Glennon R. A. (1990a) Serotonin receptors: Clinical implications. *Neurosci. Biobehav. Rev.* **14**, 35–47.

Glennon R. A. (1990b) Do hallucinogens act as 5-HT$_2$ agonists? *Neuropsychopharmacology* **3**, 509–517.

Glennon R. A., Rosecrans J. A., and Young R. (1982) The use of the drug discrimination paradigm for studying hallucinogenic agents: A review, in *Drug Discrimination: Applications in CNS Pharmacology*, (Colpaert F. C. and Slangen J. L., eds.), Elsevier Biomedical, Amsterdam, pp. 69–96.

Glennon R. A., Rosecrans J. A., and Young R. (1983a) Drug induced discrimination: A description of the paradigm and a review of its specific application to the study of hallucinogenic agents. *Med. Res. Rev.* **3**, 289–340.

Glennon R. A., Young R., and Rosecrans J. A. (1983b) Antagonism of the effects of the hallucinogen DOM and the purported serotonin agonist quipazine by 5-HT$_2$ antagonists. *Eur. J. Pharmacol.* **91**, 189–192.

Glennon R. A. and Lucki I. (1988) Behavioral aspects of serotonin agonists, in *Serotonin Receptors* (Sanders-Bush E., ed.), Humana Press, New Jersey, pp. 253–294.

Glennon R. A., Titeler M., and Lyon R. A. (1988) A preliminary investigation of 4-bromo-2,5-dimethoxyphenethylamine: A potential drug of abuse. *Pharmacol. Biochem. Behav.* **30**, 597–601.

Glennon R. A., Titeler M., and McKenney J. D. (1984) Evidence for 5-HT$_2$ involvement in the mechanism of action of hallucinogenic agents. *Life Sci.* **35**, 2505–2511.

Glennon R. A., Rosecrans J. A., Young R., and Gaines J. J. (1979) Hallucinogens as discriminative stimuli: Generalization of a 5-OMe DMT stimulus to DOM. *Life Sci.* **24**, 993–997.

Glennon R. A., Darmani N., and Martin B. R. (1991) Multiple populations of serotonin receptors may modulate the behavioral effects of serotonergic agents. *Life Sci.* **48**, 2493–2498.

Gold L. H., Koob G. H., and Geyer M. A. (1988) Stimulant and hallucinogenic behavioral profiles of 3,4-methylenedioxymethamphetamine and N-ethyl-3,4-methylenedioxyamphetamine in rats. *J. Pharmacol. Exp. Ther.* **247**, 547–555.

Goodwin G. M., Green A. R., and Johnson P. (1984) 5-HT$_2$ receptor characteristics in frontal cortex and 5-HT$_2$ receptor-mediated head-twitch behavior following antidepressant treatment in mice. *Br. J. Pharmacol.* **83**, 234–242.

Green A. R. (1984) 5-HT mediated behavior. *Neuropharmacology* **23**, 1521–1528.

Green A. R. and Heal D. J. (1985) The effects of drugs on serotonin-mediated behavioural models, in *Neuropharmacology of Serotonin* (Green A. R., ed.), Oxford University Press, Oxford, pp. 326–365.

Green A. R., Johnson P., and Nimgaonkar V. L. (1983) Increased 5-HT$_2$ receptor number in brain as a probable explanation for the enhanced 5-hydroxytryptamine-mediated behavior following repeated electroconvulsive shock administration to rats. *Br. J. Pharmacol.* **80,** 172–177.

Gudelsky G. A., Koenig J. I., and Meltzer H. Y. (1988) Involvement of serotonin receptor subtypes in thermoregulatory responses, in *5-HT Agonists as Psychoactive Drugs* (Rech R. H. and Gudelsky G. A., eds.), NPP, Ann Arbor, pp. 127–142.

Hale A. R. and Reed A. F. (1962) Prophylaxis of frequent vascular headache with methysergide. *Am. J. Med. Sci.* **243,** 92.

Handley S. L. and Singh L. (1986) Neurotransmitters and shaking behavior—more than a "gut-bath" for the brain? *Trends Pharmacol. Sci.* **7,** 324–328.

Heym J. and Jacobs B. L. (1988) 5HT$_2$ agonist activity as a common action of hallucinogens, in *5-HT Agonists as Psychoactive Drugs* (Rech R. H. and Gudelsky G. A., eds.), NPP, Ann Arbor, MI pp. 95–106.

Hirschhorn I. A. and Winter J. C. (1971) Mescaline and lysergic acid diethylamide (LSD) as discriminative stimuli. *Psychopharmacologia (Berl.)* **22,** 64–71.

Hughes R. N. (1973) Effects of LSD on exploratory behavior and locomotion in rats. *Behav. Biol.* **9,** 357–365.

Irwin S. (1968) Comprehensive observational assessment: A systematic, quantitative procedure for assessing the behavioral and physiologic state of the mouse. *Psychopharmacologia* **13,** 222–257.

Isbell H., Miner E. J., and Logan C. R. (1959) Relationships of psychotomimetic to antiserotonin potencies of congeners of lysergic acid diethylamide (LSD-25). *Psychopharmacologia* **1,** 20–27.

Jacob J. and Lafille C. (1963) Caracterisation et detection pharmacologiques des substances hallucinogenes. *Arch. Int. Pharmacodyn.* **145,** 528–545.

Jacobs B. L. (1976) An animal behavior model for studying central serotonergic synapses. *Life Sci.* **19,** 777–786.

Jacobs B. L. and Trulson M. E. (1976) An animal behavior model for studying the actions of LSD and related hallucinogens. *Science* **194,** 741–743.

Jacobs B. L. and Trulson M. E. (1978) An animal behavioral model for studying the actions of LSD and related hallucinogens, in *The Psychopharmacology of Hallucinogens* (Stillman R. C. and Willette R. E., eds.), Pergamon, New York, pp. 220–240.

Jacobs B. L., Trulson M. E., and Stern W. C. (1977) Behavioral effects of LSD in the cat: Proposal of an animals behavior model for studying the actions of hallucinogenic drugs. *Brain Res.* **132,** 301–314.

Jarbe T. U. C. (1986) State-dependent learning and drug discriminative control of behavior: An overview. *Acta Neurol. Scand.* **74 (Suppl. 109),** 37–59.

Johanson C. E. and Balster R. B. (1978) A summary of the results of a drug self-administration study using substitution procedures in rhesus monkeys. *Bull. Narcotics* **30,** 43–54.

Kabes J., Fink Z., and Roth Z. (1972) A new device for measuring spontaneous motor activity—Effects of lysergic acid diethylamide in rats. *Psychopharmacologia* **23,** 75–85.

Kulkarni A. S. (1973) Scratching response induced in mice by mescaline and related amphetamine derivatives. *Biol. Psychiat.* **6,** 177–180.

Leysen J. E., Niemegeers C. J. E., van Nueten J. M., and Laduron P. M. (1982) [³H]Ketanserin (R 41 468), a selective ³H-ligand for serotonin₂ receptor binding sites. *Mol. Pharmacol.* **21,** 301–314.

Lucki I. and Minugh-Purvis N. (1987) Serotonin-induced headshaking behavior in rats does not involve receptors located in the frontal cortex. *Brain Res.* **420,** 403–406.

Lucki I., Nobler M. S., and Frazer A. (1984) Differential actions of serotonin antagonists on two behavioral models of serotonin receptor activation in the rat. *J. Pharmacol. Exp. Ther.* **228,** 133–139.

Malick J. B., Doren E., and Barnett A. (1977) Quipazine-induced head twitch response. *Pharmacol. Biochem. Behav.* **6,** 325–329.

Marini J. L. and Sheard M. H. (1981) On the specificity of a cat behavior model for the study of hallucinogens. *Eur. J. Pharmacol.* **70,** 479–487.

Marini J. L., Jacobs B. L., Sheard M. H., and Trulson M. E. (1981) Activity of a non-hallucinogenic ergoline derivative, lisuride, in an animal behavior model for hallucinogens. *Psychopharmacology* **73,** 328–331.

Martin W. R. and Eades C. G. (1970) The action of tryptamine on the dog spinal cord and its relationship to the agonist actions of LSD-like psychotogens. *Psychopharmacologia* **17,** 242–257.

Martin W. R. and Sloan J. W. (1977) Pharmacology and classification of LSD-like hallucinogens, in *Drug Addiction II. Amphetamine, Psychotogen, and Marihuana Dependence* (Martin W. R., ed.), Springer-Verlag, Berlin, pp. 305–368.

Martin W. R. and Sloan J. W. (1986) Relationship of tryptaminergic processes and the action of LSD-like hallucinogens. *Pharmacol. Biochem. Behav.* **24,** 393–399.

Martin W. R., Vaupel D. B., Sloan J. W., Bell J. A., Nozaki M. and Bright L. D. (1978a) The mode of action of LSD-like hallucinogens and their identification, in *The Psychopharmacology of Hallucinogens* (Stillman R. C. and Willette R. E., eds.), Pergamon, New York, pp. 118–125.

Martin W. R., Vaupel D. B., Nozaki M., and Bright L. D. (1978b) The identification of LSD-like hallucinogens using the chronic spinal dog. *Drug Alcohol Depend.* **3,** 113–123.

Matthews W. D. and Smith C. D. (1980) Pharmacological profile of a model for central serotonin receptor activation. *Life Sci.* **26,** 1397–1403.

Mawson C. and Wittington H. (1970) Evaluation of the peripheral and central antagonist activities against 5-hydroxytryptamine of some new agents. *Br. J. Pharmacol.* **39,** 223P.

Mittman S. M. and Geyer M. A. (1989) Effects of 5-HT$_{1A}$ agonists on locomotor and investigatory behaviors in rats differ from those of hallucinogens. *Psychopharmacology* **98,** 321–329.

Myers R. D. and Waller M. B. (1978) Thermoregulation and serotonin, in *Serotonin in Health and Disease* (Essman W. ed.), Spectrum, New York, pp. 1–67.

Naranjo C. (1973) *The Healing Journey.* Pantheon Books, New York.

Nichols D. E., Pfister W. R., and Yim G. K. W. (1978) LSD and phenethylamine hallucinogens: New structural analogy and implications for receptor geometry. *Life Sci.* **22,** 2165–2170.

Nielsen E. B., Lee T. H., and Ellison G. (1980) Following several days of continuous administration, D-amphetamine acquires hallucinogen-like properties. *Psychopharmacology* **68,** 197–200.

Ortmann R., Bischoff S., Radeke E., Buech O., and Delini-Stula A. (1982) Correlations between different measures of antiserotonin activity of drugs. *Naunyn-Schmiedeberg's Arch. Pharmacol.* **321,** 265–270.

Otis L. S., Pryor G. T., Marquis W. J., Jensen R., and Peterson K. (1978) Preclinical identification of hallucinogenic compounds, in *The Psychopharmacology of Hallucinogens* (Stillman R. C. and Willette R. E., eds.), Pergamon, New York, pp. 126–149.

Overton D. A. (1984) State dependent learning and drug discrimination. *Handbk. Psychopharmacol.* **18,** 59–127.

Overton D. A. and Batta S. K. (1977) Relationship between abuse liability of drugs and their degree of discriminability in the rat, in *Predicting Dependence Liability of Stimulants and Depressant Drugs* (Thompson T. and Unna K. R., eds.), University Park Press, Baltimore, 125–135.

Parati E. A., Zanardi P., Cocchi D., Caraceni T., and Muller E. E. (1980) Neuroendocrine effects of quipazine in man in health state or with neurological disorders. *J. Neural Trans.* **47,** 273–297.

Peroutka S. J., Lebovitz R. M., and Snyder S. H. (1981) Two distinct serotonin receptors with distinct physiological functions. *Science* **212,** 827–829.

Pranzatelli M. R. (1988) 5-HT$_2$ receptor regulation in rat frontal cortex: Putative selective agonist and antagonist studies. *Soc. Neurosci. Abstr.* **14,** 609.

Rech R. H. and Commissaris R. L. (1982) Neurotransmitter basis of the behavioral effects of hallucinogens. *Neurosci. Biobehav. Rev.* **6,** 521–527.

Rech R. H. and Mokler D. J. (1986) Disruption of operant behavior by hallucinogenic drugs. *Psychopharmacol. Bull.* **22,** 968–972.

Rech R. H., Commissaris R. L., and Mokler D. J. (1988) Hallucinogenic 5-hydroxytryptamine agonists characterized by disruption of operant behavior, in *5-HT Agonists as Psychoactive Drugs* (Rech R. H. and Gudelsky G. A., eds.), NPP, Ann Arbor, M7, pp. 185–215.

Rusterholz D. B., Spratt J. L., Long J. P., and Barfknecht C. F. (1977) Evaluation of substituted-amphetamine hallucinogens using the cat limb flick model. *Commun. Psychopharmacol.* **1,** 589–592.

Sadzot B., Baraban J. M., Glennon R. A., Lyon R. A., Leonhardt S., Jan C. R., and Titeler M. (1989) Hallucinogenic drug interactions at human brain 5-HT2 receptors: Implications for treating LSD-induced hallucinogenesis. *Psychopharmacology* **98,** 495–499.

Schlemmer R. and Davis M. (1983) A comparison of three psychotomimetic-induced models of psychosis in non-human primate social colonies, in *Ethopharmacology: Primate Models of Neuropsychiatric Disorders* (Miczek E., ed.), Alan R. Liss, New York, pp. 33–78.

Schneider C. (1968) Behavioral effects of some morphine antagonists and hallucinogens in the rat. *Nature (Lond.)* **220,** 586,587.

Schneider C. W. and Chenoweth M. B. (1970) Effects of hallucinogens and other drugs on the nest-building behavior of mice. *Nature (Lond.)* **225,** 1262,1263.

Shulgin A. T., Shulgin A., and Jacob P. (1986) A protocol for the evaluation of new psychoactive drugs in man. *Methods Findings Exp. Clin. Pharmacol.* **8,** 313–320.

Shulgin A. T. and Shulgin A. (1991) *Phikal,* Transform Press, Berkley, California.

Silbergeld E. K. and Hruska R. E. (1979) Lisuride and LSD: Dopaminergic and serotonergic interactions in the "serotonin syndrome." *Psychopharmacology* **65,** 233–239.

Silva M. T. A. and Calil H. M. (1975) Screening hallucinogenic drugs: Systematic study of three behavioral tests. *Psychopharmacologia* **42,** 163–171.

Sloviter R. S., Drust E. G., and Connor J. D. (1978) Specificity of a rat behavioral model for serotonin receptor activation. *J. Pharmacol. Exp. Ther.* **206,** 339–346.

Sloviter R.S., Drust E. G., Damiano B. P., and Connor J. D. (1980) A common mechanism for lysergic acid, indolealkylamine, and phenethylamine hallucinogens: Serotonergic mediation of behavioral effects in rats. *J. Pharmacol. Exp. Ther.* **214,** 231–239.

Standridge R. T., Howell H. G., Gylys J. A., Partyka R. A., and Shulgin A. T. (1976) Phenalkylamines with potential psychotherapeutic utility. *J. Med. Chem.* **19,** 1400–1404.

Stoff D. M., Gillin J. C., and Wyatt R. J. (1978) Animal models of drug-induced hallucinations, in *The Psychopharmacology of Hallucinogens* (Stillman R. C. and Willette R. E., eds.), Pergamon, New York, pp. 259–267.

Stolerman I. P. (1989) Trends in drug discrimination research analysed with a cross-indexed bibliography. *Psychopharmacology* **98,** 1–19.

Tadano T., Satoh S.-E., and Kisara K. (1986) Head-twitches induced by p-hydroxyamphetamine in mice. *Japan. J. Pharmacol.* **41,** 519–523.

Taylor W. A. and Sulser F. (1973) Effects of amphetamine and its hydroxylated metabolites on central noradrenergic mechanisms. *J. Pharmacol. Exp. Ther.* **185,** 620–632.

Tilson H. A., Chamberlain J. H., and Gylys J. A. (1977) Behavioral comparisons of R-2-amino-1-(2,5-dimethoxy-4-methylphenyl) butane (BL-3912A) with R-DOM and S-amphetamine. *Psychopharmacology* **51**, 169–173.

Titeler M., Lyon R. A., and Glennon R. A. (1988) Radioligand binding evidence implicates the brain 5-HT$_2$ receptor as a site of action for LSD and phenylisopropylamine hallucinogens. *Psychopharmacology* **94**, 213–216.

Todes C. J. (1986) At the receiving end of the lisuride pump. *Lancet* **July 5**, 36–37.

Tricklebank M. D. (1985) The behavioral response to 5-HT receptor agonists and subtypes of the 5-HT receptor. *Trends Pharmacol. Sci.* **5**, 403–507.

Trulson M. E. and Jacobs B. L. (1977) Usefulness of an animal behavioral model in studying the duration of action of LSD and the onset and duration of tolerance to LSD in the cat. *Brain Res.* **132**, 315–326.

Trulson M. E. and Jacobs B. L. (1979) Effects of 5-methoxy-*N*,*N*-di-methyl-tryptamine on the behavior and raphe unit activity in freely moving cats. *Eur. J. Pharmacol.* **54**, 43–50.

Trulson M. E., Brandstetter J. W., Crisp T., and Jacobs B. L. (1981a) Behavioral effects of quipazine in the cat. *Eur. J. Pharmacol.* **78**, 295–305.

Trulson M. E., Heym J., and Jacobs B. L. (1981b)Dissociations between the effects of hallucinogenic drugs on behavior and raphe unit activity in freely moving cats. *Brain Res.* **215**, 275–293.

Uyeno E. T. and Mitoma C. (1969) The relative effectiveness of several hallucinogens in disrupting maze performace by rats. *Psychopharmacologia* **16**, 73–80.

Vetulani J., Bednarczyk B., Reichenberg K., and Rokosz A. (1980) Head twitches induced by LSD and quipazine: Similarities and differences. *Neuropharmacology* **19**, 155–158.

Wallach M. B., Friedman E. and Gershon S. (1972) 2,5-Dimethoxy-4-methylamphetamine (DOM), a neuropharmacological examination. *J. Pharmacol. Exp. Ther.* **182**, 145–154.

White F. J. and Appel J. B. (1982) Lysergic acid diethylamide (LSD) and lisuride: Differentiation of their neuropharmacological actions. *Science* **216**, 535–537.

White F. J., Holohean A. M., and Appel J. B. (1981) Lack of specificity of an animal behavior model for hallucinogenic drug action. *Pharmacol. Biochem. Behav.* **14**, 339–343.

White F. J., Holohean A. M., and Appel, J. B. (1983) Antagonism of a behavioral effect of LSD and lisuride in the cat. *Psychopharmacology* **80**, 83–84.

Winter J. C. (1979) Quipazine-induced stimulus control in the rat. *Psychopharmacology* **60**, 265–269.

Winters W. D. and Wallach M. B. (1970) Drug induced states of CNS excitation: A theory of hallucinosis, in *Psychotomimetic Drugs* (Efron D. H., ed.), Raven, New York, pp. 193–214.

Woolley D. W. and Shaw E. N. (1957) Evidence for the participation of serotonin in mental processes. *Ann. NY Acad. Sci.* **66,** 649–667.

Yamamoto T. and Ueki S. (1975) Behavioral effects of 2,5-dimethoxy-4-methylamphetamine (DOM) in rats and mice. *Eur. J. Pharmacol.* **31,** 156–162.

Yim G. K. W., Prah T. E., Pfister W. R., and Nichols D. E. (1979) An economical screen for phenethylamine-type hallucinogens: Mouse ear scratching. *Commun. Psychopharmacol.* **3,** 173–178.

Young R., Glennon R. A., and Rosecrans J. A. (1981) Discriminative stimulus properties of the hallucinogenic agent DOM. *Commun. Psychopharmacol.* **4,** 501–505.

Animal Models for Caffeine Exposure in the Perinatal Period

Ronnie Guillet

1. Introduction

Caffeine is a widely used compound found both in drug preparations and in commonly consumed foods and beverages. Most human exposure occurs primarily through consumption of caffeinated beverages (e.g., coffee, tea, and soft drinks) such that, across all age groups, the average daily dose of caffeine ranges from <1 mg/kg in infants to approx 3 mg/kg in adults, including pregnant women (Lachance, 1982; Neims and von Borstel, 1983; Sobotka, 1989; Benowitz, 1990). Because caffeine is a pharmacologically and behaviorally active chemical, exposure of the developing organism to caffeine may have far-reaching, long-term consequences. Such developmental exposure may occur *in utero* as the result of consumption of caffeine by the pregnant woman or postnatally as the result of therapeutically administered caffeine to premature infants with apnea of prematurity. This chapter will deal with animal models for such perinatal caffeine exposure. In order to facilitate extrapolation of results from experimental animals to humans, the factors that must be taken into account in the design and interpretation of these studies will be presented. The neurochemical mechanisms whereby caffeine exposure may affect subsequent neural development and behavior/function will also be discussed.

From: *Neuromethods, Vol. 24: Animal Models of Drug Addiction*
Eds: A. Boulton, G. Baker, and P. H. Wu ©1992 The Humana Press Inc.

1.1. Developmental Exposure to Neuroactive Drugs: Experimental Considerations

Drugs elicit their effects on organisms through combination with specific receptors (Langley, 1906). Thus, the existence of specific receptors in developing brains could render the developing brain vulnerable to the presence of specific exogenous compounds. Vulnerability could simply mean that the developing brain is sensitive to the compounds, but that once a response is elicited, there will be no lasting effect. On the other hand, because there is a high degree of plasticity in the central nervous system, vulnerability could imply that generation of a response in the developing brain could influence the continued course of neural development, with functional consequences that might be expressed throughout the life-span of the organism.

Precisely how a drug will influence developing nervous systems cannot be predicted, however, based simply on the presence of recognition sites for the drug. Receptor function depends not only on the presence of a specific, high-affinity recognition site, but also on the existence of effector mechanisms that translate the interaction of a drug with its binding site into a response in the receptive cell. Studies on the development of transmitter receptors in the brain indicate that some early appearing sites are linked to active effector mechanisms, whereas other early appearing receptors are unable to mediate appropriate effector responses (*see* Miller et al., 1987 for review). Additionally, effector mechanisms present during development may differ from those present in the adult.

The full expression of drug interaction with neural receptors during development will also depend on the neural interactions that receptive cells make with other neurons and on the maturational state of specific neural circuitry. Because neuroactive drugs can influence chemical neurotransmission, the developmental profile of specific neurotransmitters may be an important determinant of the effect of early developmental drug exposure. The processes underlying chemical transmission in specific neural populations, such as the catecholamine-containing neurons, are known to appear early in fetal rat brain (Coyle, 1977).

These early developing transmitter systems respond in a predictable manner to many neuroactive drugs (Coyle and Henry, 1973). Furthermore, these early appearing systems have been implicated in the regulation of cell proliferation, in differentiation, and in organization of specific neural circuitry (Lauder and Krebs, 1986). Thus, interference with specific transmitter function during development could have profound consequences on brain function.

The interpretation of the consequences of drug exposure in the developing brain of an experimental animal must include consideration of several important factors: brain developmental period, species- and age-specific drug pharmacokinetics, and potential confounding experimental variables (*see* Kellogg and Guillet, 1988 for review). The sequence of events during brain development (i.e., appearance of neurotransmitter-containing cells, followed by proliferation of neuron axons and terminals, followed by synaptogenesis, for example) is similar across species. However, the degree of maturation achieved by an arbitrary time-point (e.g., birth) in a particular species may differ from the degree of maturation achieved in another species. Thus, the developmental events taking place in the brain of a rat during the third week of gestation are not the same as those taking place in the human brain during the third trimester (*see* Miller et al., 1987 for a discussion of brain development).

The pharmacokinetics of any given compound will depend, in part, on the species exposed, and on the age and sex of the subject under investigation. The plasma half-life, elimination kinetics, and specific metabolites of a drug differ among species. In addition, the relative importance of either peak drug concentrations or area under the concentration-time curve ("total exposure") may vary with the species under investigation (Nau, 1986). Similarly, these factors will vary in fetuses vs infants vs adults of a given species because of progressive increases in hepatic, microsomal, and drug-metabolizing enzymes over this time span. Additional physiological factors, such as pregnancy or liver dysfunction, may also impact on the compound's pharmacokinetics. Therefore, simply giving the dosage of a drug in terms of mg/kg body wt may be misleading. In some instances, dosages may be more accurately compared on the basis of surface area-to-body wt ratios. Mechanical factors, such as route of administration of a given compound, can also influence the plasma levels attained. Thus, the actual measurement of plasma and/or tissue drug concentrations is essential to comparative studies.

In studies of the effects of perinatal exposure to drugs, it is also essential to determine whether the drug under investigation is acting directly on the fetus or newborn, or indirectly through effects on the placenta, pregnant dam, or lactating female. For example, developmental changes seen in the newborn exposed *in utero* to cocaine may be a result of cocaine acting directly on the developing organism (via interference with dopamine and norepinephrine reuptake at the postsynaptic junction, leading to a direct toxicity to the developing central nervous system). Alternatively, cocaine may act indirectly by causing intrauterine hypoxia and malnutrition induced by placental vasoconstriction (*see* Chasnoff, 1991 for review).

1.2. Caffeine: Pharmacokinetics and Mechanisms of Action

Caffeine is a pharmacologically active drug whose effects depend on the dose and duration of exposure of the organism. It affects the central nervous, cardiovascular, and respiratory systems; muscles; gastrointestinal secretion; rate of urine formation; basal metabolic rate; and DNA repair. At moderate doses, caffeine will increase alertness, physical activity, coordination, and attention span, but may also increase unsteadiness and nervousness, and disrupt sleep patterns. At higher doses, tachycardia, insomnia, headache, tinnitus, and tremor may ensue. Neurochemical changes that occur following exposure to caffeine include a decrease in serotonin and dopamine turnover, and an increase in norepinephrine turnover. In addition, caffeine will stimulate the medullary respiratory center, relax bronchiolar smooth muscle, and vasodilate pulmonary arterioles. At toxic levels, caffeine will cause clonic convulsions and respiratory failure. For a discussion of caffeine's mutagenic, teratogenic, and carcinogenic effects, *see* the reviews by Lachance (1982) and Sobotka (1989).

Caffeine is rapidly and completely absorbed from the gastrointestinal tract, with peak levels achieved within minutes to 1–2 h (Latini et al., 1980; Bonati et al., 1982), and there is no first-pass effect through the liver. Caffeine is distributed through the body water, such that similar caffeine concentrations are reached throughout the different tissues of the organism. This widespread, passive distribution includes the brain (i.e., the blood–brain barrier is permeable to caffeine) and occurs across the placenta to the fetus.

Caffeine is excreted by the kidney. In the newborn human, >80% is excreted unchanged in the urine; in the adult, <2% is excreted unchanged. Biotransformation takes place in the liver and is species-specific. For example, in the adult, nonsmoking human, elimination follows first-order kinetics, such that the plasma half-life is 5–6 h with a clearance rate of ~80 mL/kg/h (Neims and von Borstel, 1983; Bonati et al., 1982; Stavric and Gilbert, 1990). However, in the rat, elimination follows saturation (not first-order) kinetics and at low-moderate doses, the "half-life" is approx 1.5–2 h (Neims and von Borstel, 1983; Latini et al., 1980). At higher doses (>40 mg/kg), clearance is slower. Clearance of caffeine is age-dependent as well. Both premature and full-term human newborns eliminate caffeine slowly, because biotransformation in the liver is immature. For the first several weeks to months, the half-life is 3–4 d (clearance rate of 8–9 mL/kg/h), such that several days are required for an infant to eliminate a single dose of caffeine (Neims and von Borstel, 1983; Stavric and Gilbert, 1990). As a result, the infant will also not experience the peaks and troughs of concentration seen in the adult, but may accumulate caffeine to a much higher steady-state concentration with repeated dosing. In the newborn rat, elimination is slower than in the mature rat, but faster than in the newborn human. Throughout pregnancy in the rat (Nakazawa et al., 1985) and during the second and third trimesters in humans (Stavric and Gilbert, 1990), there is an increase in caffeine half-life and a decrease in clearance. By the end of the third trimester in humans, the half-life of caffeine increases to about 18 h, and its clearance decreases to about 23 mL/kg/h. There is a rapid return to nonpregnant values in the perinatal period (review: Neims and von Borstel, 1983).

Additional information on caffeine metabolism in different species is available (Neims and von Borstel, 1983; Stavric and Gilbert, 1990; for a review of metabolism in the rat, *see* Khanna et al., 1972; Latini et al., 1978; in the mouse, *see* Burg and Werner, 1972; in the monkey, *see* Gilbert et al., 1985). Metabolic pathways vary among species, and metabolites of caffeine may be either more or less pharmacologically active than caffeine. Different subspecies and even different individuals may exhibit varying clearance rates for different metabolites. Many physiological conditions, such as age, illness, pregnancy, and so on, may also affect the metabolism of caffeine. Thus, it is extremely important to incorporate metabolic and pharmacokinetic information in experimental protocols, particularly those in which the data obtained will be extrapolated to humans.

The biologic actions of caffeine in vivo are believed to be owing primarily to competitive antagonism of endogenous adenosine at extracellular adenosine receptors (Daly et al., 1981; Neims and von Borstel, 1983; Snyder and Sklar, 1984; Benowitz, 1990). That caffeine fits classic criteria for a compound that acts by binding to a receptor has been demonstrated by studies showing that binding is specific and of high affinity (nanomolar to micromolar range), that the displacing potency of various agonists and antagonists of specific radioligand binding matches the biological potency of caffeine, that the regional distribution of the receptor sites matches the regional distribution of endogenous adenosine, and that there is a correspondence between the in vitro receptor occupancy and the production of a biologically relevant response (e.g., modulation of locomotor activity) (Dunwiddie, 1985). Although methylxanthines inhibit phosphodiesterase in vitro, in vivo caffeine's relative low potency as an inhibitor of the enzyme contrasts with its robust potency in modulating physiological processes. Even high doses of caffeine generally fail to increase tissue cAMP levels in intact animals. Thus, it is unlikely that phosphodiesterase inhibition accounts for the stimulant action of caffeine (Snyder and Sklar, 1984). Similarly, although in vitro caffeine has effects on calcium storage and translocation, it is unlikely that these effects are directly responsible for caffeine's cardiac and CNS stimulant properties. Caffeine does augment the twitch response of skeletal muscle by releasing calcium from the sarcoplasmic reticulum, but millimolar concentrations are required for this action (Snyder and Sklar, 1984).

In the CNS, where caffeine administration results in stimulation (decreased drowsiness and fatigue, increased alertness), the mechanism of action of caffeine is believed to be the result of competitive antagonism of the tonic depressant effect of endogenous adenosine. Most of the pharmacologic effects of adenosine can be antagonized by caffeine in low concentrations—concentrations that have no apparent effects on cAMP metabolism or on intracellular calcium dynamics. There is neither an increase in cerebral cAMP nor a reduction in the specific activity of cAMP phosphodiesterase in the brain. The behavioral depression induced by adenosine analogs can also be reversed by low doses of caffeine. In addition, the brain levels of caffeine administered at behaviorally effective doses are consistent with its affinity at the adenosine receptor (Snyder et al., 1981).

1.3. Adenosine Receptors:
CNS Distribution and Ontogeny

Adenosine has been shown to play an important role in the functioning of not only the CNS but also the pulmonary, cardiovascular, renal, and immunological systems (Daly, 1985). In the CNS, adenosine modulates synaptic transmission, has inhibitory effects on neurotransmitter release, depresses spontaneous and evoked potentials, and modulates the activity of adenylate cyclase (*see* review by Dunwiddie, 1985). The modulation of adenylate cyclase activity is thought to be mediated by at least two subtypes of extracellular adenosine receptors, A1 adenosine receptors and A2 adenosine receptors (Lee and Reddington, 1986; vanCalker et al., 1979; Daly, 1985; Dunwiddie, 1985; Bruns, 1988). This distinction between the subtypes is based on their pharmacological profiles. Whereas adenosine A1 receptors and adenosine A2 receptors are probably discrete proteins (the first interacting with the Ni subunit of adenylate cyclase, and the second with the Ns subunit of adenylate cyclase), their recognition sites for agonists and antagonists show only quantitative differences. Thus, there are no truly specific agonists, and most antagonists of the xanthine class are virtually nonselective (Daly, 1985). However, although their selectivity is not absolute, several A1-selective agonists are commonly used to identify putative adenosine A1 receptors (Bruns et al., 1980; Marangos et al., 1983). Adenosine A2 receptor identification and localization have been hampered by the lack of selective ligands until very recently (Bruns et al., 1986; Parkinson and Fredholm, 1990; Barrington et al., 1989; Jarvis et al., 1989; Lupica et al., 1990; Wan at al., 1990). A third receptor subtype, called the P site, is located intracellularly. Because methylxanthines do not interact with the P site, it will not be considered further here.

The adenosine A1 receptor has been more thoroughly studied compared with the A2 receptor, primarily because of the availability of radioligands that are quite specific for this subtype of receptor and because of the higher affinity with which such ligands bind to it. The adenosine A1 receptor inhibits adenylate cyclase, acting via a specific Gi-protein, whereas the adenosine A2 receptor stimulates adenylate cyclase, acting via a specific Gs-protein. Both receptor subtypes are distributed in the CNS and in the periphery (heart, endocrine glands, vasculature, blood elements, liver, fat cells, muscles, kidney, and so forth) (for review, *see* Daly, 1985). This discussion will be limited to adenosine receptor systems in the CNS only.

The regional distribution of adenosine receptors has been studied using both tissue homogenate assays and receptor autoradiography. Using the adenosine A1-selective agonist ^{3}H-cyclohexyladenosine (CHA), Bruns et al. (1980) identified A1 receptors in crude brain membranes. They also demonstrated that A1 receptors are most highly concentrated in the cerebellum, whereas A2 receptors are most highly concentrated in the striatum (Bruns et al., 1980). The advent of quantitative receptor autoradiography allowed more extensive descriptions of adenosine receptor localization in a variety of species (Lee and Reddington, 1986; Jarvis and Williams, 1988; Deckert et al., 1988; Etzel and Guillet, 1990). In all these species, A1 receptors are most highly concentrated in the molecular layer of the cerebellum, and in the CA-1 and CA-3 regions of the hippocampus. Moderate receptor densities are generally found in the cerebral cortex, striatum, and thalamus, with little or no specific radioligand binding in the brainstem (cf Jarvis, 1988).

The distribution of adenosine A2 receptors is markedly different from that of adenosine A1 receptors. A2 adenosine receptor sites are located exclusively in the striatum and olfactory tubercle (Jarvis et al., 1989; Parkinson and Fredholm, 1990; Jarvis, 1988). Their localization was demonstrated using either an A2-specific ligand (^{3}H-CGS 21680) or ^{3}H-*N*-ethylcarboxamidoadenosine (NECA), which binds to both A1 and A2 receptors, in the presence of cyclopentyladenosine (CPA) to block ^{3}H-NECA binding to A1 receptors.

The appearance of adenosine A1 binding sites in the rat brain is gradual, regionally specific, and follows a course similar to neuronal differentiation (Marangos et al., 1982; Guillet and Kellogg, 1991). Adult receptor densities are attained by 4–6 wk in the cerebellum and by 3–5 wk in the cortex. This gradual increase in adenosine A1 receptor sites during the first postnatal month is similar to that seen for high-affinity muscarinic cholinergic, dopamine, opiate, and GABA receptors, and is in contrast to the rapid onset of benzodiazepine receptor appearance (cf Marangos et al., 1982). The increase in radioligand binding seen with age is the result of an increase in the number of binding sites, not a change in binding affinity (Marangos et al., 1982; Guillet and Kellogg, 1991). Preliminary examination of adenosine A2 receptor ontogeny suggests that adult receptor densities are attained no later than 2 wk of age in the rat (Guillet, unpublished observations). It is during their evolution that receptor-effector systems are most likely to be affected by exogenous influences (Boer et al., 1988). Thus, the adenosine receptor system, particularly the A1 receptor complex, would be susceptible to change during the perinatal period.

1.4. Clinical Correlates

1.4.1. Coffee Consumption
by the Pregnant and Lactating Woman

Exposure to caffeine during pregnancy is ubiquitous. Caffeine is consumed daily by approx 80–90% of adults in many parts of the world (Benowitz, 1990). The average consumption of caffeine in the US and Canada is about 2.4 mg/kg/d for adults and 1.1. mg/kg/d for children 5–18 yr old (cf Benowitz, 1990). One study documented that during pregnancy, 95% of mothers ingested caffeine in the form of coffee, tea, cola, or in analgesic and antihistaminic preparations (reviewed by Soyka, 1979). Because caffeine is distributed uniformly in proportion to tissue water content, it is measurable in all tissues, including fetal tissues, within about 5 min of ingestion. Both caffeine and theophylline can also be found in the breast milk of nursing mothers who consume these drugs. Approximately 1–10% of the maternal dose will be present in breast milk (Lawrence, 1980).

1.4.2. Therapeutic Use of Caffeine
in the Premature Human Neonate

There have been relatively few studies of the pharmacological therapies prescribed for human newborns, particularly premature newborns, or to pregnant women in the immediate antenatal period. One such study done by the Italian Collaborative Group on Preterm Delivery (1988) describes the pattern of drug usage in preterm newborns admitted to 20 neonatal intensive care units in Italy. All drugs prescribed from the admission of the mother through the first week of the neonate's life were compiled and analyzed. Of the 706 monitored premature infants, over 84% received at least one drug during the perinatal period; 29.3% received 5–19 different drugs. Although the most commonly administered drugs were antibiotics and vitamins (approx 100% and 80% of patients in therapy, respectively), methylxanthines (aminophylline and caffeine) were given to almost 40% of the neonates who received drugs.

Methylxanthines are currently prescribed for the management of apnea of prematurity. Neonatal apnea, defined as cessation of breathing for greater than 20–30 s, with or without bradycardia and/or cyanosis, occurs in approx 25% of neonates weighing <2500 g at birth and approx 84% of neonates weighing <1000 g at birth. Methylxanthines have been shown to regularize breathing in the majority of infants to whom they are given. Aminophylline, theophylline, and caffeine have all been used successfully in this regard. In the human newborn, theophylline is metabolized to caffeine, and this caffeine may, in fact, contribute to the therapeutic effect. In many nurseries, caffeine is now the drug of choice because it has a wider therapeutic index (i.e., a greater CNS stimulant effect with fewer peripheral side effects) (Aranda and Turmen, 1979; Fuglsang et al., 1989).

When used in the treatment of apnea of prematurity, caffeine is administered to the neonate as a loading dose of 10 mg/ kg of caffeine (equivalent to 20 mg/kg of caffeine citrate) followed 24 h later by a single daily dose of 2.5 mg/kg (equivalent to 5 mg/kg of caffeine citrate). This dose regimen will result in plasma concentrations of caffeine of approx 8–16 mg/L over the 24-h period. Because of the slow elimination of caffeine, plasma levels will not fluctuate widely over the day. Both iv and oral administration are similarly effective in achieving the desired plasma concentrations because of the high bioavailability of caffeine (Aranda and Turmen, 1979). Thus, the human neonate being treated with caffeine for apnea of prematurity may experience caffeine concentrations five to ten times that in an adult moderate coffee drinker for periods of up to 8–12 wk. Minimal acute side effects are observed, and to date, there is no evidence that such exposure is deleterious to the infant. However, data on the long-term effects of methylxanthines are virtually nonexistent. The presence of multiple confounding variables limits an investigation on the premature infants themselves. Such factors as presence or absence of intracranial hemorrhage, sepsis, and fetal drug exposure, along with other perinatal risk factors, may act independently or interact to result in compromise of the developing nervous system, thus making it difficult to determine any consequences attributable solely to the early caffeine exposure. What studies do exist suggest that long-term caffeine treatment does not adversely influence growth parameters, at least during the first year of life (Le Guennec et al., 1990), and does not appear to adversely affect gross measures of growth and development over the first 3 yr of life (Gunn et al., 1979). These very general measures may not detect more specific long-term effects. However, to elicit behavioral changes that may result from earlier exposure to caffeine, testing should be directed at areas that are affected by acute exposure. For example, caffeine given acutely increases locomotor activity. Thus, behaviors including both spontaneous activity and drug-induced activity could be tested at a time remote from initial caffeine exposure. Similarly, when given in large doses, caffeine is a convulsant. Thus, the incidence of seizures could be examined in subjects exposed during the perinatal period to caffeine.

In addition, the clinical studies that have been performed were of limited duration of followup. There is precedent in animal studies for the effects of drug exposure to become manifest only after puberty (Kellogg et al., 1991; Miranda et al., 1990a,b; Simmons et al., 1984a,b).

2. Experimental Exposure to Caffeine During the Perinatal Period: The Models

2.1. Exposure of the Fetus and Neonate via the Dam

Various models of perinatal caffeine exposure have been employed by numerous investigators (Table 1). Regimens have included exposure during gestation only (Gilbert and Pistey, 1973; Driscoll et al., 1990; West et al., 1986; Swenson et al., 1990; Hughes and Beveridge, 1990; Sinton, 1989; Enslen et al., 1980), gestation and lactation (Schneider et al., 1990; Concannon et al., 1983; Yazdani et al., 1988), and prior to conception, through gestation and lactation (Dunlop and Court, 1981; Tanaka et al., 1987; Butcher et al., 1984). In the majority of these studies, caffeine was administered to the dam via her diet or in drinking water. As a result, caffeine dosages were estimated based on the amount consumed by each animal. In many instances, the pregnant females were not disturbed for daily weighing. Plasma caffeine levels were measured in the dams to ascertain the magnitude of the exposure in only three of the above-cited studies (Gilbert and Pistey, 1973; Driscoll et al., 1990; Tanaka et al., 1987). An attempt was made in some of the studies to compare the estimated amount of caffeine consumed by rats to the number of cups of coffee consumed by a pregnant woman. However, as mentioned above (Section 1.2.), because of differences in caffeine metabolism between species (i.e., rat and human), relative amounts consumed will not necessarily be proportional to relative circulating concentrations of caffeine.

These studies of perinatal caffeine exposure do provide valuable information concerning the potential effects on the fetal and newborn rat, but the conclusions cannot be easily extrapolated to humans. In order to make that extrapolation, additional pharmacokinetic data specific to the exposure regimen are needed, especially more precise data on circulating caffeine (and caffeine metabolites) in both the animal model and in pregnant women. Without this data, the applicability of the results remains purely speculative.

2.2. Exposure of the Developing Neonate

Few studies have examined methylxanthine exposure limited to the neonatal period only (Fuller et al., 1982; Fuller and Wiggins, 1981; Quinby and Nakamoto, 1984; Guillet, 1990a,b; Guillet and Kellogg, 1991). One such study employed an animal model to document effects on brain growth of long-term exposure to theophylline in the presence or absence of malnutrition (Quinby and Nakamoto, 1984). Lactating dams were fed a diet consisting of either 6 or 20% protein, with or without added theophylline (added in amounts calculated to provide daily doses of approx 20 mg/kg body wt), from days 3–15 after parturition. The added variable of nutritional status is very pertinent to the clinical situation in which premature neonates, whose nutritional status is not always optimal, receive therapeutic theophylline or caffeine. However, neither the dose of theophylline received by the nursing pups nor the resultant circulating drug level (neither theophylline or caffeine) was documented. In addition, the rationale for the choice of duration of exposure was not stated. The 2-wk period over which the rat pups were treated corresponds to a brain developmental period much longer than that

Table 1
Models of Perinatal Caffeine Exposure

Reference	Species	Methylxanthine	Route	Magnitude of exposure	Duration of exposure	Additional parameters
Butcher et al., 1984	Rat	Caffeine, coffee	Water	20–110 mg/kg/d	Prepreg-weaning	[C] measured
Concannon et al., 1983	Rat	Caffeine	Water	10–50 mg/kg/d	E1-P25	
Driscoll et al., 1990	Rat	Caffeine	Diet	20 mg/kg/d	E7-birth	±High-protein
Dunlop and Court, 1981	Rat	Caffeine	Water	10 mg/kg/d	E0-P21	Five successive pregnancies
Enslen et al., 1980	Rat	Caffeine	Diet	3–22 mg/d	E1-birth	Second generation tested
Fuller et al., 1982	Rat	Caffeine	Gavage	20–80 mg/kg/d	P2-17	
Fuller and Wiggins, 1981	Rat	Caffeine, aminophylline or theophylline	Gavage	40, 80 mg/kg/d	P2-20	
Gilbert and Pistey, 1973	Rat	Caffeine	ip	1–16 mg/d	E0-20	Divided dose, q6h
Guillet, 1990	Rat	Caffeine	Gavage	15–20 mg/kg/d	P2-6	[C] measured
Guillet and Kellogg, 1991	Rat	Caffeine	Gavage	15–20 mg/kg/d	P2-6	[C] measured
Holloway, 1982	Rat	Caffeine	sc	5–80 mg/kg	P1 or 10	
		Caffeine	Water	25–95 mg/kg/d	E10-birth	+ Acute C or T (20 mg/kg, d1 or 10)
Holloway and Thor, 1982	Rat	Caffeine	sc	20–120 mg/kg	P1, 10, or 15	
Hughes and Beveridge, 1990	Rat	Caffeine	Water	28, 36 mg/kg/d	E0-birth	
Marangos et al., 1984	Mouse	Caffeine	Diet	40–60 mg/kg/d	E1-P23	±Malnutrition
Mori et al., 1983	Rat	Caffeine	Diet	20 mg/kg/d	P1-15	±Malnutrition
Mori et al., 1984	Rat	Caffeine	Diet	20 mg/kg/d	E13-birth	

Reference	Species	Agent	Route	Dose	Days	Comments
Nakamoto et al., 1989	Rat	Caffeine	Diet	20 mg/kg/d	P1-15	±Malnutrition
Quniby and Nakamoto, 1984	Rat	Theophylline	Diet	20 mg/kg/d	P1-15	±Malnutrition
Quinby et al., 1985	Rat	Caffeine	Gavage	10 mg/kg q.o.d.	P3-15	±Malnutrition
Schneider et al., 1990	Rat	Caffeine	Diet	10 mg/kg/d	E9-P22	
Sinton, 1989	Mouse	Caffeine	Water	60 mg/kg/d	E0-birth	Two generations tested
Sobotka et al., 1979	Rat	Caffeine	Water	23–92 mg/kg/d 37–138 mg/kg/d	E7-P0 P1-22	
Swenson et al., 1990	Rat	Caffeine	sc	60 mg/kg/d	E13-19	
Tanaka et al., 1987	Rat	Caffeine	Water	0.04, 0.02% Solution	prepreg-E21	[C & T] measured
West et al., 1986	Rat	Caffeine	Gavage	5–75 mg/kg/d	E3-19	
Yazdani et al., 1988	Rat	Caffeine	Diet	10 mg/kg/d	P1-43	

[C] = caffeince concentration
[T] = theophylline concentration
E = embryologic day
P = postnatal day

over which most premature neonates are treated. The brain development of a 15-d-old rat pup is more comparable to that of a preschool-aged child than that of a neonate in the nursery being treated for apnea of prematurity (Dobbing, 1979). Therefore, extrapolation of the results to the human clinical situation is limited.

The model developed in our laboratory circumvents some of the above limitations to interpretation. Rat pups were treated orally with caffeine over postnatal days 2–6. Pups received caffeine, 20 mg/kg on day 2 and 15 mg/kg on days 3–6, administered by gavage once daily. These doses of caffeine were shown to result in blood caffeine levels of 5–14 mg/L over a 24-h period (Guillet, 1990b). Thus, rat pups were exposed to caffeine at a time during which their brain development was comparable to that of the 26- to 38-wk postconception human infant and at levels comparable to those achieved therapeutically. Limitations remain, however. The metabolic rate of caffeine is not the same in rats and humans. Because rat neonates metabolize caffeine more rapidly than do human neonates, circulating caffeine levels will vary more over the course of a day in rats. In addition, rat and human neonates do not metabolize caffeine identically; thus, circulating metabolites will vary between the two species. These were not quantified. It is unknown whether any of the metabolites of caffeine are of importance in the effects observed following exposure.

Although none of the animal models of perinatal methylxanthine exposure duplicate the exposure experienced by the human fetus and newborn, the models do serve to direct further research in the area. Any model that is employed must be used with its limitations in mind.

3. Effects of Perinatal Caffeine Exposure on Growth

Chronic caffeine exposure of animals in the fetal and/or neonatal period has been shown to affect fetal weights (West et al., 1986), body weights (Butcher et al., 1984; Quinby and Nakamoto, 1984; Fuller and Wiggins, 1981; Fuller et al., 1982;

Holloway, 1982), and brain weights (Yazdani et al., 1988), delay various physical developmental milestones (including eye opening, tooth eruption, and so on) (West et al., 1986), and alter behavioral parameters (sleep patterns, activity, suckling, homing, exploration, passive and active avoidance, and so on) (Butcher et al., 1984; Concannon et al., 1983; Kaplan et al., 1989; Sobotka et al., 1979; West et al., 1986) (*see also* review of *in utero* effects of caffeine by West et al., 1986). These multiple effects have been produced using a variety of regimens in terms of magnitude, duration, and timing of caffeine exposure. A brief review of some specific studies follows (*see also* Table 2).

Based on the dependent variables measured, the studies that have been performed can be categorized into three general types: those that examined primarily physical parameters, those that examined primarily brain growth and composition, and those that examined primarily behavioral indices of change. The studies reviewed here will be limited to those done in mammals (especially rodents). Studies that involved administration of excessive amounts of caffeine will also not be considered here.

Of the cited studies that reported caffeine effects on fetal and neonatal growth, only one documented an alteration in litter size as a result of prenatal caffeine exposure. The study by Gilbert and Pistey (1973) utilized a dosing protocol of four daily ip injections of caffeine (1–4 mg/dose) to the pregnant dam from day 0 to day 20 of gestation. Even at the lowest dose studied (approx 20 mg/kg/d), Gilbert and Pistey found a 20% reduction in litter size, with a corresponding increase in number of resorptions. Average fetal and placental weights were also decreased in a dose–dependent manner. West et al. (1986) administered caffeine by gavage to pregnant rats over days 2–19 of gestation at dosages of 5, 25, 50, and 75 mg/kg/d. Results of that study demonstrated no effect on litter size, but a significant decrease in birthweight in neonates from dams that received 75 mg/kg/d. These offspring were persistently lighter for at least 9 wk after birth. In another study, Dunlop and Court (1981) administered caffeine (20 mg/kg/d) in drinking water throughout the day to pregnant females. They did not observe a change in litter size and only a 1.7% decrease in birthweight. Without measurements

Table 2
Effects of Perinatal Caffeine Exposure

Reference	Testing procedures	Results
Butcher et al., 1984	Physical parameters, behavioral testing	Decreased birthweight; delayed physical development; no change in learning, memory, motoric activity
Concannon et al., 1983	Physical parameters, activity, catecholamine levels, cAMP, and cGMP levels	Hypoactivity, no effect on catecholamine levels, decreased cAMP, increased cGMP
Driscoll et al., 1990	Mandible development	Effects of caffeine on development modified by protein intake
Dunlop and Court, 1981	Physical parameters	Growth retardation in successive pregnancies; no teratogenesis
Enslen et al., 1980	Sleep patterns, catecholamines	Increased paradoxical sleep in prenatally caffeine-exposed rats; decrease in dopamine levels in locus ceruleus to two generations
Gilbert and Pistey, 1973	Gross physical parameters	Fetal loss, decreased birthweight, no fetal malformations
Guillet, 1990	Locomotor activity	Change in developmental pattern of sensitivity to acute challenge with adenosine receptor ligands
Guillet and Kellogg, 1991	Adenosine A1 receptor binding, locomotor activity	Increased B_{max}, no change in affinity; change in developmental pattern of sensitivity to challenge with adenosine receptor ligands–dose/response
Holloway, 1982	Behavioral testing	Acute methylxanthine results in increased activity; alteration of response to acute caffeine following chronic exposure
Hughes and Beveridge, 1990	Behavior, adrenal weights	Low doses of caffeine increase activity and decrease emotionality; heavier adrenal glands in females, but not males at 8 mo
Marangos, 1984	Adenosine A1 receptor binding	Increased specific binding
Mori et al., 1983	Brain composition	Malnutrition may exaggerate effect of caffeine on brain protein
Mori et al., 1984	Brain composition	Effects of prenatal caffeine and malnutrition interact

Reference	Measure	Finding
Nakamoto et al., 1989	Brain composition	Altered sensitivity of brain tissue to postnatal caffeine in malnourished state
Quinby and Nakamoto, 1984	Brain composition	Postnatal caffeine differentially affects brain cell number and size
Quinby et al., 1985	Brain composition	Differential effect of caffeine on brain cell number and size depends on nutritional state
Schneider et al., 1990	Mandible and femur growth	Impairment of bone development
Sinton, 1989	Passive avoidance	Increased passive avoidance latency through two generations
Sobotka et al., 1979	Physical parameters, brain neurochemistry, behavioral testing	Nominal effects on physical development; no effect on neurochemistry; increased exploratory activity
Swenson et al., 1990	Passive avoidance	Enhanced retention of response in female, but not male offspring
Tanaka et al., 1987	Placental, body and cerebral growth	Disproportionate decrease in cerebral weight
West et al., 1986	Physical parameters, locomotor activity, passive avoidance, active avoidance	Delayed postnatal development and altered avoidance performance
Yazdani et al., 1988	Brain composition	Decreased total brain weight, DNA and RNA content; differential effects on different brain regions

of circulating caffeine levels, neither the maximum nor the total daily caffeine exposure in the cited studies can be compared as a function of dosing regimen. The significant reduction in litter size and fetal weight seen by Gilbert and Pistey and the persistent effect on growth seen by West et al. may suggest that wide swings in caffeine concentration, with marked peaks and troughs, and/or higher serum caffeine concentrations predispose to greater developmental effects than do lower, more consistent caffeine levels throughout the 24-h period.

In addition to the dosing regimen, the growth parameter measured is also important in ascertaining the presence or absence of fetal effects of prenatal caffeine exposure on growth. Whereas Schneider et al. (1990) and Driscoll et al. (1990) found no effect of prenatal caffeine exposure on litter size or neonatal birthweight, they did observe a significant reduction in bone weight and calcium content. The dosing regimens utilized for these studies included supplementation of the diet with 10 and 20 mg/kg/d caffeine, respectively. Serum caffeine levels were not measured in the first study; levels averaged approx 1–4 mg/L in the second study. Although this was a protocol in which dramatic variations in serum caffeine concentrations would not be expected over the course of a day, no information is provided in that regard.

Human fetuses and premature infants, in addition to being exposed to caffeine in the perinatal period, may simultaneously be subjected to other environmental conditions that have potential for affecting growth and development. Nutrition is an important variable that has been addressed alone (Slob et al., 1973; Smart et al., 1989; Dobbing, 1979; Resnick and Morgane, 1983; Smart, 1990) and in conjunction with caffeine exposure. For example, Driscoll et al. (1990) examined the effects of protein deficiency, and Nakamoto and Shaye (1986) examined the effects of protein excess on fetal mandible development. Both extremes of protein intake interacted with the effects of caffeine on bone weight, calcium content, and phosphatase activity.

Quinby and Nakamoto (1984) fed lactating female rats either a standard (20% protein) diet or a protein-deficient (6% protein) diet. One-half of each of these groups received, in addition, approx 20 mg/kg/d of theophylline in the chow beginning on the day of delivery. The pups were weighed every third day, and at 15 d of age, brain weight, DNA, RNA, and protein content were determined. The results of this study indicated that, with respect to the above parameters, the pups' response to theophylline depended on their nutritional state. For example, in the malnourished pups, theophylline exposure resulted in a further decrease in body wt by the end of the study period, whereas in the normally nourished pups, theophylline exposure resulted in an increased body wt. There was, however, no effect of theophylline exposure on brain weight in either nutritional group. The amount of DNA in the brains of malnourished pups was not affected by theophylline exposure; however, in normally nourished pups, theophylline exposure resulted in a significant decrease in brain DNA content. Quinby and Nakamoto (1984) interpreted the changes in amount of DNA (both per brain and per mL brain) as indicative of a change in cell number, such that in normally nourished pups there was a decrease in the number of brain cells in the pups that received theophylline. Without histological examination of the brains, it is impossible to determine if neuronal (neurons or glia) or other (capillaries and so on) elements were affected. Brain protein:DNA, used by the authors as an estimate of cell size, was affected by diet and by theophylline exposure. Malnutrition alone resulted in an increase in cell size. The effect of theophylline on cell size was dependent on the nutritional status of the pup: Cell size decreased in the malnourished pups given theophylline and increased in normally nourished pups given theophylline compared with their respective nontheophylline controls. These conclusions concerning the interaction of nutritional state and theophylline exposure are based on several assumptions that may not be justified. For example, because the brain is not comprised of a homogeneous collection of cells and exogenous influences may interact with different cell types differently, examining the brain as a whole may obscure or accentuate cell type- or region-specific effects resulting in erroneous conclusions.

Similar studies were performed in the same laboratory using caffeine exposure in normally nourished and malnourished rat pups (Mori et al., 1983,1984; Quinby et al., 1985; Nakamoto et al., 1989). In each of these studies, the dams were fed a control or protein-deficient diet either during the last week of gestation (Mori et al., 1984) or during the first two postpartum weeks (Mori et al., 1983; Quinby et al., 1985; Nakamoto et al., 1989). In addition, half of the pups were exposed to caffeine either via the dam's milk (20 mg/kg caffeine to the dam in her diet) (Mori et al., 1983,1984; Nakamoto et al., 1989) or directly (10 mg/kg to the pup by gavage) (Quinby et al., 1985) over the same experimental period. In all of these studies, the effects of caffeine exposure and nutritional status were interrelated. Differential effects of caffeine on brain weight and brain DNA, RNA, and protein content were observed depending on the nutritional status of the dam. As discussed above, the interpretation of changes in these end points relies on several assumptions. In addition, the differential effects may be owing in part to either an altered metabolism of caffeine or to an altered sensitivity of brain tissue to caffeine in the malnourished vs the normally nourished state. Thus, the nutritional status of the experimental model (and perhaps the human fetus or infant) exposed to methylxanthines may be a critical factor in determining the effects of such exposure.

4. Effects of Perinatal Caffeine on Behavior

Prenatal exposure to caffeine (Holloway, 1982; West et al., 1986; Hughes and Beveridge, 1990; Swenson et al., 1990; Concannon et al., 1983; Butcher et al., 1984; Sobotka et al., 1979) and postnatal exposure to caffeine (Concannon et al., 1983; Holloway, 1982; Guillet, 1990b) both affect various behavioral parameters in both preweanling and postweanling (adolescent/adult) offspring. These effects are often subtle, suggesting that caffeine exposure modifies discrete neuronal subsystems that function to modulate the organism's response to environmental stimuli (Sobotka, 1989). The behavioral parameters tested, the age of the organism studied, and the dosing protocol may be crucial variables that differ from experiment to experiment, resulting in an apparent lack of consistency of effects across studies.

Holloway (1982) examined the effects of both chronic *in utero* and postnatal caffeine exposure. Pregnant dams were given either a 0.0125 or 0.05% caffeine solution as their source of drinking water over gestational days 10 to parturition. Pups were then tested on day 1 following acute exposure to 0, 20, 40, or 80 mg/kg caffeine or theophylline, sc. Activity testing on day 1 consisted of tabulating movements of the head and/or legs over a 2-min testing period. Alternatively, pups from unexposed dams were injected sc with 20 mg/kg/d caffeine over postnatal days 1–9. These pups were then tested on day 10 after acute exposure to 0, 5, 20, or 80 mg/kg caffeine. During a 3-min testing period, latency to enter the "home area," seconds in the "home area," and total number of squares entered in the testing chamber were recorded. One-day-old rats whose mothers had received tap water during gestation had a linear increase in activity with increasing doses of caffeine. The pups whose mothers received 0.0125% caffeine in their drinking water had a linear, but attenuated response to caffeine. The rats whose mothers had 0.05% caffeine in their water during gestation exhibited a large increase in activity following a 20 mg/kg injection and very small increases at the higher doses. (Similar decreases in activity at high caffeine doses have been observed in adult rats [Snyder et al., 1981; Kaplan et al., 1990]). Rats exposed to caffeine on days 1–9 after birth did not increase their activity in response to caffeine administration on day 10, unlike pups that received vehicle injections on days 1–9. Thus, long-term exposure to caffeine either during gestation or the first 9 d of lactation moderately increased the pups' activity levels and altered the activity increase observed following an acute caffeine challenge. Although chronic caffeine exposure affected these measures of nondirected behaviors, such caffeine exposure had no effect on directed behaviors, such as nipple attachment and home orientation. Of note, however, is the fact that all of the testing was performed within 24 h of the last dose of caffeine. Since no serum caffeine determinations were done on either dams or pups, it is not known how long caffeine persisted in the neonate's system and therefore whether, at 24 h after the final dose of caffeine, persistence of or withdrawal from caffeine influenced the testing.

Some information on persistence of caffeine effects may be gleaned from the following studies. Concannon et al. (1983) exposed dams to approx 14 mg/kg/d caffeine in drinking water throughout gestation and/or approximately 48 mg/kg/d caffeine in drinking water during lactation (to day 25). Offspring were tested for their development of spontaneous activity over the first 35 d of life. In this study, activity was assessed as present or absent once a minute for 1 h, thus generating 60 measures for each animal. Results demonstrated that both prenatal and postnatal caffeine exposure contributed to relative hypoactivity from day 15 on. Although this relative hypoactivity was seen from day 15 to day 25, during which time the pups were still being exposed daily to caffeine either in mothers' milk or in the water bottle, the effect persisted 5–10 d after withdrawal of the caffeine from the drinking water.

Of particular interest is the observation that, in many other instances, behavioral changes were seen well after the caffeine exposure had been terminated (Hughes and Beveridge, 1990; West et al., 1986; Swenson et al., 1990; Sobotka et al., 1979; Guillet, 1990b). Sobotka et al. (1979) tested offspring that had been exposed to caffeine beginning on gestational day 7 (23, 49, or 92 mg/kg/d via drinking water to pregnant dams) and continuing to postnatal day 22 (37, 75, or 138 mg/kg/d via drinking water to lactating dams). Testing of adolescent males revealed an alteration in performance in progressive fixed-ratio schedule testing as a function of caffeine exposure. In addition, open field testing revealed an increase in amount of rearing and in number of squares crossed during the testing interval. This hyperactivity was in contradistinction to the hypoactivity observed by Concannon et al. (1983). Although mode of delivery of caffeine and exposure periods were similar in the two studies, the magnitude of exposure differed. Whether an approximate doubling of the dose of caffeine used can alone account for the discrepancy in effects is unknown. The age at testing for locomotor activity may also be crucial. Concannon et al. (1983) observed the relative hypoactivity in males from 15–25 d of age; Sobotka et al. (1979) found relative hyperactivity in males ages 28–50 d of age. The period from approx day 12–18 is one in which spontaneous motor activity is undergoing characteristic changes (Guillet, 1990b). Rats typically exhibit relative locomotor hyperactivity over days 14–16 compared with the periods immediately preceding and following. Early developmental caffeine exposure may alter locomotor behaviors or their ontogeny, making age at testing an important variable.

West et al. (1986) exposed pregnant dams to 0, 5, 25, 50, or 75 mg/kg/d caffeine via gavage over days 3–19 of gestation. To rule out the possibility that exposure of the dams resulted in maternal-pup interactions that might alter effects of caffeine seen in the offspring, dams were tested through gestation and lactation. Although some physical and neurobehavioral effects were found in the dams, maternal influences on behavior of the offspring, based on correlations, were extremely limited. Offspring were tested over the first nine postnatal weeks of life. Observations revealed that at least some neurobehavioral effects were found at all doses of caffeine. Auditory startle in males occurred significantly earlier than control at 5 mg/kg, but was significantly delayed at 75 mg/kg. A similar trend was observed for female rats, but the results did not attain statistical significance. Locomotor activity was measured on lactational days 9, 12, 15, 18, and 21 and on postweaning days 41–44, and was significantly reduced on day 12 in offspring of dams exposed to 50 and 75 mg/kg/d caffeine. In this study, caffeine exposure did not affect locomotor activity on any other postnatal day. Thus, results of activity testing in perinatally caffeine-exposed pups appear to be specific to the age at which the testing is performed (Concannon, 1983; Sobotka, 1979; West, 1986). *In utero* exposure to caffeine did not significantly alter the increased activity induced by an acute challenge with 5 mg/kg of caffeine at 9 wk of age.

The effects seen by West et al. (1986) of *in utero* caffeine exposure on passive avoidance in males tested during the seventh postnatal week were dose-dependent. The proportion of animals that learned to avoid during the 24-h test session was significantly reduced at 5 and 25 mg/kg, and was significantly increased at 50 mg/kg. Exposure to 60 mg/kg/d caffeine, sc, over the same gestational period was shown by Swenson et al. (1990) to result in enhanced retention of a passive avoidance response in female offspring tested as adults. No such effects were found for male offspring. These results suggest that the effects of *in utero* exposure to caffeine may be both dose- and sex-dependent. There may be sexually dimorphic effects of perinatal caffeiene exposure on learning.

Hughes and Beveridge (1990) exposed pregnant dams to 0, 28, or 36 mg/kg/d throughout gestation via their drinking water. Subsequent testing of the offspring at 1, 2, 4, 6, and 8 mo of age on open-field locomotor and rearing activity and on emergence latency revealed an inverted-U-shaped prenatal caffeine dose–response curve for ambulation, frequency of walking, and rearing. There was an opposite trend for emergence latency. Thus, a lower dose of prenatal caffeine appeared to increase activity and decrease emotionality, whereas a higher dose appeared to have the opposite effect.

Even brief periods of exposure to caffeine in the postnatal period can affect subsequent behavior. Guillet (1990a,b) exposed rat pups to 15–20 mg/kg/d caffeine, po, over postnatal days 2–6. Resultant blood caffeine concentration ranged from 5–15 mg/L over the 24-h period between administrations. Pups were then tested for locomotor activity at 12, 15, 18, or 28 d of age. These ages were chosen to represent times of characteristic locomotor hyperactivity (days 14–16), 2 d prior (day 12), 2 d post (day 18), and stable adult level activity (28 d). Fifteen minutes prior to testing, animals were injected ip with saline, 15 mg/kg caffeine citrate (an adenosine receptor antagoinist), 10 μmol/kg D-phenyl-isopropyladenosine (D-PIA) (an adenosine receptor agonist), or left uninjected. Activity was then recorded for 30 min. Analysis revealed that baseline locomotor activity increased as a function of age: baseline activity was increased approx 35% in 12-d-old caffeine-exposed pups compared with controls; however, at 15, 18, and 28 d there were no differences in baseline activity as a function of neonatal exposure. This again demonstrates the age specificity for observation of effects of previous caffeine exposure. Drug-induced activity in caffeine-exposed pups also differed from that in control pups at specific ages. At 28 d, a 52–67% decrease in locomotor activity in response to acute challenge with D-PIA and a 70–93% increase in activity to response to acute caffeine were demonstrated in all treatment groups. (These responses were comparable to those seen in adult rats similarly challenged [Bruns et al., 1983; Choi et al., 1988; Daly et al., 1981; Kaplan et al., 1989; Latini et al., 1980; Neims and von Borstel, 1983; Snyder and Sklar, 1984].) However, the developmental pattern for achievement of the responses differed between the control and caffeine-exposed groups. At 12 d of age, D-PIA decreased activity over all groups, but acute caffeine was without effect. At 15 d of age, D-PIA again decreased activity over all groups; acute caffeine decreased activity as well. Analysis at 18 d demonstrated a neonatal exposure effect, an acute treatment effect, and an interaction of the two. Acute caffeine challenge significantly increased activity over baseline in control pups, but not in 18-d-old pups neonatally exposed to caffeine. Thus, neonatal caffeine exposure developmentally delayed the stimulatory effect of caffeine on locomotor activity. Of note also was the observation that the locomotor response to D-PIA, an adenosine agonist, seems to develop earlier than the locomotor response to caffeine, an adenosine antagonist.

In order to determine whether the above results were dose-dependent (i.e., whether an effective dose of caffeine could be found that produced increased locomotor activity in 18-d-old caffeine-exposed pups or in 12- or 15-d-old control pups), dose–response studies were then conducted (Guillet, 1991). Pups were again tested at 12, 15, 18, and 28 d of age for locomotor response 15 min following acute challenge with caffeine citrate (15, 30, or 60 mg/kg, approx 100, 200, or 400 μmol/kg, ip), D-PIA (0.038 or 0.38 mg/kg, approx 1 or 10 μmol/kg, ip), or saline. To determine baseline activity, an uninjected group was tested at each age. Analyses of the data presented in Fig. 1 demonstrated that there was a significant effect of age ($p < 0.0001$), neonatal exposure ($p < 0.03$), and acute dose of caffeine ($p < 0.0001$) on locomotor activity. In addition, there was a significant interaction of age and acute dose of caffeine ($p < 0.05$) as well as an interaction of age, neonatal exposure and acute dose of caffeine ($p < 0.02$). Thus, in contrast to 18-d-old control pups that responded to 15 mg/kg caffeine with a significant increase in locomotor activity, caffeine-exposed pups showed little response

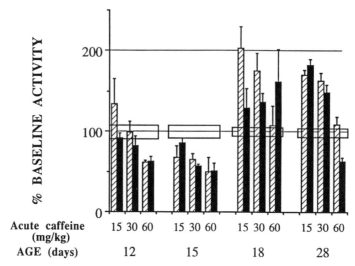

Fig. 1. Dose–response study of acute caffeine treatment as a function of age and neonatal caffeine exposure: Responsiveness expressed as percent baseline activity (n = 5–13 rats at each age, neonatal exposure, acute caffeine combination). Mean baseline activity indicated by box depicting 100% ±SEM (n = 25–35 rats at each age). At 18 d of age, unlike control pups, neonatally caffeine-exposed pups did not increase their activity in response to acute caffeine injection even at high doses of caffeine. Neonatal exposure: ▨ control; ■ caffeine.

to acute caffeine at that age even at 30 mg/kg. At a dosage of 60 mg/kg, neither control nor caffeine-exposed pups showed an increase in locomotor activity at 18 d of age. By 28 d of age, both control and caffeine-exposed groups demonstrated a significant increase in locomotor activity at both 15 and 30 mg/kg acute caffeine (p < 0.0001).

Responses to acute injection of D-PIA are shown in Fig. 2. Not only were there main effects of age (p < 0.0001) and acute D-PIA dose (p < 0.0002), but there was also an interaction of age and neonatal exposure (p < 0.02). Caffeine-exposed pups demonstrated a depressant effect of 10 μmol/kg D-PIA as early as 12 d of age vs 15 d in control pups, and of 1 μmol/kg D-PIA as early as 15 d vs 28 d in controls.

The results of the above experiments suggest that, in the naive rat, there is a developmental change in the adenosine

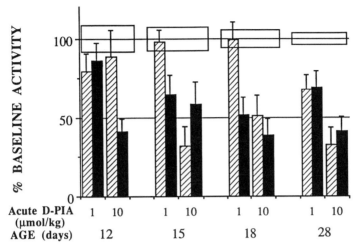

Fig. 2. Dose–response study of acute D-PIA treatment as a function of age and neonatal caffeine exposure: Responsiveness expressed as percent baseline activity (n = 5–12 rats at each age, neonatal exposure, acute D-PIA combination). Mean baseline activity indicated by box depicting 100% ± SEM (n = 25–35 rats at each age). Neonatally caffeine-exposed pups demonstrated the locomotor depressant effect of 10 µmol/kg D-PIA as early as 12 d vs 15 d in control pups and of 1 µmol/kg D-PIA as early as 15 d vs 28 d in controls. Neonatal exposure: ▨ control; ■ caffeine.

receptor system occurring over at least the first three postnatal weeks of life. Over this time frame, the rat becomes sensitive to the effects on locomotor activity of acute exposure to adenosine receptor agonists and antagonists. Moreover, this development of sensitivity is not synchronous for the two categories of ligands. Further, early caffeine exposure in some way disrupts the chronology, such that the animal becomes sensitive to the effects of acute D-PIA earlier and to the effects of acute caffeine at a later age. The etiology of these developmental patterns remains unknown, but probably involves the ontogeny of other components of the adenosine receptor system (e.g., G-protein or receptor/G-protein coupling).

These experiments illustrate, in part, the potential complexities of any receptor system. In order to evaluate fully effects of earlier exposures, not only must the receptor be available to bind the appropriate ligands, but also the rest of the receptor complex and neural network involved in mediating responses to the ligand must be present at the time of testing. Thus, the effects of chronic exposure to any given chemical will be apparent at a given stage of development only if the entire apparatus required to observe a particular response is functional. In other words, the behaviors tested must be age-appropriate tasks. Otherwise, effects of chronic exposures may appear to be "latent." By the same token, behavioral testing must be tailored to the receptor system under investigation, because specific disruptions caused by the agent under investigation may or may not be manifested in gross measures of development. The receptor system to which the drug binds to effect its responses must be identified, and the involvement of the receptor system in specific behaviors delineated.

5. Effects of Perinatal Caffeine on Brain Neurochemistry

Long-term exposure to methylxanthines has been shown to have multiple effects on the CNS in adult animals (Zielke and Zielke, 1987; Goldberg et al., 1982; Boulenger et al., 1983; Kirch et al., 1990; Ramkumar et al., 1988; Green and Stiles, 1986; Fredholm, 1982; Hawkins et al., 1988; Lupica et al., in press). Using dosages of approx 50 and 100 mg/kg/d caffeine added to the diets of mice for 12, 26, or 40 d, Boulenger et al. (1983) demonstrated a statistically significant increase in adenosine receptor agonist (^{3}H-CHA) binding in whole brain that was attributable to an increase in receptor density. No effect on binding affinity $(1/K_d)$ was seen. The increased binding was greatest after 12 d of exposure and not significant after 40 d of exposure. Similar results were obtained by Zielke and Zielke (1987) using silastic tubing containing either caffeine or theophylline. After 14 d of exposure to an average plasma level of 1.2 μg/mL theophylline or 7.1 μg/mL caffeine, there was a significantly increased maximal binding density (B_{max}) measured in cerebellum, a marginally significant increase in B_{max} measured in cortex, and no change in K_d. Hawkins et al. (1988) exposed adult rats to 75 mg/kg/d for 12 d, at which time they, too, found increases in B_{max} in cortex and cerebellum without concomitant effects on K_d. In addition, they examined binding to hippocampal membranes and found no effect of caffeine exposure in that region. In another study (Fredholm, 1982), exposure to caffeine for periods as short as 1 wk at 20 mg/kg/d, ip produced similar upregulation of adenosine receptors.

Ramkumar et al. (1988) and Green and Stiles (1986) demonstrated that chronic caffeine exposure in adult rats affects the A1 adenosine receptor-adenylate cyclase system of cerebral cortex at multiple levels. In both studies involving exposing rats to 600–1000 mg/L caffeine in drinking water for periods of 12–28 d, there was an increase in the proportion of receptors in the high-affinity state, an increase in coupling of the receptor with the associated G-protein, and an increase in the quantity of adenosine receptor-specific G-protein. These results suggest an up-regulation of different components of the adenosine receptor-adenylate cyclase system after chronic caffeine exposure.

The effects of chronic caffeine exposure on the CNS include more than its interaction with the adenosine receptor system. There is evidence for interaction with the β-adrenergic system also. Goldberg et al. (1982) observed an increase in norepinephrine (NE) utilization, but no change in basal levels of NE in rat forebrain following three doses of 50 mg/kg caffeine over the course of 24 h. There was also a decrease in forebrain β-adrenergic receptor numbers with no change in binding affinity (Goldberg et al., 1982; Green and Stiles, 1986). This effect may, however, be regionally specific and, for example, occur in frontal cortex and not cerebellum or striatum (Kirch, 1990).

The effects on brain neurochemistry of chronic caffeine exposure during the perinatal period have been incompletely examined at this time. Enslen et al. (1980) examined first and second generation offspring of dams exposed to approx 2.9, 5.7, or 21.5 mg caffeine/d throughout gestation. They found no effect on NE levels, but a decrease in dopamine (DA) levels in the locus ceruleus, the decrease being attenuated in the second generation. Concannon et al. (1983) exposed rats to 14 mg/kg/d caffeine throughout gestation and approx 48 mg/kg/d caffeine during lactation (to day 25). They found regional effects on cyclic nucleotide concentrations in whole brain and in cerebellum, but no effect on either brain NE or DA levels. β-Adrenergic receptor binding was not examined in either of these two studies.

The effects of chronic caffeine exposure on the developing adenosine receptor system itself are similar to those seen in adult animals with mature adenosine receptor systems, in that there is an apparent regionally specific increase in A1 adenosine receptor densities without a change in binding affinity (Marangos et al., 1984; Guillet and Kellogg, 1991). Marangos et al. (1984) exposed pregnant rats to approx 40–60 mg/kg/d caffeine throughout gestation and lactation. After weaning at 18 d of age, pups were continued on the same caffeine-containing diets as their dams. Regional binding in cortex, cerebellum, brain stem, thalamus, and hippocampus was measured at 16 and 23 d of age. There was an increase in specific binding to membranes derived from cerebellum and brain stem on day 16, and from all regions except hippocampus on day 23. Saturation binding experiments on whole brain suggested that the increase in binding was the result of an increase in B_{max} from postnatal day 11 onward. Thus, increased receptor numbers and no change in binding affinity were most likely responsible for the regional changes as well.

Similar effects were obtained following a more limited caffeine exposure. Examining at weekly intervals from 14–90 d of age, Guillet and Kellogg (1991a) found an increase in specific binding in membrane preparations from cortex, cerebellum, and hippocampus, but not brain stem or hypothalamus, after only 5 d of exposure (postnatal days 2–6) to 15–20 mg/kg/d caffeine. Saturation analysis of ^{3}H-CHA binding in cortex, cerebellum, and hippocampus indicated that the increase in specific binding measured in these regions at these ages was owing to an increase in maximal binding density and not to a change in binding affinity (Guillet and Kellogg, 1991a). Interestingly, however, in pups at 18 d of age, an age at which behavioral data suggested that changes were taking place in the adenosine receptor system, there was a significant decrease in binding affinity in cortex from caffeine-exposed rats (K_d control = 0.78 ± 0.07; K_d caffeine-exposed = 3.63 ± 0.20). This effect was limited to the cortex, and was not observed in cerebellar or hippocampal tissue (Guillet and Kellogg, 1991b).

Whether the apparent changes in adenosine A1 receptor densities seen following perinatal caffeine exposure represent "upregulaton" in the face of blockade of the receptors or a long-term change in the development of the receptors is not clear. Rather than merely increasing receptor densities, early caffeine exposure may instead accelerate the development of the adenosine receptor system with attainment of adult densities at earlier ages. At the present time, there is little information available concerning the ontogeny of the rest of the receptor system (i.e., G-proteins, adenosine receptor/G-protein coupling, adenylate cyclase activity) and no information on the effect of chronic caffeine exposure thereon. Thus, it is not known if there are any effects on these components that would suggest an altered ontogeny. Preliminary data (Guillet, unpublished observations) indicate an absence of an effect of early neonatal caffeine exposure on adenosine A2 receptor binding in rat striatum, which is consistent with the lack of effect of chronic theophylline exposure on this receptor population in adult animals as well (Lupica et al., in press). The A2 adenosine receptor and its associated G-protein may therefore be less susceptible to exogenous influences. Further study is needed to determine the maturity of these components during the period of caffeine exposure in order to reach any conclusions about susceptibility during development.

Because of the complexities inherent in the adenosine receptor systems (A1 and A2), a simple correlation between behavioral and neurochemical changes brought about by chronic perinatal caffeine exposure is not apparent. Until there is more information on the ontogeny of the various components of this receptor system and on how caffeine may affect each component, correlations will be difficult to make. Furthermore, because the behaviors in which the receptor system is involved are complex and may involve many other neural systems, assigning a direct cause-and-effect role to changes in any one component of the adenosine receptor system may be impossible. However, by perturbing the development of the adenosine receptor system by exposure to caffeine, facets of its normal ontogeny may become clearer.

6. Summary and Conclusions

Early developmental exposure to exogenous chemicals has been shown to result in delayed and long-lasting effects on receptor-effector systems. For example, despite the apparent absence of measurable effects in the prepubertal animal, exposure to diazepam during the third week of gestation in the rat affects metabolic, biochemical, and behavioral parameters in rats tested in adulthood (Kellogg and Guillet, 1988; Simmons et al., 1984b). Thus, appropriate development of later neural and behavioral functions may well depend upon orderly sequential development of earlier systems. The use of caffeine, both in the infant with apnea of prematurity and by women during the course of pregnancy, is sufficiently widespread and of such magnitude as to constitute a potential risk to the developing nervous system in the fetus and newborn. Evidence suggests that effects of caffeine exposure would be mediated by the adenosine receptor system. In order to identify the locus of these effects, appropriate animal models must be developed that reflect both the magnitude and developmental period over which the human exposure occurs. Similarly, the subsequent testing using these models must be appropriate with respect to neurochemical, physiological, and behavioral properties of the neural systems believed to be involved in these effects.

This chapter has dealt with some of the factors that must be taken into consideration when designing experiments intended to examine potential effects of perinatal caffeine exposure. The effects are dependent on timing and magnitude of exposure, and on timing and utilization of appropriate provocative testing. Factors that influence either exposure parameters (e.g., drug pharmacokinetics or developmental stage of the subject) or testing performance (e.g., age or sex of the subject) must be carefully considered in the interpretation of results.

The importance of well-designed studies on the effects of perinatal caffeine exposure goes beyond the acquisition of knowledge concerning alterations in the adenosine system caused by such exposure. Similarities among different receptor systems exist (e.g., involvement of G-proteins in signal transduction) and what is learned about a particular receptor complex may be applicable to another. The developing brain is at risk for the effects of centrally acting drugs administered during periods of rapid evolution of its various receptor systems, regardless of the route or intent of exposure. The resultant functional alterations will depend on the particular drug exposure and the brain region(s) in which the appropriate receptors are evolving at the time of such exposure. Furthermore, by experimentally using drugs to perturb a developing receptor system, one may learn more about the normal ontogeny of that system.

The use of animal models for perinatal drug exposure will allow a more focused investigation of the effects of such exposure in humans. As the normal ontogeny of the adenosine receptor system and the neurochemical bases of the effects of caffeine exposure on the adenosine and other neural systems become elucidated in animal models, the most appropriate indices of function to be examined in the clinical situation will emerge. Long-term consequences of methylxanthine treatment for apnea of prematurity can then be sought in a more rational, orderly, and cost- and time-effective manner.

References

Aranda J. V. and Turmen T. (1979) Methylxanthines in apnea of prematurity. *Clin. Perinatol.* **6**, 87–108.

Aranda J. V., Cook C. E., Gorman W., Collinge J. M., Loughnan P. M., Outerbridge E. W., Aldridge A., and Neims A. H. (1979) Pharmacokinetic profile of caffeine in the premature newborn infant with apnea. *J. Pediatr.* **94**, 663–668.

Barrington W. W., Jacobson K. A., Hutchison A. J., Williams M., and Stiles G. L. (1989) Identification of the A2 adenosine receptor binding subunit by photoaffinity crosslinking. *Proc. Natl. Acad. Sci.* **86**, 6572–6576.

Benowitz N. L. (1990) Clinical pharmacology of caffeine. *Ann. Rev. Med.* **41**, 277–288.

Boer G. J., Feenstra M. G. P., Mirmiran M., Swaab D. F., VanHaanen F., Chen-Pelt W., and Eikelboom T. (1988) *Neurochemistry of Functional Neuroteratology: Permanent Effects of Chemicals on the Developing Brain.* Elsevier, Amsterdam.

Bonati M., Latini R., Galletti F., Young J. F., Tognoni G., and Garattini S. (1982) Caffeine disposition after oral doses. *Clin. Pharmacol. Ther.* **32**, 98–106.

Boulenger J. P., Patel R. M., Parma A. M., and Marangos P. J. (1983) Chronic caffeine consumption increases the number of brain adenosine receptors. *Life Sci.* **32,** 1135–1142.

Bruns R. F. (1988) Adenosine receptor assays, in *Adenosine Receptors* (Dermot M. F. and Cooper C. L., eds.), A. R. Liss, New York, pp. 41–62.

Bruns R. F., Daly J. W., and Snyder S. H. (1980) Adenosine receptors in brain membranes: Binding of N^6-cyclohexyl[^{3}H]adenosien and 1,3-diethyl-8-[^{3}H]phenylxanthine. *Proc. Natl. Acad. Sci. USA* **77,** 5547–5551.

Bruns R. F., Katims J. J., Annau Z., Snyder S. H., and Daly J. W. (1983) Adenosine receptor interactions and anxiolytics. *Neuropharmacology* **22,** 1523–1529.

Bruns R. F., Lu G. H. and Pugsley T. A. (1986) Characterization of the A2 adenosine receptor labeled by [^{3}H]NECA in rat striatal membranes. *Mol. Pharmacol.* **29,** 331–346.

Burg A. W. and Werner E. (1972) Tissue distribution of caffeine and its metabolites in the mouse. *Biochem. Pharmacol.* **21,** 923–936.

Butcher R. E., Vorhees C. V., and Wootten V. (1984) Behavioral and physical development of rats chronically exposed to caffeinated fluids. *Fund. Appl. Toxicol.* **4,** 1–13.

Chasnoff I. J. (1991) Cocaine and pregnancy: Clinical and methodologic issues. *Clin. Perinatol.* **18,** 113–124.

Choi O. H., Shamim M. T., Padgett W. L., and Daly J. W. (1988) Caffeine and theophylline analogues: Correlation of behavioral effects with activity as adenosine receptor antagonists and as phosphodiesterase inhibitors. *Life Sci.* **43,** 387–398.

Concannon J. T., Braughler J. M., and Schechter M. D. (1983) Pre- and postnatal effects of caffeine on brain biogenic amines, cyclic nucleotides and behavior in developing rats. *J. Pharmacol. Exp. Ther.* **226,** 673–679.

Coyle J. T. (1977) Biochemical aspects of neurotransmission in the developing brain. *Int. Rev. Neurobiol.* **20,** 65–103.

Coyle J. T. and Henry D. (1973) Catecholamines in fetal and newborn rat brain. *J. Neurochem.* **18,** 2061–2075.

Daly J. W. (1985) Adenosine receptors, in *Advances in Cyclic Nucleotide and Protein Phosphorylation Research,* vol. 19 (Cooper D. M. F. and Seamon K. B., eds.) Raven Press, New York, pp. 29–46.

Daly J. W., Bruns R. F., and Snyder S. H. (1981) Adenosine receptors in the central nervous system: Relationship to the central actions of methylxanthines. *Life Sci.* **28,** 2083–2097.

Deckert J., Morgan P. F., and Marangos P. J. (1988) Adenosine uptake site heterogeneity in the mammalian CNS? Uptake inhibitors as probes and potential neuropharmaceuticals. *Life Sci.* **42,** 1331–1345.

Dobbing J. (1979) Prenatal nutrition and neurological development, in *Early Malnutrition and Mental Development* (Cravido J., Hambraeus L., and Vahlquist B., eds.) Almquist and Wiksell, Stockholm, pp. 96–110.

Driscoll P. G., Joseph F., Jr., and Nakamoto T. (1990) Prenatal effects of maternal caffeine intake and dietary high protein on mandibular development in fetal rats. *Br. J. Nutrition* **63**, 285–292.

Dunlop M. and Court J. M. (1981) Effects of maternal caffeine ingestion on neonatal growth in rats. *Biol. Neonate* **39**, 178–184.

Dunwiddie T. V. (1985) The physiological role of adenosine in the central nervous system. *Int. Rev. Neurobiol.* **27**, 63–139.

Enslen M., Milon H., and Wurzner H. P. (1980) Brain catecholamines and sleep states in offspring of caffeine-treated rats. *Experientia* **36**, 1105–1106.

Etzel B. A. and Guillet R. (1990) The ontogeny of adenosine A1 receptors in rat brain using receptor autoradiography. *Soc. Neurosci. Abstr.* **16**, 697.

Fredholm B. B. (1982) Adenosine actions and adenosine receptors after 1 week treatment with caffeine. *Acta Physiol. Scand.* **115**, 283–286.

Fuglsang G., Nielsen K., Nielsen L. K., Sennes F., Jakobsen P., and Thelle T. (1989) The effects of caffeine compared with theophylline in the treatment of idiopathic apnea in premature infants. *Acta Paediatr. Scand.* **78**, 786–788.

Fuller G. N. and Wiggins R. C. (1981) A possible effect of the methylxanthines caffeine, theophylline and aminophylline on postnatal myelination of the rat brain. *Brain Res.* **213**, 476–480.

Fuller G. N., Divakaran P., and Wiggins R. C. (1982) The effect of postnatal caffeine administration on brain myelination. *Brain Res.* **249**, 189–191.

Gilbert E. F. and Pistey W. R. (1973) Effect on the offspring of repeated caffeine administration to pregnant rats. *J. Reprod. Fert.* **34**, 495–499.

Gilbert S. G., Stavric B., Klassen R. D., and Rice D. C. (1985) The fate of chronically consumed caffeine in the monkey (Macaca fascicularis). *Fund. Appl. Toxicol.* **5**, 578–587.

Goldberg M. R., Curatolo P. W., Tung C.-S., and Robertson D. (1982) Caffeine down-regulates β adrenoreceptors in rat forebrain. *Neurosci. Lett.* **31**, 47–52.

Green R. M. and Stiles G. L.(1986) Chronic caffeine ingestion sensitizes the A1 adenosine receptor-adenylate cyclase system in rat cerebral cortex. *J. Clin. Invest.* **77**, 222–227.

Guillet R. (1990a) Neonatal caffeine exposure alters adenosine A1 receptor kinetics during a critical developmental period. *Soc. Neurosci. Abstr.* **16**, 66.

Guillet R. (1990b) Neonatal caffeine exposure alters adenosine receptor control of locomotor activity in the developing rat. *Dev. Pharmacol. Ther.* **15**, 94–100.

Guillet R. and Kellogg C. K. (1991) Neonatal exposure to therapeutic caffeine alters the ontogeny of adenosine A1 receptors in brain of rats. *Neuropharmacol.* **30**, 489–496.

Guillet R. and Kellogg C. K. (1991) Neonatal caffeine exposure alters develpmental sensitivity to adenosine receptor ligands. *Pharmacol. Biochem. Behav.* **40,** 811–817.

Gunn T. R., Metrakos K., Riley P., Willis D., and Aranda J. V. (1979) Sequelae of caffeine treatment in preterm infants with apnea. *J. Pediatr.* **94,** 106–109.

Hawkins M., Dugich M. M., Porter N. M., Urbanic M., and Radulovacke M. (1988) Effects of chronic administration of caffeine on adenosine A1 and A2 receptors in rat brain. *Brain Res. Bull.* **21,** 479–482.

Holloway W. R. (1982) Caffeine: Effects of acute and chronic exposure on the behavior of neonatal rats. *Neurobehav. Toxicol Teratol.* **4,** 21–32.

Holloway W. R., Jr. and Thor D. H. (1982) Caffeine sensitivity in the neonatal rat. *Neurobehav. Toxicol. Teratol.* **4,** 331–333.

Hughes R. N. and Beveridge I. J. (1990) Sex- and age-dependent effects of prenatal exposure to caffeine on open-field behavior, emergence latency and adrenal weights in rats. *Life Sci.* **47,** 2075–2088.

Italian Collaborative Group on Preterm Delivery (1988) Early neonatal drug utilization in preterm newborns in neonatal intensive care units. *Dev. Pharmacol. Ther.* **11,** 1–7.

Jarvis M. F. (1988) Autoradiographic localization and characterization of brain adenosine receptor subtypes, in *Receptor Localization: Ligand Autoradiography,* Alan R. Liss, Inc., New York, pp. 95–111.

Jarvis M. F. and Williams M. (1988) Differences in adenosine A-1 and A-2 receptor density revealed by autoradiography in methylxanthine-sensitive and insensitive mice. *Pharmacol. Biochem. Behav.* **30,** 707–714.

Jarvis M. F., Schulz R., Hutchison A. J., Do U. H., Sills M. A. and Williams M. (1989) [^{3}H]CGS 21680, a selective A2 adenosine receptor agonist directly labels A2 receptors in rat brain. *J. Pharmacol. Exp. Ther.* **251,** 888—893.

Kaplan G. B., Tai N. T., Greenblatt D. J., and Shader R. I. (1990) Caffeine-induced behavioural stimulation is dose- and concentration-dependent. *Br. J. Pharmacol.* **100,** 435–440.

Kaplan G. B., Greenblatt D. J., Leduc B. W., Thompson M. L. and Shader R. I. (1989) Relationship of plasma and brain concentrations of caffeine and metabolites to benzodiazepine receptor binding and locomotor activity. *J. Pharmacol. Exp. Ther.* **248,** 1078–1083.

Kellogg C. K., Primus R. J., and Bitran D. (1991) Sexually dimorphic influence of prenatal exposure to diazepam on behavioral responses to environmental challenge and on γ-aminobutyric acid (GABA)-stimulated uptake in the brain. *J. Pharmacol. Exp. Ther.* **256,** 259-265.

Khanna K. L., Rao G. S., and Cornish H. H. (1972) Metabolism of caffeine-^{3}H in the rat. *Toxicol. Appl. Pharmacol.* **23,** 720–730.

Kirch D. G., Taylor T. R., Gerhardt G. A., Benowitz N. L., Stephen C., and Wyatt R. J. (1990) Effect of chronic caffeine administration on monoamine and monoamine metabolite concentrations in rat brain. *Neuropharmacology* **29,** 599–602.

Lachance M. P. (1982) The pharmacology and toxicology of caffeine. *J. Food Safety* **4**, 71–112.

Langley J. N. (1906) On nerve-endings and on special excitable substances in cells. *Proc. R. Soc. Lond. (Biol.)* **78**, 170–194.

Latini R., Bonati M., Marzi E., Tacconi M. T., Sadurska B., and Bizzi A. (1980) Caffeine disposition and effects in young and one-year-old rats. *J. Pharm. Pharmacol.* **32**, 596–599.

Latini R., Bonati M., Castelli D., and Garattini S. (1978) Dose-dependent kinetics of caffeine in rats. *Toxicol. Lett.* **2**, 267–270.

Lauder J. M. and Krebs H. (1986) Do neurotransmitters, neurohumors, and hormones specify critical periods? in *Developmental Neuropsychobiology* (Greenough W. T. and Junaska J. M., eds.), Academic Press, New York, pp. 119–174.

Lawrence R. A. (1980) *Breastfeeding: A Guide for the Medical Profession*. Mosby, St. Louis.

Lee K. S. and Reddington M. (1986) Autoradiographic evidence for multiple CNS binding sites for adenosine derivatives. *Neurosci.* **19**, 535–549.

Le Guennec J. C., Sitruk F., Breault C., and Black R. (1990) Somatic growth in infants receiving prolonged caffeine therapy. *Acta Pediatr. Scand.* **79**, 52–56.

Lupica C. R., Jarvis M. F., and Berman R.F. Chronic theophylline treatment in vivo increases high affinity adenosine A1 receptor binding and sensitivity to exogenous adenosine in the in vitro hippocampal slice. *Brain Res.* (in press).

Lupica C. R., Cass W. A., Zahniser N. R., and Dunwiddie T. V. (1990) Effects of the selective adensoine A2 receptor agonist CGS 21680 on in vitro electrophysiology, cAMP formation and dopamine release in rat hippocampus and striatum. *J. Pharmacol. Exp. Ther.* **252**, 1134–1141.

Marangos P. J., Boulenger J.-P. and Patel J. (1984) Effects of chronic caffeine on brain adenosine receptors: Regional and ontogenetic studies. *Life Sci.* **34**, 899–907.

Marangos P. J., Patel J., and Stivers J. (1982) Ontogeny of adenosine binding sites in rat forebrain and cerebellum. *J. Neurochem.* **39**, 267–270.

Marangos P. J., Patel J., Martino A. M., Dilli M., and Boulenger J. P. (1983) Differential binding properties of adenosine receptor agonists and antagonists in brain. *J. Neurochem.* **41**, 367–374.

Miller R. K., Kellogg C. K., and Saltzman R. A. (1987) Reproductive and perinatal toxicology, in *Fundamentals of Toxicology* (Berndt W. O. and Haley W., eds.), Hemisphere Pub. Corp., Washington, DC, pp. 159–309.

Miranda R., Ceckler T., Guillet R., and Kellogg C. K. (1990a) Early developmental exposure to benzodiazepine ligands alters brain ^{31}P NMR spectra in young adult rats. *Brain Res.* **506**, 85–92.

Miranda R., Ceckler T., Guillet R., and Kellogg C. K. (1990b) Aging-related changes in brain metabolism are altered by early developmental exposure to diazepam. *Neurobiol. Aging* **11**, 117–122.

Mori M., Wilber J. F., and Nakamoto T. (1983) Influences of maternal caffeine on the neonatal rat brains vary with the nutritional states. *Life Sci.* **33**, 2091–2095.

Mori M., Wilber J. F., and Nakamoto T. (1984) Protein-energy malnutrition during pregnancy alters caffeine's effect on brain tissue of neonate rats. *Life Sci.* **35**, 2553–2560.

Nakamoto T. and Shaye R. (1986) Protein-energy malnutrition in rats during pregnancy modifies the effects of caffeine in fetal bones. *J. Nutr.* **116**, 633–640.

Nakamoto T., Hartman A. D., and Joseph F., Jr. (1989) Interaction between caffeine intake and nutritional status on growing brains in newborn rats. *Ann. Nutrition Metab.* **33**, 92–99.

Nakazawa K., Tanaka H., and Arima M. (1985) The effect of caffeine ingestion on pharmacokinetics of caffeine and its metabolites after a single administration in pregnant rats. *J. Pharmacobio.-Dyn.* **8**, 151–160.

Nau H. (1986) Species differences in pharmacokinetics and drug teratogenesis. *Env. Health Perspectives* **70**, 113–129.

Neims A. H. and von Borstel R. W. (1983) Caffeine: Metabolism and biochemical mechanisms of action, in *Nutrition and the Brain, vol. 6* (Wurtman R. J. and Wurtman J. J., eds.), Raven, New York, pp. 1–30.

Parkinson F. E. and Fredholm B. B. (1990) Autoradiographic evidence for G-protein coupled A2-receptors in rat neostriatum using [³H]-CGS 21680 as a ligand. *Naunyn-Schmiedeberg's Arch. Pharmacol.* **342**, 85–89.

Quinby G. E. Jr. and Nakamoto T. (1984) Theophylline effects on cellular response in protein-energy malnourished neonatal rat brain. *Pediatr. Res.* **18**, 546–549.

Quinby G. E., Batirbaygil Y., Hartman, A. D., and Nakamoto T. (1985) Effects of orally administered caffeine on cellular response in protein energy-malnourished neonatal rat brain. *Pediatr. Res.* **19**, 71–74.

Ramkumar V., Bumgarner J. R., Jacobson K. A., and Stiles G. L. (1988) Multiple components of the A1 adenosine receptor-adenylate cyclase system are regulated in rat cerebral cortex by chronic caffeine ingestion. *J. Clin. Invest.* **82**, 242–247.

Resnick O. and Morgan P. J. (1983) Animal models for small-for-gestational-age (SGA) neonates and infants-at-risk. *Devel. Brain Res.* **10**, 221–225.

Schneider P. E., Miller H. I., and Nakamoto T. (1990) Effects of caffeine intake during gestation and lactation on bones of young growing rats. *Res. Exp. Med.* **190**, 131–136.

Simmons R. D., Kellogg C. K., and Miller R. K. (1984a) Prenatal diazepam exposure in rats: long-lasting, receptor-mediated effects on hypothalamic norepinephrine-containing neurons. *Brain Res.* **293**, 73–83.

Simmons R. D., Miller R. K., and Kellogg C. K. (1984b) Prenatal exposure to diazepam alters central and peripheral responses to stress in adult rat offspring. *Brain Res.* **307**, 39–46.

Sinton C. M. (1989) Preliminary indications that functional effects of fetal caffeine exposure can be expressed in a second generation. *Neurotoxicol. Teratol.* **11**, 357–362.

Slob A. K., Snow C. E., and de Natris-Mathot E. (1973) Absence of behavioral deficits following undernutrition in the rat. *Dev. Psychobiol.* **6**, 177–186.

Smart J. L., Billing A. E., Duggan J. P., and Massey R. F. (1989) Effects of early life undernutrition in artificially-reared rats: 3. Further studies of growth and behavior. *Physiol. Behav.* **45**, 1153–1160.

Smart J. L. (1990) Vulnerability of developing brain to undernutrition. *Upsala Med. Sci.* **Suppl 48**, 21–41.

Snyder S. H. and Sklar P. (1984) Behavioral and molecular actions of caffeine: Focus on adenosine. *J. Psychiatr. Res.* **18**, 91–106.

Snyder S. H., Katims J. J., Annau Z., Bruns R. F., and Daly J. W. (1981) Adenosine receptors and behavioral actions of methylxanthines. *Proc. Natl. Acad. Sci. USA* **78**, 3260–3264.

Sobotka T. J. (1989) Neurobehavioral effects of prenatal caffeine, in *Prenatal Abuse of Licit and Illicit Drugs* (Hutchings D. E., ed.), Ann. NY Acad. Sci. vol. 562, NY Acad. Sci., New York, pp. 327–339.

Sobotka T. J., Spaid S. L., and Brodie R. E. (1979) Neurobehavioral teratology of caffeine exposure in rats. *Neurotoxicol.* **1**, 403–416.

Soyka L. F. (1979) Effects of methylxanthines on the fetus. *Clin. Perinatol.* **6**, 37–51.

Stavric B. and Gilbert S. G. (1990) Caffeine metabolism: A problem in extrapolating results from animal studies to humans. *Acta Pharm. Jugosl.* **40**, 475–489.

Swenson R. R., Beckwith B. E., Lamberty K. J., and Krebs S. J. (1990) Prenatal exposure to AVP or caffeine but not oxytocin alters learning in female rats. *Peptides* **11**, 927–932.

Tanaka H., Nakazawa K., and Arima M. (1987) Effects of maternal caffeine ingestion on the perinatal cerebrum. *Biol. Neonate* **51**, 332–339.

vanCalker D., Müller M., and Hamprecht B. (1979) Adenosine regulates via two different types of receptors, the accumulation of cyclic AMP in cultured brain cells. *J. Neurochem.* **33**, 999–1005.

Wan W., Sutherland G. R., and Geiger J. D. (1990) Binding of the adenosine A2 receptor ligand [^{3}H]CGS 21680 to human and rat brain: Evidence for multiple affinity sites. *J. Neurochem.* **55**, 1763–1771.

West G. L., Sobotka T. J., Brodie R. E., Beier J. M., and O'Donnell M. W., Jr. (1986) Postnatal neurobehavioral development in rats exposed in utero to caffeine. *Neurobehav. Toxicol. Teratol.* **8**, 29–43.

Yazdani M., Hartman A. D., Miller H. I., Temples T. E., and Nakamoto T. (1988) Chronic caffeine intake alters the composition of various parts of the brain in young growing rats. *Dev. Pharmacol. Ther.* **11**, 102–108.

Zielke C. L. and Zielke H. R. (1987) Chronic exposure to subcutaneously implanted methylxanthines: Differential elevation of A1-adenosine receptors in mouse cerebellar and cerebral cortical membranes. *Biochem. Pharmacol.* **36**, 2533–2538.

Index